Advances in Intelligent Systems and Computing

Volume 1411

The series "Advances in Intelligent Systems and Computing" contains publications on theory, applications, and design methods of Intelligent Systems and Intelligent Computing. Virtually all disciplines such as engineering, natural sciences, computer and information science, ICT, economics, business, e-commerce, environment, healthcare, life science are covered. The list of topics spans all the areas of modern intelligent systems and computing such as: computational intelligence, soft computing including neural networks, fuzzy systems, evolutionary computing and the fusion of these paradigms, social intelligence, ambient intelligence, computational neuroscience, artificial life, virtual worlds and society, cognitive science and systems, Perception and Vision, DNA and immune based systems, self-organizing and adaptive systems, e-Learning and teaching, human-centered and human-centric computing, recommender systems, intelligent control, robotics and mechatronics including human-machine teaming, knowledge-based paradigms, learning paradigms, machine ethics, intelligent data analysis, knowledge management, intelligent agents, intelligent decision making and support, intelligent network security, trust management, interactive entertainment, Web intelligence and multimedia.

The publications within "Advances in Intelligent Systems and Computing" are primarily proceedings of important conferences, symposia and congresses. They cover significant recent developments in the field, both of a foundational and applicable character. An important characteristic feature of the series is the short publication time and world-wide distribution. This permits a rapid and broad dissemination of research results.

Indexed by DBLP, INSPEC, WTI Frankfurt eG, zbMATH, Japanese Science and Technology Agency (JST).

All books published in the series are submitted for consideration in Web of Science.

More information about this series at https://link.springer.com/bookseries/11156

Gaurav Gupta · Lipo Wang · Anupam Yadav ·
Puneet Rana · Zhenyu Wang
Editors

Proceedings of Academia-Industry Consortium for Data Science

AICDS 2020

Springer

Editors
Gaurav Gupta
School of Mathematical Sciences, College of Science and Technology
Wenzhou Kean University
Wenzhou, China

Lipo Wang
Department of Electrical and Electronic Engineering
Nanyang Technological University
Singapore, Singapore

Anupam Yadav
Department of Mathematics
Dr. B. R. Ambedkar National Institute of Technology Jalandhar
Jalandhar, Punjab, India

Puneet Rana
School of Mathematical Sciences, College of Science and Technology
Wenzhou Kean University
Wenzhou, China

Zhenyu Wang
Newford Research Institute of Advanced Technology
Wenzhou, China

ISSN 2194-5357 ISSN 2194-5365 (electronic)
Advances in Intelligent Systems and Computing
ISBN 978-981-16-6886-9 ISBN 978-981-16-6887-6 (eBook)
https://doi.org/10.1007/978-981-16-6887-6

This Springer imprint is published by the registered company Springer Nature Singapore Pte Ltd.
The registered company address is: 152 Beach Road, #21-01/04 Gateway East, Singapore 189721, Singapore

Preface

Academia-Industry Consortium for Data Science (AICDS 2020) was an event to bring researchers, academicians, industries, and government personnel together to share and discuss the various aspects of intelligent systems. It is organized during December 19–20, 2020. The consortium was sponsored by Wenzhou Education Bureau. The consortium aimed to enforce the interaction between academia and industry, leading to innovation in both fields. During this event, significant advances in data science were presented, bringing together prominent figures from business, science, and academia to promote the use of data science in industry. AICDS also encourages industrial sectors to propose challenging problems where academician can provide insight and new ideas.

This volume is a curated collection of the articles which are presented during the conference. This book focuses on the current and recent developments in data science, machine learning, data security and IoT, medical imaging, deep learning, convolutional neural networks, decision making, watermarking, air quality analysis models, data hiding in encrypted images, image encryption algorithms, and various other aspects of machine learning.

In conclusion, the edited book comprises papers on diverse aspects of intelligent systems with many real-life applications.

Wenzhou, China — Gaurav Gupta
Singapore, Singapore — Lipo Wang
Jalandhar, India — Anupam Yadav
Wenzhou, China — Puneet Rana
Wenzhou, China — Zhenyu Wang

Contents

About the Editors

Dr. Gaurav Gupta is an Assistant Professor and Department Chair of Mathematics at the Wenzhou-Kean University, Wenzhou, China. He is having 13 years of teaching and research experience. From 2007–2010, he worked for Indian Space Research Organization (ISRO). He obtained more than $120,000 in grant funding from Wenzhou-education bureau. His research focus is in the area of Data Analytics, Image Processing, Computer vision and soft Computing. Dr. Gupta has published 31 research papers in reputed journals and conferences. He has guided two PhD students, 9 Masters Dissertations and 3 undergraduate Projects. He has participated and contributed in many conferences and workshops as keynote speaker, technical committee and session chair.

Dr. Lipo Wang received a Bachelor degree from National University of Defense Technology (China) and PhD from Louisiana State University (USA). He is presently on the faculty of the School of Electrical and Electronic Engineering, Nanyang Technological University, Singapore. His research interest is an artificial intelligence with applications to image/video processing, biomedical engineering, and data mining. He has 340+ publications, a U.S. patent in neural networks and a patent in systems. He has co-authored 2 monographs and (co-)edited 15 books. He has 8,800+ Google Scholar citations, with H-index 43. He was a keynote speaker for 36 international conferences. He is/was Associate Editor/Editorial Board Member of 30 international journals, including 4 IEEE Transactions, and guest editor for 10 special journal issues. He was a member of the Board of Governors of the International Neural Network Society, IEEE Computational Intelligence Society (CIS), and the IEEE Biometrics Council. He served as CIS Vice President for Technical Activities and Chair of Emergent Technologies Technical Committee, as well as Chair of Education Committee of the IEEE Engineering in Medicine and Biology Society (EMBS). He was President of the Asia-Pacific Neural Network Assembly (APNNA, recently renamed as APNNS – "Society") and received the APNNA Excellent Service Award. He was founding Chair of both the EMBS Singapore Chapter and CIS Singapore Chapter. He serves/served as chair/committee members of over 200 international conferences.

Dr. Anupam Yadav, Assistant Professor, Department of Mathematics, Dr. BR Ambedkar National Institute of Technology Jalandhar, India. His research area includes numerical optimization, soft computing, and artificial intelligence, he has more than ten years of research experience in the areas of soft computing and optimization. Dr. Yadav has done a Ph.D. in soft computing from the Indian Institute of Technology Roorkee and he had worked as a research professor at Korea University. He has published more than twenty-five research articles in journals of international repute, has published more than fifteen research articles in conference proceedings. Dr. Yadav has authored a textbook entitled "An introduction to neural network methods for differential equations". He has edited three books which are published by AISC, Springer Series. Dr. Yadav was the General Chair, Convener and member of the steering committee of several international conferences. He is a member of various research societies.

Dr. Puneet Rana is working as an Assistant Professor of Mathematics at the Wenzhou-Kean University, Wenzhou, China. He has more than ten years of teaching and research experience. He has received his PhD degree in the field of applied mathematics with specialization in "nanofluids" from Indian Institute of Technology Roorkee, INDIA in 2013. His research interest includes nanotechnology, soft computing, numerical methods, thermal stability analysis etc. He was awarded "research fellowship" for pursuing higher education by CSIR, INDIA and travel grant for attending various international conferences by INSA and DST, INDIA. He has also guided two Ph. D. students in the field of "nanotechnology" and published more than 60+ research papers in international journal and conferences of high impact with more than 1800+ Google and Scopus citation. He was also invited to the school of mathematical sciences, Universiti Sains Malaysia, Penang, Malaysia for collaborative research work. One of his paper is recognized by Science-Direct as a most cited paper published in Communications in Nonlinear Science and Numerical Simulation, Elsevier in 2012. He has participated and contributed in many conferences and workshops as a keynote speaker and technical committee member. He served as a reviewer of several Top ISI journals including Scientific reports, Physics of fluids, IJHMT, IJTS and many more.

Dr. Zhenyu Wang is the Director of the NRIAT, Wenzhou. He obtained his DPhil Degree from the Department of Computer Science, University of Oxford, and he was a member of Linacre College. Dr. Wang graduated from East China Teaching University, holding a BSc in Computer Science and Technology. He received his Msc in Natural Computation from the School of Computer Science at University of Birmingham. His current research areas include fuzzy rule-based systems, ensemble Learning, evolutionary algorithms, natural inspired optimizations, applications of intelligent system in cancer diagnosis and bioinformatics.

Neural Artistic Style Transfer Using Deep Neural Networks

Bharath Gupta, K. Govinda, R. Rajkumar, and Jolly Masih

Abstract Neural style transfer (NST) is a category of software algorithm that uses digital videos and images, adopting the look or viewing style of another video or image. Neural style transfer techniques use deep neural networks to perform image transformation. NST is a high-performance visualization process that allows us to use and thus replicate the style content of another style. It takes two pictures: image content and style reference, and we combine both together to get the resulting image, thus making it look like it was painted in style image style, despite keeping the main elements of the content image. In this paper, the proposed activity uses a convolutional neural network to transmit neural style.

Keywords VGG Net · Error · Loss · Content · CNN

1 Introduction

The process of transferring text from one image to another is called neural transfer, the purpose of which is to integrate text from the image source to identify and store the content of the target image. A photograph taken by a street photographer can provide the content of the image, and van Gogh's artwork, 'The Starry Night' could be a reference style image. The result of the photo shoot can be a stand-alone image that looks like the creation of a real van Gogh. Motivation is to produce a new style from two different images that look like a style image.

After discovering a popular paper by Gatys et al., demonstrating the work of style transfer, as well as a variety of research articles based on improving the beauty of the transfer style, improving speed, video styling, and analyzing parameter tuning, such as working on communication and informal information. This chapter is to create amazing style transfer effects and apply a new style to an image while retaining its original content [1].

B. Gupta · K. Govinda (✉) · R. Rajkumar · J. Masih
School of Computer Science and Engineering, VIT University, Tamil Nadu, Vellore 014, India
e-mail: kgovinda@vit.ac.in

G. Gupta et al. (eds.), *Proceedings of Academia-Industry Consortium for Data Science*, Advances in Intelligent Systems and Computing 1411,
https://doi.org/10.1007/978-981-16-6887-6_1

The input to the neural style transfer includes

- A generated image—the image that contains the final output which is **the only trainable variable**
- A style image—the image from which we want to transfer the style that is to be incorporated in the content image
- A content image—the image on which we want to perform the style transfer.

1.1 The Steps in Neural Style Transfer

We take an input image and a style image and then resize them to equal dimensions.

1.1.1 Then download the pre-trained convolutional neural network (CNN) such as VGG19.

1.1.2 We can distinguish CNN's content layers (image-related elements) and style-based layers (basic shapes, colors, etc.), and we can separate CNN's various layers to work independently in style and content.

1.1.3 The following should work on the problem of performance that we are trying to reduce:

- Loss of content (distance between output and input images—we strive to retain content)
- Loss of style (distance between output and style photos—striving to apply a new style)
- Complete loss of contrast (regularity—slippery location to remove sound from photo)

1.1.4 Set the gradients and do the right thing.

The VGG19 network was upgraded during The ImageNet Large Scale Visual Recognition Challenge (ILSVRC)-2014, for image segmentation, with an error rate of 7.3%. VGG19 is a 19-layer deep convolutional neural network. We can download a pre-trained network version of over a million images from the ImageNet database. A pre-trained network can split images into 1000 objects, such as a keyboard, mouse, pencil, and many other animals. As a result, the network has read rich presentations of various images. The network has 224-by-224 image input size. We can use the separation to separate new images using the VGG19 network [3].

As suggested in the paper by Gatys et al., I have used the VGG19 network for this project. VGG19 allows us to issue feature and style maps, content and custom images. VGG19 is a simpler CNN model compared to Inception and ResNet. VGG19 feature maps work best on neural art transfer. VGG19 has five layers of convolutional—conv1 to conv5 layers—each of which has two to four layers called conv1_2… conv5_4, which are also followed by fully connected layers that form a separation. Convolutional layers perform element extraction by detecting a variety of geometric shapes and complex patterns of continuity while fully connected layers act as perceptrons and separate objects based on the conditions in the image. Therefore, a pre-trained VGG network with completely separated layers will act as a pure pattern and output element [1].

1.2 Content and Style Representations

In order to find both the content and style of our image, we will look at some of the inner layers within our model. The middle layers represent the feature maps that start to get more organized as you go deeper. In this project, the construction of the VGG19 network, which is a pre-trained image editing network, is being used. These middle layers are needed to define content and style representation from our images. We will try to match the corresponding style and content in these layers in the middle of the installation image. We will download our pre-trained photo editing network. After that, we hold layers of interest. We then describe the model by setting the model input to the image and the effects of the style and output of the content. In other words, we have created a model that will take the image of the installation and extract the content and style of the middle layers [3].

1.3 Intermediate Layers

These are the results of the center of our pre-trained image editing network that enables us to define style and representation of content. For the network to perform image classification, it must understand the image. This includes taking a green image as pixels to insert and create internal representations and transformations that convert green pixels into a complex understanding of the elements in an image. This is also why CNN is able to do so well: It is able to take the supernatural and explain the elements in the studies that believe that God is not known by background noise and other issues. Therefore, somewhere in the middle of where the green image is inserted and the partition label is removed, the model acts as a complex element output, which is why by finding the middle layers, we are able to define the content and style of input images [3].

The manuscript is arranged as follows in paragraph 2 describing the use of style transfer; the proposed method is discussed in Sect. 3, and its implementation and implementation are discussed in Sect. 4 followed by the conclusion and indications.

2 Applications

Neural artistic style transfer finds its much-needed program in various video and image editing software. From sharing custom-made images to enhancing user-generated movies and music videos, and more, the ability to incorporate popular art styles into photos and video clips carries a new level of attraction to the user base and adds unprecedented power to innovative tools in line with NST. NST art transfer style models can be easily installed on edge devices—for example, mobile phones—allowing a variety of applications to convert photos and video, after processing, in real time. It thus enables the right image and video editing tools to be easily usable and accessible.

2.1 Commercial Art

Neural artistic style transfer is one of the few ways computer vision makes AI-enabled art possible. Whether it is emerging artists opening their way to sharing their world-class beauty art or auctioned art, neural artistic style transfer promises to transform the real meaning, the ways we think about art and how we represent art in real time. Neural artistic style transfer can be used to create high-quality printing and reproduction of large advertising campaigns or office buildings. The transfer of neural artistic style can drastically change our view of the effects of commercial art.

2.2 Neural Artistic Style Transfer on Mobile

Applications to use neural artistic style transfer are not limited to those working in cloud or servers. We can create neural artistic style transfer models that are small and fast enough to run directly on a variety of platforms, thus opening up many opportunities, including creative tools, powerful video editors and images, etc.

From user engagement and storage to product reliability, and more, using neural artistic on-device style transfer has the potential to attract users in a variety of new and lasting ways, while reducing cloud costs and maintaining user data privacy.

2.3 Gaming

Back in 2019, during a press conference at Game Developers Conference'19, Google launched Stadia, its powerful video game streaming service. One of Stadia's most important features is the in-game style transfer features that enable the automatic redesign of the visual world with color palettes and graphics embedded in an unlimited range of art styles.

According to Google, efforts like these are aimed at empowering an existing artist within all developers. Therefore, gambling, like any other neural art transfer app, will make the creation of art accessible to those who are not traditionally recognized as artists.

2.4 *Virtual Reality*

Virtual reality is like playing. Immersed VRs represent the anchor of the user experience and explore various possibilities for the transfer of neural art style. Various applications for the transfer of neural art style to virtual reality are still in the research phase with promising and exciting opportunities. Facebook has been actively testing the ability to transfer neural art style to completely change the way virtual reality developers present visual stories through their films, apps, games, etc., neural artistic style transfer [6–8].

3 Proposed Method

In the proposed work, the L-BFGS optimization algorithm (limited memory Broyden–Fletcher–Goldfarb–Shanno algorithm) is used and is widely used in machine learning to measure parameters. Content submissions and style representations may be separated from CNN and read during computer viewing (e.g., photo manipulation or visual function). NST uses a pre-trained convolution neural network (CNN) to transfer styles from a given image to another. This is done by defining a lost function that attempts to reduce the contrast between a style image and a content image, by combining both seamlessly to create a structured image.

Two networks, i.e., an already trained feature extractor and a transmission network, are required to train the transfer model (where to avoid the use of training data in pairs, using a pre-trained feature extractor). The trend of individual CNN-trained image classification layers can be used to focus on understanding specific aspects of the image. Layers can learn to extract content and image texture. The style transfer conveys these two images with the pre-trained NN when comparing the similarity of the pre-trained network release to many different layers. Images that produce the same output on one of the pre-trained model layers may have the same content while having the same style and the corresponding output on the other layer. The transmission network has an encoder–decoder structure that helps to create a custom-made image from a pre-trained model only that allows us to compare the content and style of both images.

At the beginning of the training, one or more styles are developed with the pre-trained feature keyboard, and the results of the various style layers are retained for

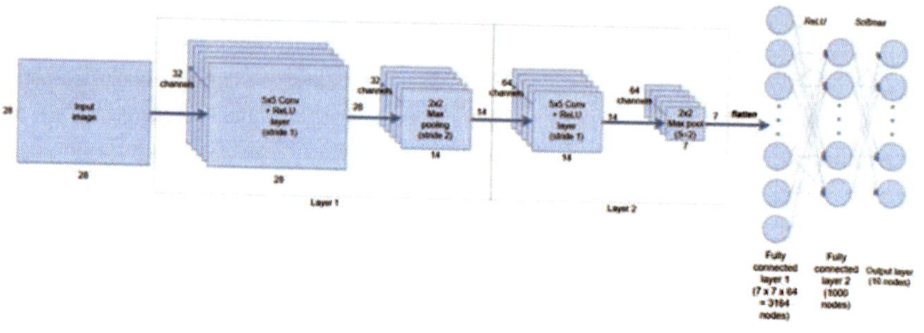

Fig. 1 Convolutional neural network architecture

later comparison. Content images are then included in the program. Each content image is accompanied by a pre-trained feature extractor, in which different layers of content are extracted. Image content is then passed through a transfer network, which provides a customized image. The styled image is also processed with a feature button, and the effects on the content and layers of style are saved. Custom loss function and our intentions both content and style define the quality of the styled image. The output elements of the styled image are compared to the original image. Output style features are compared to the style elements present in the index style images. After each step, only the transfer network is updated. The weights of a pre-trained trailer remain constant throughout. By measuring the different terms of the lost work, we can train models to produce images by producing a simple or sophisticated style.

Neural artistic style transfer model consists of two sub-models:

1. **Style transform model** is a neural network that creates a stylized image after taking an image and then applying a style bottleneck vector to the content image.
2. **Style prediction model** is a neural network that takes a style image as input to a 100-dimension style bottleneck vector.

In Fig. 1, in the first layer, using 32 filters, a convoluted neural network (CNN) can capture simple patterns, for example, a horizontal line or a straight line, which may not make sense to us but is very important to CNN. Slowly as you descend into the second layer, with 64 filters, CNN begins capturing more complex features that it could have by using its previous layer. It could be the face of a cat or a dog or a person or an inanimate object. Feature representation is the process of capturing various simple and complex aspects. The important thing to note here is that CNN even though they do not know what a particular image is can learn to encode in that image. The code type of convolutional neural networks (CNNs) assists us in the transmission of neural artistic style.

The VGG19 network, CNN with 19 layers, is widely used in neural artistic style transmission and is trained in millions (almost) images from the ImageNet database, as it has the ability to detect high-quality features in images. CNN's 'coding environment' is the key to neural artistic style transfer. Our first step will be to start a sound image, which will be our image created or output (G). The second step is to calculate the similarity of this image and style and image content in any particular layer of our VGG19 network. We calculate the loss of the created image (G) in relation to the content (C) and style (S) because we want our output/created image (G) to contain content and image style (C) and style (S), respectively.

3.1 Content Loss

How similar is the random image created randomly in the content image? We achieve this by calculating the loss of content using the proposed method below. Obviously, a hidden layer (L) is selected from the pre-trained CNN network (i.e., VGG network) to calculate the loss. Taking P to be the first image and F to be the generated image and F [I] and P [I] have both become part of the image of those images, loss of layer L content can be defined as given below.

$$\mathcal{L}_{\text{content}}(\vec{p}, \vec{x}, l) = \frac{1}{2} \sum_{i,j} \left(F_{ij}^{l} - P_{ij}^{l}\right) \tag{1}$$

3.2 Content Cost Function

The content cost function ensures that the content contained in the content image can be seen in the generated image. L {content} captures the square root error between using the generated image and the content image.

Different Feature maps in layer L

In Fig. 2, different channels/feature maps/filters at a particular chosen layer L can be seen. We will calculate how correlated these filters are to each other to capture

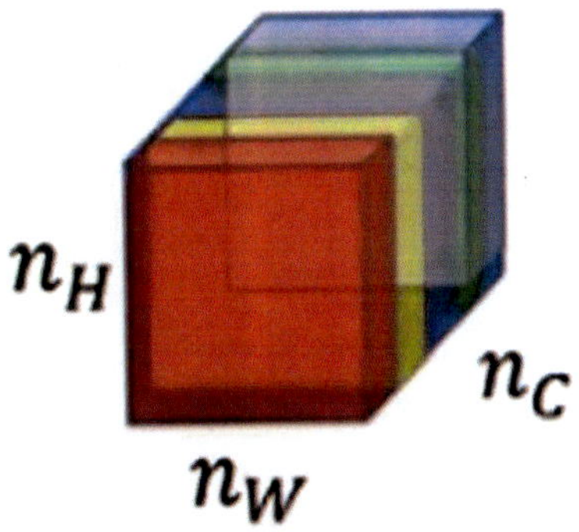

Fig. 2 Image depicting different feature maps in layer L

the style of an image. The distribution of features of a set of feature maps in a given layer is captured by the gram matrix where we match the distribution of features of two images by attempting to minimize the style loss between the two images [1, 2].

Estimating whether the two images are correlated or not

1. The channels are correlated if the dot product between the activation of two filters is large.
2. The channels are un-correlated if the dot product between the activation of two filters is small.

Gram matrix of style image (S): k and k' are the variables representing the different filters of the layer L. We can assume it to be Gkk″ [l][S].

$$J_{\text{style}} = \sum_{l}^{H} \sum_{j}^{W} \left(A_{ijk}[l][G] - A_{ijk'}[l][G]\right) \tag{2}$$

Gram matrix for generated image (G): The variables k and k' represent different filters or channels of the layer L. We can assume it to be Gkk′ [l][G].

$$J_{\text{style.generated}} = \sum_{l}^{H} \sum_{j}^{W} \left(A_{ijk}[l][G] - A_{ijk'}[l][G]\right) \tag{3}$$

Style loss: The cost function between the generated image and the style image is the square of the difference between the style matrix or gram matrix of the generated image and the style matrix or gram matrix of the style image.

$$J(S, G) = \frac{1}{\left(2H^{l}W^{l}C^{l}\right)} \sum_{k} \sum_{k'} \left(Gkk'^{[l][S]} - Gkk'^{[l][G]}\right) \tag{4}$$

3.3 *Style Cost Function*

Loss of function: All CNN layers are used to extract style information from the VGG network. Style details can be measured as the number of intersections between object maps in a particular layer. The difference at the crossroads between feature maps combined using image and built-in style is called loss. We are trying to calculate the style matrix for the made and style images. The root means the square difference between the two style metrics called the loss of style.

Total Loss Function for Neural Style Transfer
Total loss function is the sum of the cost of the style image and the content image which can be mathematically expressed as follows:

$$\mathcal{L}_{\text{total}}(\vec{p}, \vec{\alpha}, \vec{x}) = \alpha \mathcal{L}_{\text{content}}(\vec{p}, \vec{x}) + \beta \mathcal{L}_{\text{style}}(\vec{\alpha}, \vec{x}) \quad (5)$$

We can weigh content and style cost using alpha and beta, respectively, from the above equations which define the weightage of both costs in the output image generated. These are user-defined hyper-parameters which can control the amount of content and style to be put in the input image from the style and content images. The loss can be minimized using back-propagation that will optimize the random image into a meaningful art piece, given that we calculate the loss.

The neural style algorithms which were developed by Leon A. Gatys, Matthias Bethge, and Alexander S. Ecker have been implemented in this project that allows us to reproduce a selected image with a new artistic style. Three images are taken (input, content, and style), and the input is changed to resemble the content and style of the content and style images which are chosen, respectively [8, 9].

Underlying Principle

The underlying principle is simple to understand. Two distances need to be defined, namely content (DC) and style (DS). DC measures the difference in content between the two images while DC measures the difference in their styles. A third image is then taken along with the input and transformed to minimize the content distance with content image and style distance with style image. The necessary packages can then be imported to begin the neural transfer [1].

3.4 *Content Loss*

Content loss is a function that represents a version that weighs the content of each layer. We will install the content loss module directly after the convolution layers used to calculate the content range. This way every time an integrated neural network is inserted with an input image, content loss will be calculated in the desired layers, and due to auto grad, all gradients will be calculated. In order to make the content loss layer visible, we need to define a way forward that will calculate the content loss and restore the layer input. Combined losses will be saved as module parameters.

3.5 *Gradient Descent*

I used the L-BFGS algorithm to make a gradient drop. I want to train the input image to minimize content/style loss, as opposed to network training. So I will build a PyTorch L-BFGS optimizer. L-BFGS transfers the installation image to it as a well-built helper.

Fig. 3 Content image

Limited memory BSFGS L-BFGS or LM-BFGS is basically an optimization algorithm that uses a limited amount of computer memory to measure the Broyden–Fletcher–Goldfarb–Shanno (BFGS) algorithm. The L-BFGS algorithm is best known for measuring parameters in the field of machine learning.

4 Implementation and Results

In the proposed work, a 19-layer deep convolutional neural network (CNN), called VGG19 (VGG—visual geometric group), with 16 CONV layers and three FC layers is used. Stanford University's Standford Vision Laboratory pre-trained VGG19 on the ImageNet dataset used for implementation. Pooling reduces the spatial volume of feature vectors, thereby reducing the number of computations.

Neural style transfer allows us to blend two images: content image (Fig. 3) and style image (Fig. 4) together to create new piece of art as output image (Fig. 5) from the given input image.

5 Conclusion

In the proposed work demonstrated how neural style transfer works for images, the same approach can be extended to video. The neural style transfer has lot of applications in real world, and the accuracy of the model can be increased by using more perceptron.

Fig. 4 Style image

Fig. 5 Output image

References

1. Gatys LA, Ecker AS, Bethge M (2016) Image style transfer using convolutional neural networks. In: Proceedings of the IEEE conference on computer vision and pattern recognition, pp 2414–2423
2. Gatys LA, Ecker AS, Bethge M (2015) A neural algorithm of artistic style. arXiv preprint arXiv:1508.06576
3. Jing Y, Yang Y, Feng Z, Ye J, Yu Y, Song M (2019) Neural style transfer: A review. IEEE transactions on visualization and computer graphics 26(11):3365–85
4. Luan F, Paris S, Shechtman E, Bala K. Deep photo style transfer (2017) In: Proceedings of the IEEE conference on computer vision and pattern recognition, pp 4990–4998

5. Gatys LA, Ecker AS, Bethge M, Hertzmann A, Shechtman E (2017) Controlling perceptual factors in neural style transfer. In: Proceedings of the IEEE Conference on Computer Vision and Pattern Recognition, pp 3985–3993
6. Ulyanov D, Vedaldi A, Lempitsky V (2016) Instance normalization: The missing ingredient for fast stylization. arXiv preprint arXiv:1607.08022
7. Johnson J, Alahi A, Fei-Fei L (2016) Perceptual losses for real-time style transfer and super-resolution. In: European conference on computer vision, pp 694–711. Springer, Cham
8. Huang X, Belongie S. Arbitrary style transfer in real-time with adaptive instance normalization (2017) In: Proceedings of the IEEE International Conference on Computer Vision, pp 1501–1510
9. Ghiasi G, Lee H, Kudlur M, Dumoulin V, Shlens J (2017) Exploring the structure of a real-time, arbitrary neural artistic stylization network. arXiv preprint arXiv:1705.06830
10. Dumoulin V, Shlens J, Kudlur M (2016) A learned representation for artistic style. arXiv preprint arXiv:1610.07629
11. Gatys LA, Bethge M, Hertzmann A, Shechtman E (2016) Preserving color in neural artistic style transfer. arXiv preprint arXiv:1606.05897

Choice-Based One-Time Pad Approach Using Hybrid Cellular Automata

G. Kumaresan, C. N. Umadevi, and N. P. Gopalan

Abstract Cloud computing is a service-based technology which is accessed through the Web. Due to this reason, security and robustness are the important task in a public cloud. However, the present authentication mechanism does not give adequate protection in cloud technologies. So, this article tries to propose a choice- based one-time pad (OTP) encryption method using hybrid cellular automata for the open cloud. This proposed work assures confusion, diffusion, and avalanche requirements. It is identified that by consolidating checksum and hybrid cellular automata technique, a strong final key in the cloud could be produced. The executed results confirm that the proposed technique prevents chosen plaintext and brute force attacks. It has sufficient key space along with accessing time also improved in comparison with earlier methods. The theoretical analysis certified that the proposed hybrid cellular automata-based one-time pad key generation solution provides better security.

Keywords One-time pad · Cellular automata · Public cloud

1 Introduction

Today, cloud computing is a Web-based computing technology that is runtime accessible, virtualized, and the resources that are accessed over the World Wide Web. Anywhere, anytime, and anyplace, the cloud users can be accessed through the Internet. As per the NIST [1] description, "Cloud computing as a model for enabling ubiquitous, convenient, on-demand network access to shared pool of configurable computing resources that can be rapidly provisioned and released with minimal management

G. Kumaresan (✉)
School of Computer Science and Applications, REVA University, Bangalore 560064, India

C. N. Umadevi
Research and Development Centre, Bharathiar University, Coimbatore 641046, India

N. P. Gopalan
Department of Computer Applications, National Institute of Technology, Tiruchirappalli 620015, India

G. Gupta et al. (eds.), *Proceedings of Academia-Industry Consortium for Data Science*, Advances in Intelligent Systems and Computing 1411,
https://doi.org/10.1007/978-981-16-6887-6_2

effort or service provider interaction." Cloud computing consists of three typical services, namely platform as a service (PaaS), software as a service (SaaS), and infrastructure as a service (IaaS), and its claim can be processed to four deployment models. Figure 1 shows the private cloud, public cloud, hybrid cloud, and community cloud [2–5]. In addition, it has some important characteristics such as data storage, disaster recovery, scalability, pay-per-use, and application development and virtualization. Secrecy of the customer's information is very crucial one. So, cloud service provider (CSP) takes in charge to give protection for customer's data. Also, it is dense with many services which can help users to access their information from on-demand services [6]. Nowadays, cryptographic block encryption for key generation plays a vital role in an open cloud. Generally, the key generation methods are used to prevent the cloud user's secret information over the Web. Hence, conventional encryption technique does not give sufficient protection in public cloud [7–10]. So, there is a need for an efficient encryption method in Web applications. As stated in Kerckhoff's [11] principle, "secrecy of the key only provides security though the cryptographic system is more complicated." Hereafter, Shannon was reformed [12] as "enemy knows the system," Due to this reason, many researchers developed different techniques and a new approaches for key generation in open cloud [22], but with advanced attackers are out, there is knowledge to launch a successful attack against any security model. Thus, there is requiring for secret keys to strengthening in distributed cloud.

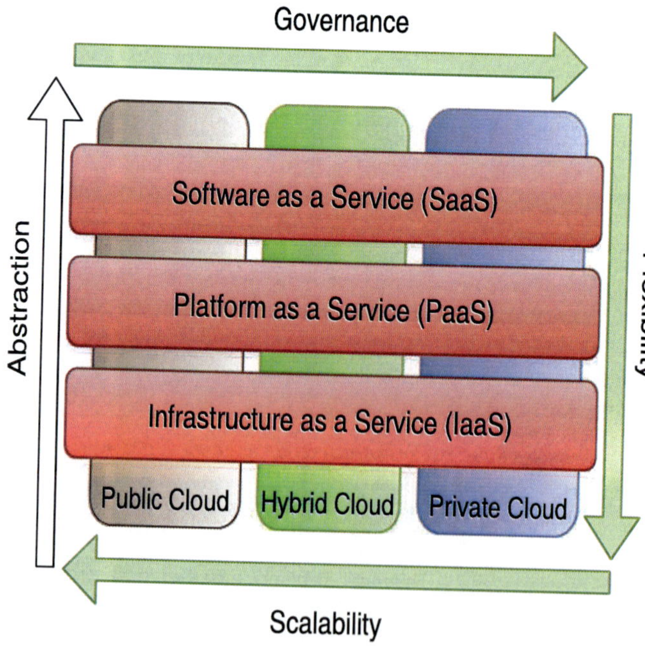

Fig. 1 Cloud computing model

1.1 *Cellular Automata*

A cellular automaton is all about a regular grid of cells that depends on neighborhood cells on its right, its left, and the cell itself [13]. As per the linear transformation method, in case if it is global map is returnable, then it is reversible one which established in two ways is called as encryption. For each reversible rule, consider as an inverse rule in the CA method. By executing CA rule in both ways like forward and backward and then iterating for n times, the original structures can be obtained which is shown in Fig. 2. For example, Rule 15 has an inverse to Rule 85. Cellular automata have become the vital technique to developing high-speed encryption-based hardware and software. One of the significant skin tones of using reversible cellular automata is the capability to generate highly suitable random numbers that are appropriate for high-speed applications [14], especially in the open cloud.

2 Related Work

Information security is the major concern in public cloud because of its openness, and it is subject to many kinds of attacks. The existing literatures are studied, many different type of risks are found, and they also address different solutions. The cryptographic researchers had presented many innovative methods to protect data of public cloud. Ruj et al. proposed [15] techniques to authenticate cloud data anonymously and a decentralized access control. In that work, the authors propose to authenticate the cloud users without any identity, and hence, authenticated users can only decrypt the data. The cloud knows the access policy, so the replay attacks are prevented. A performance improvement over this scheme was found in the work of Siddiqui et al. a three-factor, dynamic cloud-based user authentication system for telecare medical information system [16].

In this work, the smart phone of the user acts as a unique identity to access the services; thus, the DOS attack and password guessing attacks are efficiently prevented.

Rule 15	111	110	101	100	011	010	001	000
	0	0	0	0	0	0	0	0
N	1	0	1	1	1	0	0	0
N+1	1	0	1	0	0	0	1	1
N+2	0	0	1	0	1	1	1	0

Rule 85	111	110	101	100	011	010	001	000
	0	1	0	1	0	1	0	1
N	0	0	1	0	1	1	1	0
N+1	1	0	1	0	0	0	1	1
N+2	1	0	1	1	1	0	0	0

Fig. 2 Rule 15 has an inverse to Rule 85

But it suffers from high cost and increased computational complexity. A two-factor authentication system was found in the literature work of Nitin et al. using M-Pin server for secure cloud computing environment [17]. A crypt user name is used as the first factor, and the second factor is a password with ATM pin called M pin; thus, the replay attack and brute force attacks are prevented, but the system does not suit for multicloud environment. A dynamic two-stage authentication framework was proposed by Kumaresan et al. [18] to enhance the security in public cloud providing services to educational platforms. The major advantage of this scheme is less access time compared to the above-discussed schemes. The work found in [19, 20] provides cellular automata-based authentication techniques, and they are the recent one. The authentication methods using multi-factor authentication techniques suffer from severe computational complexity, time consuming and are subject to chosen plain-text and statistical attacks. The cellular automata-based user authentication technique proposed in this paper overcome these drawbacks and provides efficient techniques over the multi-cloud as well as public cloud.

3 Proposed Work

The major plan for the proposed work is to reinforce user protection in open cloud. The proposed model involves two levels such as registration and authentication levels which are shown in Fig. 3.

3.1 Registration Part

An authenticated cloud user's needs to register themselves by using personal computer, laptop, and smart phone which is shown in Fig. 3. The registered user details are stored as well as monitored by the cloud service provider who takes responsibility to create a unique identification number for each registered user and stored in the local server database. After activated of registration, an acknowledgment will be sent to the registered user via smart phone or email.

3.2 Authentication Part

The proposed technique gives admittance to the authenticated users only. Likewise, a choice-based process using hybrid cellular automata is proposed to reinforce the security in public cloud. Initially, the eight best cellular automata rules are stored in the present stack which is shown in Fig. 3.

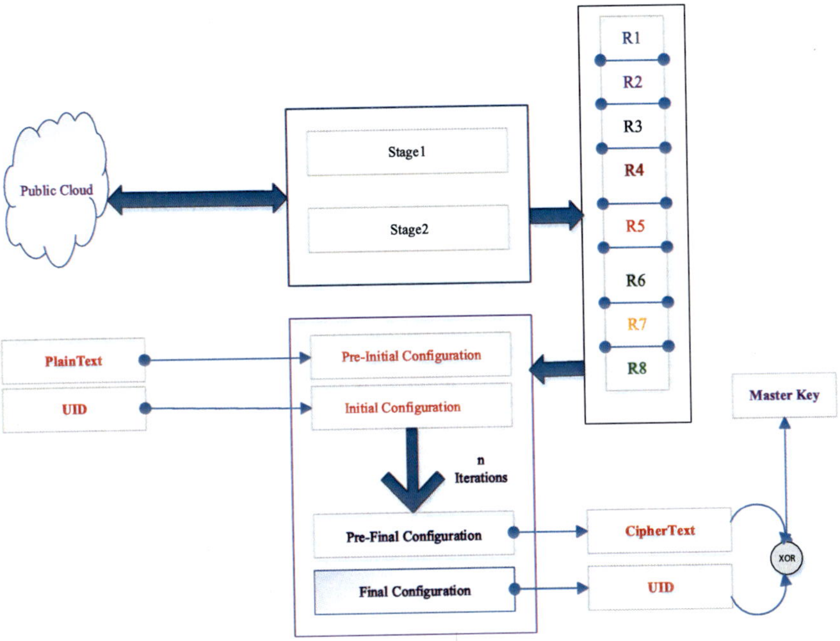

Fig. 3 Process of encryption

User Level Selection In the first level, the corresponding cloud user is randomly selecting a HCA rule from the present stack which is stored in the CA initial configuration.

CSP Machine Level Selection Similarly in the second level, the CSP machine is randomly choosing a HCA rule from rule vector matrix. Here, first iteration result is input of the second selection rule.

CSP Level Selection In the third level, the CSP randomly is choosing a HCA rule from rule vector matrix. Here, second iteration result is input of the third selection rule. After iterating 2128 times, it produces the output structure.

Content Owner Level Selection In the fourth level, the content owner randomly choosing a HCA rule from rule vector matrix. Here, third iteration result is input of the fourth selection rule. After iterating 2^{128} times, it produces the output structure.

3.3 *Working Method*

The corresponding cloud user to enter their username and password to the cloud service access page. Hence, the password (plaintext) data block takes as an input of the pre-initial CA configuration, and the corresponding unique identification number (UID) of the particular user takes as an input of the initial CA configuration. Then, the

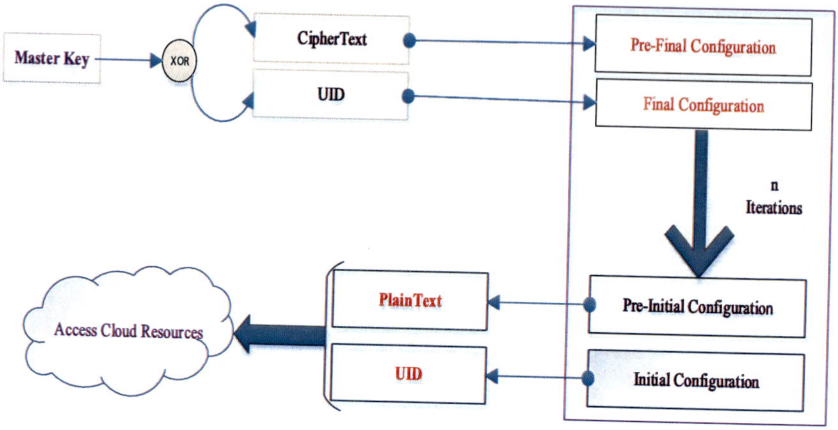

Fig. 4 Process of decryption

selected rules are iterated in 2^{128} times, and it gives the two output configurations such as pre-final and final configurations. These two output configurations are XORed and are added to checksum values as secret key. In the end, the result is captured in the form of binary values denoted as a master key which is converted into decimal and sent to the registered users. In case, the authenticated user's has entered the receiving key, then they can allow accessing the cloud resources or else the request will be rejected. The decryption process is like encryption step which is shown in Fig. 4. The proposed model achieves better security and improves response time for accessing data, in comparison with the earlier method.

4 Mathematical Analysis

4.1 Correctness of Proposed Key Generation

Let the logistic function

$$a_{n+1} = \delta \times a_n(1 - a_n)$$

be the key seed value for the proposed methodology with the interval from $0 < a_n < 1$ and $3.99 < \delta < 4$. Here, δ represents the growth rate and a_n is a population at time n. Hence, proposed method obtained the following matrix:

$$\mathbf{K}_{P\times P} = \begin{bmatrix} 0.817 & 0.33. & 0.151 & 0.752 \\ 0.569 & 0.177 & 0.262 & 0.359 \\ 0.424 & 0.391 & 0.165 & 0.454 \\ 0.719 & 0.823 & 0.119 & 0.658 \end{bmatrix}$$

Hence, every value in the matrix multiplies with 255.

$$\mathbf{K}_{P\times P} = \begin{bmatrix} 208\ 85\ 39\ 192 \\ 145\ 45\ 67\ 92 \\ 108\ 100\ 42\ 116 \\ 183\ 210\ 30\ 168 \end{bmatrix}$$

The obtained square matrix converted into binary value and is applied into 1's complement method:

$$\mathbf{K}_{P\times P} = \begin{bmatrix} \mathbf{11010000}\ 01010101\ 00100111\ 11000000 \\ 10010001\ 00101101\ 01000011\ 01011100 \\ 01101100\ 01100100\ 00101010\ 01110100 \\ 10110111\ 11010010\ 00011110\ 10101000 \end{bmatrix}$$

Hence, the proposed method takes first binary values of the matrix as input for the cellular automata structure. Therefore,

$$\text{CA1} = 11010000$$

According to the following equation, the proposed method uses to select the CA rules in the vector matrix.

$$\text{C} = |x_n \times 7| + 1$$

where a_n is obtained from logistic function.

First Choice: In case the user select rule 153 ($M_i^{t+1} = M_i^t \text{xnor} M_{i+1}^t$) from rule vector matrix the proposed method obtained the following operations:

$$\mathbf{CA1'} = 10001110$$

Where CA1′ represents first selection first iteration result.

$$\mathbf{CA1''} = 01101100$$

where CA1″ represents first selection second iteration result.

Second Choice: Similarly the CSP machine selects rule 101 ($M_i^{t+1} = M_{i-1}^t \text{nor} M_{i+1}^t$) from rule vector matrix, the following operations are obtained:

$$\mathbf{CA2} = 01101100$$

$$\mathbf{CA2'} = 00000001$$

where CA2′ represents second selection first iteration result.

$$\mathbf{CA2''} = 01111101$$

Where CA2″ represents second selection second iteration result.

Third Choice: CSP selects rule 86 ($M_i^{t+1} = [M_{i-1}^t \text{nor} M_i^t] \oplus [\text{not}(M_i^t)]$) from rule vector matrix the following operations are obtained:

$$\mathbf{CA3} = 01111101$$
$$\mathbf{CA3'} = 10000010$$

where CA3′ indicates third selection first iteration result.

$$\mathbf{CA3''} = 01000001$$

where CA3″ represents third selection second iteration result.

Fourth Choice: Content owner selects rule 105 [($M_i^{t+1} = \text{not}[M_{i-1}^t \text{xor} M_i^t \text{xor}$ $(M_{i+1}^t)]$) from rule vector matrix the following operations are obtained:

$$\mathbf{CA4} = 01000001$$
$$\mathbf{CA3'} = 10011100$$

where CA4′ indicates fourth selection first iteration result.

$$\mathbf{CA4''} = 00010100$$

where CA4″ represents fourth selection second iteration result.

4.2 Correctness of Proposed OTP Encryption

Suppose the user plaintext is "Two One Nine Two," and it becomes:

$$\mathbf{PT} = \begin{bmatrix} \mathbf{54} & 4F & 4E & 20 \\ 77 & 6E & 69 & 54 \\ 6F & 65 & 6E & 77 \\ 20 & 20 & 65 & 6F \end{bmatrix}$$

Hence, the proposed method takes first hexadecimal values of the matrix as plaintext input in the one-time process. Therefore, PT=01010100, and initial key I_k is: 00010100 which is obtained by one-time pad key $\mathbf{K}_{\text{otp}}$

$$\mathrm{I}_k = 00010100$$
$$\mathbf{K}_{\text{otp}} = 01010100 \oplus 00010100$$
$$\mathbf{K}_{\text{otp}} = 01000000$$

Add checksum into OTP key $\mathbf{K}_{otp}$. Hence, the proposed method computes the checksum value:

$$\mathbf{K}_{otp} = 0 + 1 + 0 + 0 + 0 + 0 + 0 + 0 (\bmod 2) = 1$$

Then, the proposed method adds an odd bit $\boxed{1}$ into $\mathbf{K}_{otp}$ and send to destination.

$$\mathbf{K}_{otp} = 01000000\boxed{\mathbf{1}}$$

4.3 Correctness of Proposed Decryption

After getting key K_{otp} with parity bit $\boxed{1}$, the user needs to calculate and ensure the proper transmission in the destination.

$$K_{otp} = 01000000 + \boxed{1}$$
$$K_{otp} = 0 + 1 + 0 + 0 + 0 + 0 + 0 + 0 + 1 (\bmod 2) = 0$$

The proposed technique acknowledges proper transmission after observing likelihood even result (i.e., odd bit). The registered user needs to enter the proper code and found the following mode of executions. The decryption process is similar to encryption process which is done in earlier stage of the proposed method. Therefore, the proposed techniques obtain the following initial key and plaintext.

Plaintext: 01010100

Initial Key: 00010100

4.4 Avalanche, Confusion, and Diffusion Properties

As per the security requirement, the master key should fulfill three important requirements such as diffusion, confusion, and avalanche. When an input changes a tiny and the output changes extremely, this is called avalanche criterion. Suppose that the tiny changes in the plaintext or master key should origin the amazingly change in the ciphertext. The proposed technique fulfills avalanche property using HCA rules in the RCA configurations. Confusion is the connection between ciphertext and master key; it should be complex as much as possible which fulfills this property by using RCA and it makes the complex manners. Diffusion property is the connection between ciphertext and plaintext and should be complex as essential one. As a result, the RCA method and hybrid rules fulfill the needs. Thus, the proposed technique satisfies the basic three requirements, and it prevents possible attacks in the cloud.

4.5 *Key Space Analysis*

Based on the user's requirement, the different keys are generated in the proposed method. Here, authenticated users password and unique identification number of corresponding user details are combined, and it can be used as the key initiator. Every key bears 128-bit key length, and the total key size is 2^{128} and is enough to avert all possible attacks. The CA holds 256-bit for each block and is used for encryption which is having $2^{8X256X256}$ space size. Hence, the proposed method is to avert brute force attack.

5 Result and Discussion

As per our experiment, the proposed technique has been analyzed in open virtual cloud environment. The response times are calculated based on cloud service provider database records. The obtained results are calculated using (5) and are compared with earlier methods.

$$\text{Speed (\%)} = \frac{\text{EF} - \text{PF}}{\text{EF}} \times 100 \quad (1)$$

where EF indicates existing factor and PF indicates proposed factor. The total number of records has been compared with earlier methods which are given in Table 1 and shown in Fig. 5. It has been identified that based on the number of records there is an improvement in speed percentage which is shown in Fig. 6. It is understood that as the number of records increased the improvement speed is decreased and vice versa. In the end, it protects all possible attacks in an open cloud.

It has been noticed that depending on the number of records there is an improvement in speed percentage which is shown in Fig. 6. It is understood that as the number of records increased the improvement speed is decreased and vice versa. In the end, it averts all possible attacks in the public cloud.

Table 1 Comparison of response time (*ms*)

Total records	EF [18]	EF [21]	PF	Speed (%)
19	13.9	15.1	10.2	0.26
14	9.0	10.5	7.5	0.16
10	8.7	9.9	6.0	0.31
6	3.7	5.8	2.9	0.21
2	2.6	4.4	1.5	0.42

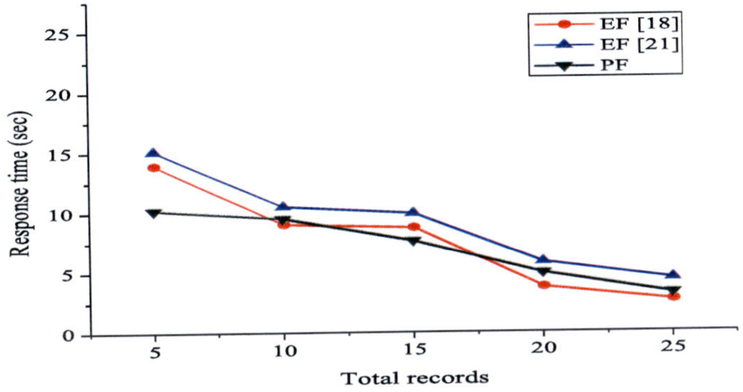

Fig. 5 Comparison of response time

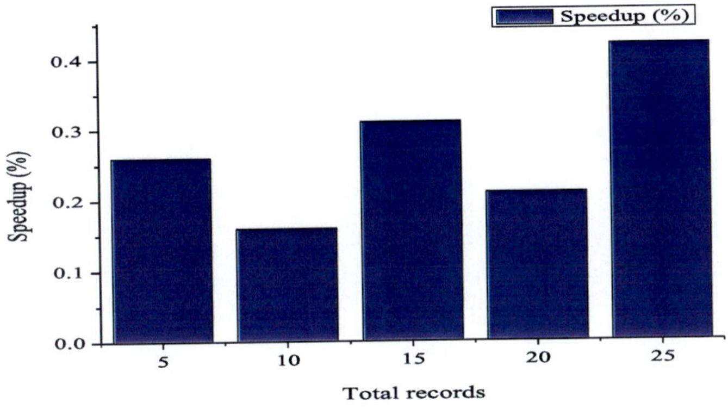

Fig. 6 Speed up percentage

6 Conclusion

In this article, hybrid cellular automata-based one-time pad key generation technique has been analyzed and proposed in the virtual cloud. The proposed technique created key prevents chosen plaintext and brute force attacks, and it is fulfilled diffusion, confusion, and avalanche requirements. In addition, the proposed technique provides enhanced security and improves accessing time in comparison with earlier techniques.

References

1. Mell P, Grance T (2009) The NIST definition of cloud computing version 15 technical report. Comput Inf Sci 53(6):1–10
2. Vaquero L, Rodero-Merino L, Caceres J, Lindner M (2009) A break in the clouds toward a cloud definition. ACM Sigcomm Comput. Commun. Rev. 39:50–55

3. Badger L, Bohn R, Chu S, Hogan M, Liu F, Kaufmann V, Jian (2011) Useful information for cloud adopters. NIST Cloud Comput Program 2(1):1–73
4. Buyya R, Yeo C, Vengopal S (2008) Market oriented cloud computing vision, hype, and reality for delivering IT services as computing utilities. In: IEEE international conference on high performance computing and communications, pp 5–13
5. Buyya R, Yeo C, Venugopal S, Broberg J, Brandic I (2009) Cloud computing and emerging IT platforms vision hype and reality for delivering computing as the 5th utility. Further Gener Comput Syst 25:599–616
6. Subashini S, Kavitha V (2011) A survey on security issues in service delivery models of cloud computing. J Netw Comput Appl 1–11
7. Viega J (2009) Cloud computing and common man. Computer 42(8):106–08
8. Wang C, Wang Q, Ren K (2009) Ensuring data storage security in cloud computing. In: Proceedings of the international workshop on quality of service (IEEE)
9. Whitman ME (2004) In defense of the realm: Understanding the threats to information security. Int J Inf Manag 24(1):43–57
10. Zissis D, Lekkas D (2012) Addressing cloud com-putting security issues. Future Gener Comput Syst 28:583–592
11. Kerckhoff's A (1883) La cryptographie militaire. J Sci Militaires 9:161–191
12. Shannon C (1949) Communication theory of secrecy systems. Bell Syst Tech J
13. Kumaresan G, Gopalan NP (2017) An analytical study of cellular automata and its applications in cryptography. Int J Comput Netw Inf Secur 9(12):45–54
14. Anghelescu P (2011) Encryption algorithm using programmable cellular automata. In: IEEE world congress on internet security, pp 233–239
15. Ruj S, Stojmenovic M, Nayak A (2014) Decentralized access control with anonymous authentication of data stored in clouds. IEEE Trans Parallel Distrib Syst 25(2):384–394
16. Siddiqui Z, Abdullah AH, Khan MK, Alghamdi AS (2014) Smart environment as a service: three factor cloud based user authentication for telecare medical information system. J Med Syst 38:9997
17. Nagar Nitin, Sumen Ugrasen (2014) Two factor authentication using M-pin server for secure cloud computing environment. Int J Cloud Appl Comput 4(4):42–54
18. Kumaresan G, Veeraragavan N, Arockiam L (2015) A dynamic two stage authentication framework to enhance security in public educloud. Int J Appl Eng Res 10(82):126–131
19. Song S, Lee J, Lee JK, Jun MS (2015) User authentication method design based on biometrics in a multi-cloud environment. J Adv Comput Sci Ubiquit Comput 373:283–288
20. Liu JK, Au MH, Huang X, Lu R, Li J (2016) Fine grained two factor access control for web based cloud computing services. IEEE Trans Inf Forensics Secur 11(3):484–497
21. Kumaresan G, Gopalan NP (2017) Educloud: a dynamic three stage authentication framework to enhance security in public cloud. Int J Eng Manufa 6:12–26
22. Kumaresan G, Gopalan NP (2020) An additive cellular automata based random key generation for one-time pad encryption using luhn method. J Appl Secur Res 15:255–269

Novel Imaging Approach for Mental Stress Detection Using EEG Signals

Swaymprabha Alias Megha Mane and Arundhati A. Shinde

Abstract The mental stress level estimation using EEG signal is challenging task. The neural network approach can provide better solution over other classical approaches. The main challenge involved in EEG signal processing is irregularity in signal data shapes as fixed data shapes are required as input to the convolutional neural networks. The image-based spectrogram processing approach can solve the shape issue and also multiparametric dependency of brain state estimation along with better efficiency in convergence capabilities of the network models. This paper contributes to show the stress detection using convolutional neural networks using images obtained from azimuthal projections of EEG signals. These projected 2D images are processed using convolutional neural network for stress detection. This performance is evaluated with varying number of convolutional layers in the model. The results obtained show 93% accuracy of mental stress detection obtained using DASPS database of EEG dataset.

Keywords Mental stress · EEG · CNN · Azimuthal projection · 2D image

1 Introduction

Brain state detection is a new area for the researchers which are used to define the treatment patterns for the people suffering from stress-related issues. Medical assistive technology using EEG signal-based analysis shows the huge scope for the researchers in signal processing domain. Prediction-based methods which make use of neural network platforms have shown significant impact in terms of improved accuracy levels. The use of neural network models has variety of approaches in which signals can be processed directly as one dimensional vectors or these signals can be converted into spectrograms and these spectrograms can be treated as images. Such images are processed using convolutional neural networks. Treating each signal spectrogram differently is possible when signals are processed in frequency domain and shows more details while classifying each signal from others in significantly

S. A. M. Mane (✉) · A. A. Shinde
Bharati Vidyapeeth (Deemed to be University), College of Engineering, Pune, India

G. Gupta et al. (eds.), *Proceedings of Academia-Industry Consortium for Data Science*, Advances in Intelligent Systems and Computing 1411,
https://doi.org/10.1007/978-981-16-6887-6_3

improved manner. While performing such classification operations, the capability of the networks in terms of accuracy and processing complexities while configuring the models are the two important aspects required to be considered. Whenever there is feasibility of obtain simple aspects of differing signal characteristics in some other possible conversions such as spectrogram images, then such approach shows more impact in terms of accuracy and reduced complexities.

Stress in particular person has variety of causes and identifying and discussing each one of them is huge work and hence not addressed here and should be considered as medical side work. The stress can be seen with the stimulation which directly depends on anxiety. The stress level of particular person directly depends on the involvement of that person in particular task, when task dependent stress is taken into consideration. Other emotional impacts in day to day life of particular person that cause increase in stress levels is not the focus of the experimentation involved in this paper. Rather the experimentation for stress detection using stimulated effects in the brain in terms of anxiety, it is considered in this work.

The work consists of various stages in which recorded EEG signals are processed to obtain 2D images, and then, these images are processed using convolutional neural network for detecting stress.

2 Related Work

The EEG recording using 10–20-based EEG recoding systems has shown significant improvement for the identification of brain activities and diseases. The noncontact methods of EEG signal capturing and then processing for characteristic feature identification with respect to disease or mental states such as alertness, dizziness, sleep state, and stress or anxiety is becoming popular in medical assistive technologies. The EEG processing methods mainly comprise stages such as preprocessing, feature extraction, and classification. The classification with the use of nonlinear classifiers such as K-nearest neighbor, support vector machine, or other machine learning-based algorithms, or deep neural network-based methods are significantly showing satisfactory performances. The brain state or activity identification using EEG signal or other medical diagnosis methods involving signal processing and neural network platforms are addressed here to form the platform toward contribution of proposed work.

These days' medical imaging-based EEG signals processing are showing huge impact in the market due to its promising results due to its virtualization technology in which EEG signals directly get affected with respect to task without actually performing it [1]. Most of the time dimensionality is the important aspect that creates problem while dealing with EEG signals and their respective characteristic features [2]. Such problems are caused by means of time intervals of recordings, presence of artifacts, nonstationary signals, and channel correlations when multichannel recording paradigms are involved. The problems associated with practical world EEG signal recordings while using motor imagery (MI) technology may involve sensor developments, signal processing techniques, robustness, reliability, and complexity of technology used to obtain actual signals [3]. For long period of time, it can be seen that researchers have preferred use of traditional neural network platforms. These networks have shown processing methods for time series data, images, and videos. But these networks shown inadequacy of the processing methodologies due to limited scopes for the improvements in accuracy as features of the input data were dependent hand crafted rules [4]. On the other hand, deep learning (DL)-based methods have shown adequate level of accuracy along with robustness compared to traditional neural networks which were time consuming processing along with proper tuning of the outcomes [5] and suffer from losing information [6]. Lee and Choi [7] used the continuous wavelet transform (CWT) to construct 2D images for training a CNN model. Deep learning involves large number of samples in the dataset along with intense floating point matrix multiplications [8–10]. The huge computational requirements due to large datasets were also been tacked with the use of graphical processing units (GPUs). Though involvement of complexities, these methods using DL have shown significant diagnosing capabilities [11]. Machine interfaced signals have been processed by large number of researchers using deep neural networks (DNNs) [12] along with wide range of applications such as controlling wheelchair or robots [13], helping paralyzed patients to walk [14], upper limb exoskeleton [15], and many others [3]. Also, MI signals have small amplitudes and mixture of artifacts degrade the performance of the systems [16, 17].

Li et al. [18] have shown the use of CNN for classification of signals by extracting temporal features from EEG signals. The CNN was adopted for extracting high-level features. The 2D image obtaining method is deduced for the use of CNNs by Alazrai et al. [1]. The energy distribution over time and frequency domain with the use of FFT techniques were possible using such images [19]. VGG-16 model of CNN was used by Xu et al. [4]. The transfer learning method was adopted for training the model. Zhao et al. [20] proposed a multibranch CNN for the classification of MI signals. In their network, the EEG data are applied to three separated CNNs in parallel fashion then quantifying the final classification result based on summing the results retained by the three layers of convolutional layers. Tabar and Halici [21] suggested a deep network that is composed of CNN and stacked auto encoders (SAE). They used the short-time Fourier transform (STFT) to construct 2D images for training their network. The features within the MI signals were extracted by the CNN then they classified using the SAE. Murugappan et al. [22] have processed ECG signals for heart disease detection with the analysis of PQRST features of the ECG in which R to tend features are focused to identify the sudden cardiac arrest from recorded ECG signals. This method is addressed here as it uses K-nearest neighbor and support vector machine classifiers for heart disease detection using ECG signals. The extreme levels of hallucination, unreal behavior type of symptoms are termed as schizophrenia condition of the brain which can be analyzed using EEG signals [23]. Palani et al. [23] have given use of radial basis function support vector machine (RBF-SVM) classifier for the detection of schizophrenia condition with processing of EEG signals. Multivariate empirical mode decomposition for obtaining intrinsic mode function signals from recorded multichannel EEG is the processing approach used for feature extraction. The f1 score obtained is almost 93% with this method. The method is suitable when distinctive features of schizophrenia condition are captured in the EEG signals.

3 Proposed Work

The proposed work of stress detection is shown in block diagram in Fig. 1.

The EEG recorded dataset is obtained from DASPS database [24]. The dataset consists of different anxiety levels obtained through psychological stimulations. The study and experimentation of EEG signals in terms of alpha, beta, and theta signals projections are obtained in the form of azimuthal equidistant projection [25] for converting signals into 2D projections for patterns analysis. The steps in this study and experimentation include:

3.1 Preprocessing

EEG signals are preprocessed for converting into 2D images which comprise following steps.

1. **Hanning Window**: First, the data were chopped up into overlapping 1-s "frames" and a Hanning window were applied.
2. **Fast Fourier Transform (FFT)**: FFT was applied to transform data for each frame from time domain to frequency domain.
3. **Frequency Binning**: FFT amplitudes were grouped into theta (4-8 Hz), alpha (8-12 Hz), and beta (12-40 Hz) ranges, giving 3 scalar values for each probe per frame.

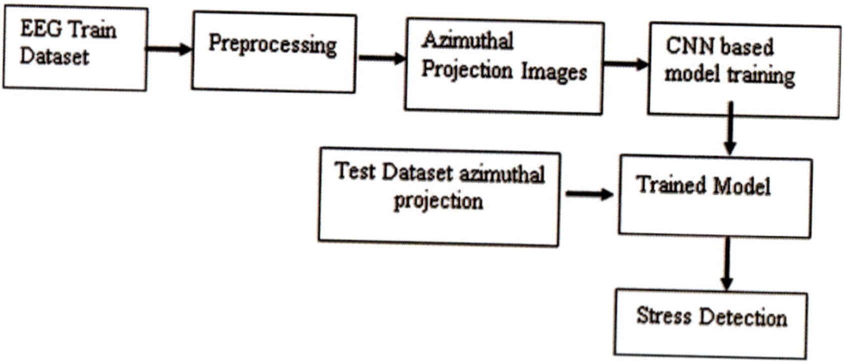

Fig. 1 Proposed system block diagram

3.2 2D Azimuthal Projection

The theta, alpha, and beta values can be represented as RGB channel values with different levels of intensity mixing phenomenon in 2D map of the head. A circular shaped canvas is considered for projecting the color mixings. The projected image is based on platform graph which shows the arranged shape of EEG wearable electrode cap over the scalp.

Azimuthal equidistant projections (AEPs) as shown in [25]. As per the concept from azimuthal projections, the projections are usually either perpendicular to the pole and equator and any other point lying intermediately. The shape of cap can be considered as approximated sphere and the same can be used to obtain projections of the signals on 2D surface as an image.

The projections are not with respect to locations of the electrodes as may incase where electrode positions-based projections are obtained. On the other hand, in our work, here, the input signals are divided in frequency bands using fast Fourier transform (FFT) and then different intensity variations are obtained in the image when projected using azimuthal projection concept. As per method by Clough-Tocher scheme (Alfeld, 1984), which interpolates the scattered power of signals over scalp region over a 32 × 32 mesh. The point here to be considered is there is consideration of covering of entire scalp region using 10–20 system electrodes which may constitute 32 electrodes and their respective signal catching. The process is continued until all the frequencies within specific band are used and projected in terms of image with color intensities. Figure 2 specifies the illustration of mental state projection using theta, alpha, and beta signals. Also, variations in signal color intensities can be easily observed when mental state is observed while performing known task causing less level of stress and then unknown task causing more stress on the brain. The point here to be considered is that when brain activity is monitored for the person involved in

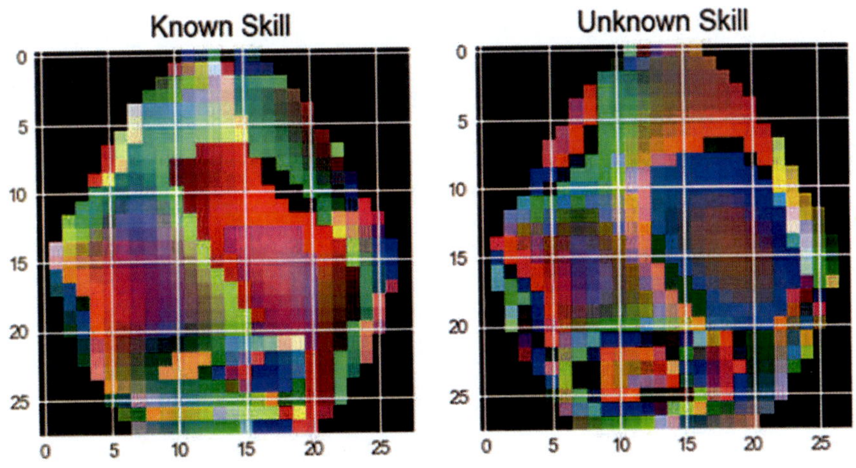

Fig. 2 Azimuthal projection of input EEG signal

known task and unknown task, it is the anxiety level of brain which results further in stress as per illustrations by medical experts. Although this particular view is required to be considered for the experimentation using neural network for classification of stress in brain, the input dataset which is recorded for anxiety level is used.

Figure 2 shows the azimuthal projection for EEG signal recorded during handling the tasks which give more and less stress obtained through known skill and unknown skill.

Every azimuthal projection output is treated as 2D image that is to be processed using convolutional network for stress classification task. The size of 2D images is set to 28 × 28.

3.3 Implementation of CNN Classifier for 2D Azimuth Projections

CNN model architecture to classify the EEG signals based on known task and unknown task and resulting stress and normal state of mind are implemented. The sequential model obtained through implementation of Python using Keras and TensorFlow as platform tools is shown in Fig. 3.

The proposed network model is configured for training using configuration parameters as given in Table 1.

4 Results and Analysis

The training process can be observed using loss rate analysis (Fig. 4).

Detection of stress on test dataset. The test dataset is prepared by splitting the total dataset in 80–20 form and 20% is used for testing purpose. Few sample results of which are shown in Table 2.

Table 3 shows the formulae for different parameters.

Where true positive (TP) stands for input with stress and detected as stress, when input is stress and detected as normal, it is considered true negative (TN). Also, false positive (FP) stands for input normal and detected as stress and finally, false negative (FN) stands for input normal and detected as normal.

4.1 Performance Evaluation

Performance evaluation is shown in Table 4 for the prediction of mental stress on test data. The CNN-based deep network shows better performance compared to all other classical classifiers.

Layer (type)	Output Shape	Param #
conv2d_1 (Conv2D)	(None, 28, 28, 32)	896
activation_1 (Activation)	(None, 28, 28, 32)	0
conv2d_2 (Conv2D)	(None, 26, 26, 32)	9248
activation_2 (Activation)	(None, 26, 26, 32)	0
max_pooling2d_1 (MaxPooling2	(None, 13, 13, 32)	0
flatten_1 (Flatten)	(None, 5408)	0
dense_1 (Dense)	(None, 10)	54090
activation_3 (Activation)	(None, 10)	0
dense_2 (Dense)	(None, 2)	22
activation_4 (Activation)	(None, 2)	0

Total params: 64,256
Trainable params: 64,256
Non-trainable params: 0

Fig. 3 Proposed CNN model for mind stress detection

Table 1 Parameter configuration

Parameter	Value
Total number of convolution layers	2, 4, 6, 8, 10, 12, 14
Number of epochs	100
Steps per epoch	10
Optimizer	Adam
Metric	Accuracy
Training-validation split	80–20

The performance of sensitivity and specificity is seen less in model with 2 convolutional layers, and hence, the increment in number of convolutional layers is done with reference to which performance is seen increased with increased processing time.

The comparative of the results is shown in Table 5 and graph in Fig. 5 for the test data.

It is seen that increment in number of convolutional layers is responsible to increment in performance up to 10 number of layers and it remains same for further

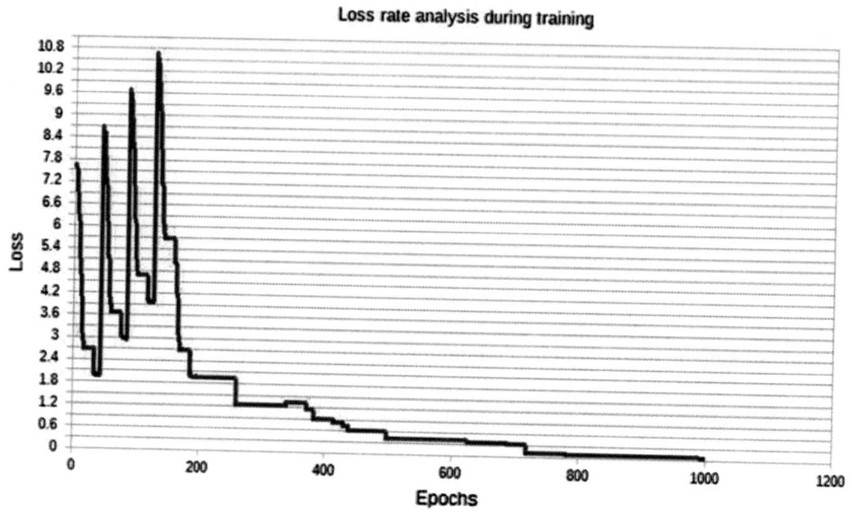

Fig. 4 Loss rate analysis

increment in count that is up to 14 layers. This analysis shows that number of convolutional layers are required to be set at optimum value for image-based analysis of EEG signals represented using azimuthal projections.

5 Conclusion

This paper contributes in terms of a novel approach for mental stress detection using EEG signal records. The stress level is stimulated using task performing works as specified in DASPS dataset. The 2D azimuthal projection shows the characteristic features appearing in the projected images and then processing these images using CNNs. The classification accuracy of stress and normal images is obtained and shows almost 93% accuracy level which is satisfactory. The stress detection era is new in the age, and hence, comparative study along with combinations of various methods in terms of more experimentation is the requirement in coming future.

Table 2 Performance based on confusion matrix

Vector name	Actual type	Detected result	Nomination
INPUT1	Stress	Stress	TP
INPUT2	Stress	Stress	TP
INPUT3	Stress	Stress	TP
INPUT4	Stress	Stress	TP
INPUT5	Stress	Stress	TP
INPUT6	Stress	Stress	TP
INPUT7	Stress	Normal	TN
INPUT8	Stress	Normal	TN
INPUT9	Stress	Stress	TP
INPUT10	Stress	Stress	TP
INPUT11	Stress	Stress	TP
INPUT12	Stress	Stress	TP
INPUT13	Stress	Stress	TP
INPUT14	Normal	Stress	FP
INPUT15	Normal	Stress	FP
INPUT16	Normal	Normal	FN
INPUT17	Normal	Normal	FN
INPUT18	Normal	Normal	FN
INPUT19	Normal	Normal	FN
INPUT20	Normal	Normal	FN
INPUT21	Normal	Stress	FP
INPUT22	Normal	Normal	FN
INPUT23	Normal	Normal	FN
INPUT24	Normal	Normal	FN
INPUT25	Normal	Normal	FN
INPUT26	Normal	Stress	FP
Performance			TP = 11 FP = 3 TN = 4 FN = 9

Table 3 Parameters formulae

Parameter	Formula
Sensitivity	TP/(TP + FN)
Specificity	TN/(TN + FP)
Accuracy	TP + FN/(TP + TN + FP + FN)
Precision	TP/(TP + FP)
Recall	TP/(TP + FN)
F1 Score	2 * ((Precision * Recall)/(Precision + Recall))

Table 4 Performance evaluation

Accuracy	Sensitivity	Specificity	F1 score
0.86	0.55	0.57	0.6285

Table 5 Performance evaluation with respect to number of convolutional layers

Number of convolutional layers	Accuracy	Sensitivity	Specificity	F1 score	Processing time level
2	0.86	0.55	0.57	0.6285	Low
4	0.875	0.64	0.63	0.67	Low
6	0.89	0.67	0.744	0.685	Medium
8	0.91	0.71	0.78	0.71	High
10	0.934	0.78	0.81	0.76	High
12	0.934	0.78	0.81	0.76	High
14	0.934	0.78	0.81	0.76	High

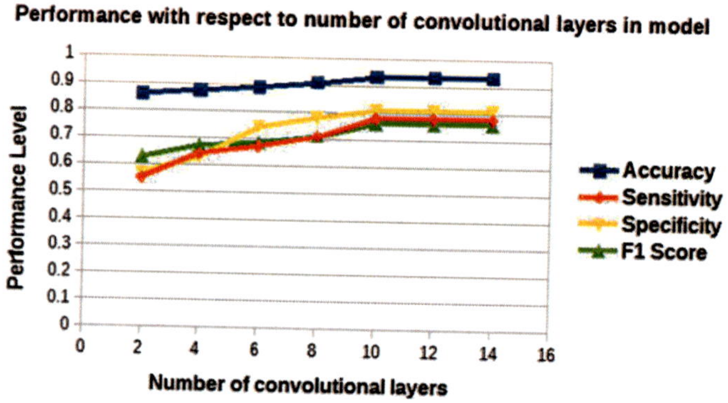

Fig. 5 Performance graph with respect to convolutional layers

References

1. Alazrai R, Abuhijleh M, Alwanni H, Daoud MI (2019) A deep learning framework for decoding motor imagery tasks of the same hand using EEG signals. IEEE Access 7:109612–109627. https://doi.org/10.1109/access.2019.2934018
2. Sakhavi S, Guan C, Yan S (2018) Learning temporal information for brain-computer interface using convolutional neural networks. IEEE Trans Neural Netw Learn Syst 29(11):5619–5629
3. Padfield N, Zabalza J, Zhao H, Masero V, Ren J (2019) EEG-based brain-computer interfaces using motor-imagery: techniques and challenges. Sens (Switz) 19(6):1–34. https://doi.org/10.3390/s19061423
4. Xu G et al (2019) A deep transfer convolutional neural network framework for EEG signal classification. IEEE Access 7:112767–112776. https://doi.org/10.1109/access.2019.2930958.
5. Litjens G et al (2017) A survey on deep learning in medical image analysis. Med Image Anal 42:60–88. https://doi.org/10.1016/j.media.2017.07.005

6. Qiao W, Bi X (2019) Deep spatial-temporal neural network for classification of EEGbased motor imagery. In: ACM international conference proceeding series, pp 265–272. https://doi.org/10.1145/3349341.3349414
7. Lee HK, Choi Y (2019) Application of continuous wavelet transform and convolutional neural network in decoding motor imagery brain-computer interface. Entropy 21(12):1–12. https://doi.org/10.3390/e21121199
8. Majidov I, Whangbo T (2019) Efficient classification of motor imagery electroencephalography signals using deep learning methods. Sens (Switz) 19(7):1–13. https://doi.org/10.3390/s19071736
9. Khan SH, Hayat M, Porikli F (2019) Regularization of deep neural networks with spectral dropout. Neural Netw 110:82–90. https://doi.org/10.1016/j.neunet.2018.09.009
10. Nurvitadhi E et al (2017) Can FPGAs beat GPUs in accelerating next-generation deep neural networks? In: FPGA 2017—proceedings of the 2017 ACM/SIGDA international symposium on field-programmable gate arrays, pp 5–14. https://doi.org/10.1145/3020078.3021740
11. Faust O, Hagiwara Y, Hong TJ, Lih OS, Acharya UR (2018) Deep learning for healthcare applications based on physiological signals: a review. Comput Methods Programs Biomed 161: 1–13. https://doi.org/10.1016/j.cmpb.2018.04.005
12. B. Rim, N. Sung, S. Min, M. Hong, Deep learning in physiological signal data: a survey, Sensors (Switzerland) 20 (4) (2020) 969, https://doi.org/10.3390/s20040969
13. Ron-angevin R, Velasco-álvarez F, Fernández-rodríguez Á, Díaz-estrella A, Blanca-mena MJ, Vizcaíno-martín FJ (2017) Brain-computer Interface application: auditory serial interface to control a two- class motor-imagery-based wheelchair. J Neuroeng Rehabil 14:1–16. https://doi.org/10.1186/s12984-017-0261-y
14. Yu H, Lu H, Wang S, Xia K, Jiang Y, Qian P (2019) A general common spatial patterns for EEG analysis with applications to vigilance detection. IEEE Access 7:111102–111114. https://doi.org/10.1109/access.2019.2934519
15. Souto DO, Cruz TKF, Coutinho K, Julio-Costa A, Fontes PLB, Haase VG (2020) Effect of motor imagery combined with physical practice on upper limb rehabilitation in children with hemiplegic cerebral palsy. NeuroRehabilitation 46:53–63. https://doi.org/10.3233/NRE-192931
16. Wang F (2020) Motor imagery classification using geodesic filtering common spatial pattern and filter- bank feature weighted support vector machine Motor imagery classification using geodesic filtering common spatial pattern and filter-bank feature weighted support vec. Rev Sci Instrum 91:034106. https://doi.org/10.1063/1.5142343
17. Ramya PS, Yashasvi K, Anjum A, Bhattacharyya A, Pachori RB (2020) Development of an effective computing framework for classification of motor imagery EEG signals for brain—computer interface. Advances in computational intelligence techniques. Springer, Singapore, pp 17–35
18. Li Y, Zhang XR, Zhang B, Lei MY, Cui WG, Guo YZ (2019) A channel-projection mixed-scale convolutional neural network for motor imagery EEG decoding. IEEE Trans Neural Syst Rehabil Eng 27(6):1170–1180. https://doi.org/10.1109/TNSRE.2019.2915621
19. Lu N, Li T, Ren X, Miao H (2017) A deep learning scheme for motor imagery classification based on restricted boltzmann machines. IEEE Trans Neural Syst Rehabil Eng 25(6):566–576. https://doi.org/10.1109/TNSRE.2016.2601240
20. Zhao X, Zhang H, Zhu G, You F, Kuang S, Sun L (2019) A multi-branch 3D convolutional neural network for EEG-Based motor imagery classification. IEEE Trans Neural Syst Rehabil Eng 27(10):2164–2177. https://doi.org/10.1109/TNSRE.2019.2938295
21. Tabar YR, Halici U (2016) A novel deep learning approach for classification of EEG motor imagery signals. J Neural Eng 14(1):16003. https://doi.org/10.1088/1741-2560/14/1/016003
22. Murugappan M, Boon H, Raj ANJ, Krishnan G, Palanisamy K (2020) ECG morphological features based sudden cardiac arrest (SCA) prediction using nonlinear classifiers. Advances in electrical and computer technologies, Springer Singapore, pp 789—801
23. Krishnan PT, Raj ANJ, Balasubramanian P, Chen Y (2020) Schizophrenia detection using multivariate empirical mode decomposition and entropy measures from multichannel EEG signal. Biocybernetics Biomed Eng 40(3):1124–1139. https://doi.org/10.1016/j.bbe.2020.05.008

24. Baghadadi A, Aribi Y, Fourati R, Halouani N, Siarry P, Alimi AM (2019) DASPS: a database for anxious states based on psychological stimulation. http://www.regim.org/publications/databases/dasps/
25. Ozdemir MA, Degirmenci M, Guren O, Akan A (2019) EEG based emotional state estimation using 2-D deep learning technique. In: 2019 medical technologies congress (TIPTEKNO), Izmir, Turkey, pp 1–4. https://doi.org/10.1109/TIPTEKNO.2019.8895158

Classification of Fundus Images for Diabetic Retinopathy Using Machine Learning: a Brief Review

Ruchika Bala, Arun Sharma, and Nidhi Goel

Abstract Diabetic retinopathy (DR) is an eye disease that is caused by damage of blood vessels in the retina due to excess sugar level. DR is a severe problem that might lead to blindness among working age people if not detected in time. As per the WHO report, in 2016, diabetes was the main reason of 1.6 million deaths. Between 2000 and 2016, there was an increment of 5% in premature deaths due to diabetes. Thus, early diagnosis of DR is desirable to safeguard human beings from the severity of DR. Analysis of blood vessel structures in retinal fundus images play a vital role in ophthalmology. However, it is challenging to perform manual analysis of the retinal fundus images by the medical fraternity as it is time-consuming, expensive, and tedious. The blood vessel structure includes variation in the size of the vessels, branching patterns, appearance, changes in thickness, etc. Many machine learning and deep learning-based automated techniques have been proposed for analyzing and detecting blood vessel structures for determining DR classification. The proposed paper aims at discussing and comparing automated processes for predicting various stages of DR by reviewing prominent and novel papers in the field of automatic classification of DR. Even though the field of automatic classification of retinal fundus images has progressed a lot, there are still some shortcomings or lacunas which need to be addressed.

Keywords Diabetic retinopathy · Fundus images · Retina · Classification · Grading · Proliferative · Non-proliferative · Microaneurysms · Heamorrhages · Exudates

R. Bala (✉) · A. Sharma · N. Goel
Indira Gandhi Delhi Technical University for Women, Delhi, India
e-mail: ruchika003phd19@igdtuw.ac.in

A. Sharma
e-mail: arunsharma@igdtuw.ac.in

G. Gupta et al. (eds.), *Proceedings of Academia-Industry Consortium for Data Science*, Advances in Intelligent Systems and Computing 1411,
https://doi.org/10.1007/978-981-16-6887-6_4

1 Introduction

Diabetic Retinopathy (DR) is a severe eye disease caused due to high sugar levels found in the population across the world. When initial screening or treatment is not started in the early stage, this may lead to blindness. High sugar levels cause several lesions in the eye and affects human vision. The lesions include microaneurysms, heamorrhages, red or white spots, leakage of fluid, breaking of blood vessels in retina, neovascularization, etc. Based on the lesions found in the retina, diabetic retinopathy severity is measured in different scales. DR is broadly classified as non-proliferative DR (NPDR) and proliferative DR (PDR). As per the outcome of Early Treatment of Diabetic Retinopathy Study (ETDRS), DR is classified as no DR, mild DR, moderate DR, severe DR, and PDR [1].

Eye specialists analyzed different types of images such as photographic retinal fundus images, Fluorescein Angiograms (FA), and Optical Coherence Tomography (OCT) to identify lesion patterns [2]. It was observed that use of FA and OCT patterns for classification are time consuming, complex and not suitable for clinical purposes. Thus, analysis of blood vessel patterns in retinal fundus images plays a vital role in detecting DR severity scale. Since the ratio of patients to eye specialists is very high, it is challenging to diagnose each individual manually; thus there is a high demand to automate the process of screening. Many data scientists and machine learning experts have proposed automated approaches for screening diabetic retinopathy. These approaches are categorized into two categories such as machine learning based approaches and deep learning based approaches. Based on the datasets available, different approaches have been used. In case of sparse data, machine learning based approaches are in use, whereas if bulk data is available, deep learning based approaches are recommended.

As shown in Fig. 1, initially, a retinal fundus image is acquired from a retinal image capturing device, followed by image preprocessing techniques. These techniques are applied to remove noise, enhance contrast, smooth or sharpen edges, resize, crop, etc. [3]. To separate background pixels from the foreground pixels, segmentation techniques are applied to the pre-processed images. The segmentation techniques are classified in four different categories such as machine learning, vessel enhancement, tracking based, and deformable models [4]. Features of the blood vessels are extracted from the segmented information. The extracted features include width, tortuosity, ridges, bifurcation, shape, size, etc. [5]. Further, the extracted features are used to classify fundus images into different DR severity scales based on machine learning and deep learning based approaches.

The review aims at analyzing all prominent automated DR classification approaches found in the literature, and discussing state-of-the-art of these approaches. Machine-learning based approaches are discussed in Sect. 2.1, and deep-learning based approaches are discussed in Sect. 2.2.

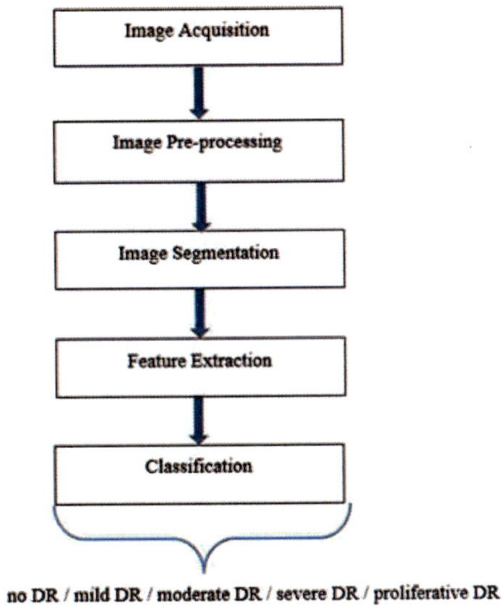

Fig. 1 Flow chart of DR classification

2 Automated Approaches

Manual analysis of color fundus images is complex and time consuming. Automated approaches make it simple and quick. The automated approaches are divided into two categories—machine learning based, and deep learning based. Tables 1 and 2 provides the summary of the DR classification based on machine learning and deep learning approaches, respectively.

2.1 Machine Learning Based Approaches

Machine-learning based approaches are broadly divided into two csategories—supervised machine learning and unsupervised machine learning. In supervised machine learning, output labels are provided to train the model and used to classify different objects, whereas in unsupervised machine learning, hidden patterns are identified and objects are clustered into different categories [3].

Machine learning based automated grade classifier for diabetic retinopathy has been proposed by Carrera et al. [6]. Retinal images undergo preprocessing steps separately, each for extracting characteristics of blood vessels, microaneurysms, and hard exudates. They have been isolated to extract features of standard deviation of the RGB components, density of blood vessels, number of microaneurysms, density of hard exudates, and green component entropy. These features are given as input to SVM classifier for determining grading of retina images. Further, DR diagnosis has been performed by removing extracted vessels and optic disk region. Candi-

Table 1 Performance summary of machine learning based approaches

Existing work	Database	Acc	Se	Sp	AUC
Carrera et al [6]	Messidor	0.85	–	–	–
Kar and Maity [7]	DRIVE, STARE, DIARETDBI, Messidor	0.0736	0.895	–	–
Yu et al [8]	Messidor, High-Resolution Fundus Image database, DIARETDB0, EyePACS	0.9523	0.9290	0.9630	0.9851
Leeza and Farooq [9]	Kaggel	–	0.9592 (SVM)	0.9890(SVM)	–
		-	0.8383(ANN)	0.9597(ANN)	–

date lesions have been detected using matched filter and LoG filter, and finally post processing is performed to remove the erroneous pixel detection [7]. Even though image preprocessing is performed to detect lesions, but the method is not efficient in detecting red lesions as it is difficult to detect red lesions from background due to illumination effects.

Illumination correction has been emphasized as a preprocessing step by Yu et al.[8] for an automatic detection of neovascularization. Non-local means filter is used for retrieving major vessels and removing fine vessels from the retinal fundus images. Multi-level Gabor wavelet filter and adaptive thresholding are employed for segmentation. 48 features are extracted based on vessel analysis, fractal dimensions, and granulometry. Relevant features are selected using PCA algorithms, and provided to SVM for classification. However, the approach detects neovascularization in OD region only and not elsewhere in the retina.

To save time, automated approach is recommended without performing preprocessing. Leeza and Farooq [9] proposed DR classification approach without preprocessing by extracting different features from the images based on LoG filters and histograms are clustered to generate a dictionary. Radial kernel space vector machine or neural network model is applied for DR classification on the dictionary. The method does not use preprocessing or postprocessing. However, it takes long training time for larger datasets and degrades the overall performance [9].

Table 2 Performance summary of deep learning based approaches

Existing work	Database	Acc	Se	Sp	Precision	Recall	AUC
Gulshan et al. [17]	EyePACS-1 Messidor-2	–	0.903 0.87	0.981 0.985	–	–	0.991 0.990
Li et al. [18]	Messidor-2	0.9349	0.9693	0.9345			0.9905
Sahlsten *et al.*[19]	Digifundus	(0.936–0.947)	–	–	–	-	–
Lam et al. [20]	Kaggel, Messidor	0.745 0.6425(w/o preprocessing)	–	–	–	–	–
Singh et al. [21]	Messidor, Kaggle	0.71(Binary) 0.56(Multi)	–	–	–	–	–
Prabhu et al. [22]	PIDD	–	–	–	1.0	0.6791	–
Momeni et al. [23]	IDRiD, Messidor	–	–	–	–	–	0.945
Tymchenko et al. [24]	IDRID, Messidor, APTOS2019	–	0.99	0.99	–	–	–
Orlando et al. [26]	DIARETDB1, e-ophtha, Messidor	–	0.9721	–	–	–	0.9347
Shreyas et al. [27]	Kaggle, Messidor, RIGA, HEI-MED	0.95(VGG16) 0.93(ResNet50)	–	–	–	–	–
Piotr et al. [28]	e-Ophtha,DIARETDB1,ROC	–	0.355	–	–	–	–
Lifeng et al. [29]	IDRiD	0.978	–	–	–	–	–
Zago et al. [30]	Kaggel,IDRiD,DDR, DIARETDB0, DIARETDB1, Messidor, Messidor-2	–	0.940	–	–	–	0.912

2.2 Deep Learning Based Approaches

In deep learning, more layers are added to the model to identify intrinsic features. These features are used to classify different objects. One of the most widely used deep learning approaches is Convolutional Neural Networks (CNN). CNN is based on Artificial Neural Network (ANN) that works on the principle of vortex used for image classification and object detection. In deep learning, when a trained model is used for other related problem domains to transfer knowledge, then it is called transfer learning. In visual recognition Imagenet competitions, several models have been trained such as AlexNet, GoogleNet, VGG, Inception, Resnet, etc. [10–14] and used for different problem domains for image classification. There is lot of work done in the area of diabetic retinopathy based on deep learning and transfer learning. [15, 16]

DR detection model based on transfer learning using Inception-v3 architecture [13] is proposed for classification of fundus images by Gulshan et al. [17]. Initially, images are preprocessed and then provided to the neural network. The network is optimized based on backpropagation and performance is tested on two databases EyePACS-1 and Messidor-2. An automatic DR detection is also proposed based on transfer learning using Inception v3 [18]. Further, referral recommendations are facilitated to patients for future investigations. However, the performance needs to be determined in real time clinical settings [17, 18].

Further, Sahlsten et al. [19] used a deep CNN based on Inception-v3 as an ensemble of five classification systems to classify retinal images into DR or macular edema. Each image is cropped into a square size containing circular area of fundus. Each of the cropped images were resized to five different standard input image sizes of 256×256, 299×299, 512×512, 1024×1024, and 2095×2095 pixels. Further, the effect of image resolution is determined with respect to the model performance. Higher model performance is achieved with the increased resolution of the fundus image with the exception of image size of 2095×2095.

Most of the research in diabetic retinopathy domain involves lot of preprocessing and data cleaning. Diagnosing DR diseases as binary, 3-ary, and 4-ary classification via CNN approaches (AlexNet, GoogeNet, and GoogleNet with preprocessing) are proposed [20]. Feature localization is performed, followed by classification and further preprocessing is performed to find diseases difficult to perceive manually. Overall, improvement in the detection of mild disease and multi-grade disease detection are shown, however, low performance is obtained when GoogleNet is employed without preprocessing. Further, automatic classification of DR into its degree of severity is proposed using deep CNN without performing image preprocessing. Fundus images are obtained from online datasets for classification. 3-layer CNN for 2-ary classification (no DR and DR) and 4-layer CNN for 5-ary classification (no DR, mild DR, moderate DR, severe DR, PDR) have been proposed [21].

Deep Belief Neural Network (DBN) is designed by following preprocessing, pretraining of deep belief neural network and fine-tuning back propagation for supervised classification [22]. In preprocessing step, max-min normalization is performed, followed by feature selection using PCA. In pre-training step, each layer of the DBN is trained as Restricted Boltzmann Machines (RBM). At last, fine-tuning is used to train the classifier to predict diabetic mellitus. Experiment is performed on Pima Indian Women Diabetes dataset (PIDD). The performance of DBN is proved to be better than conventional machine learning methods including SVM, naïve bayes, logistic regression, decision tree, and random forest.

DR classification using EfficientNet CNN on CLAHE based preprocessed images has been proposed [23]. The effect of varying various parameters such as width, height, resolution and derived different models (B0-B7), and B5 model has been used for classification. Ensemble method also proved good performance in designing the models. Combination of classification and regression models with pre-trained weights has been proposed [24]. Ensemble of EfficientNet-B4 [25], EfficientNet-B5, and SE-ResNeXt50 is proposed for classification of retinal fundus images.

Further, Orlando et al. [26] proposed random forest ensemble classifier for DR classification by extracting features based on 63-dimensional hand crafted feature vector and LeNet convolution network. Various intensity based and shape based features have been used for generating hand-crafted features.

Deep transfer learning CNN models (VGG and ResNet) have been employed on different datasets after applying CLAHE [27]. As compared to ResNet, VGG 16 model illustrated the best classification performance. Transfer learning based models has less training time as they use ImageNet weights; however transfer learning is not suitable for all kinds of datasets.

Microaneurysms (MA) is detected by applying CNN model on cleaned images based on morphological operations [28]. Since the method is very specific, other features such as heamorrhages, exudates, etc., are not detected. Further, Lifeng [29] proposed diagnosis of MA and DR detected by extracting features using matched filter with LoG filter and PCA. The extracted features are applied on deep CNN for DR classification. However, initial preprocessing impacts the speed of the model.

In some of the cases, instead of considering full image, different patches are considered for performing convolution. DR detection is proposed by employing fully patch based CNN model for lesion localization by using higher values of stride to increase the learning rate [30].

3 Conclusion

This review discussed theoretical background of the machine learning and deep learning approaches. Prominent and novel approaches to solve diabetic retinopathy classification has been discussed briefly. Based on the study, it is observed that lot of research has been done in image classification based on conventional machine learning methods, ANN, deep learning, and transfer learning. In machine-learning approaches, feature extraction step requires deep domain knowledge to retrieve the most suitable features, while in deep learning, intrinsic features of the images are extracted automatically. Deep learning is advantageous over other conventional methods as it utilizes increasing computational power (e.g., GPU) with high data availability. Deep learning approaches for fundus images classification are becoming very popular and highly recommended for image classification.

References

1. Wilkinson CP, Ferris FL, Klein RE, Lee PP, Agardh CD, Davis M, Dills D, Kampik A, Pararajasegaram R, Verdaguer JT (2003) Proposed international clinical diabetic retinopathy and diabetic macular edema disease severity scales. Ophthalmology 110(5):1677–1682. PMID: 13129861 https://doi.org/10.1016/S0161-6420(03)00475-5
2. Wu L, Fernandez-Loaiza P, Sauma J et al (2013) Classification of diabetic retinopathy and diabetic macular edema. World J Diab 4(6):290–4
3. Memari N, Ramli AR, Saripan MIB, Mashohor S, Moghbel M (2019) Retinal blood vessel segmentation by using matched filtering and fuzzy C-means clustering with integrated level set method for diabetic retinopathy assessment. J Med Biol Eng 39(5):713-731. https://doi.org/10.1007/S40846-018-0454-2
4. Mocciaa S, De Momi E, El Hadji S, Mattos LS (2018) Blood vessel segmentation algorithms—review of methods, datasets and evaluation metrics. Comput Methods Programs Biomed 158:71–91 (Elsevier)
5. Panchal P, Bhojani R, Panchal T (2016) An algorithm for retinal feature extraction using hybrid approach. In: International conference on communication, computing and virtualization (ICCCV)
6. Carrera EV, González A, Carrera R (2017) Automated detection of diabetic retinopathy using SVM. In: 2017 IEEE XXIV international conference on electronics, electrical engineering and computing (INTERCON). IEEE (2017)
7. Kar SS, Maity SP (2018) Automatic detection of retinal lesions for screening of diabetic retinopathy. IEEE Trans Biomed Eng 65(3):608–618
8. Yu S, Xiao D, Kanagasingam Y (2018) Machine learning based automatic neovascularization detection on optic disc region. IEEE J Biomed Health Inform 22(3):886–94
9. Leeza M, Farooq H (2019) Detection of severity level of diabetic retinopathy using Bag of features model. IET Comput Vis 13(5):523–530
10. Krizhevsky A, Sutskever I (2012) Imagenet classification with deep convolutional neural networks. Neural Inf Process Syst 25:1097–1105
11. Szegedy C, Liu W, Jia Y, Sermanet P, Reed S, Anguelov D, Erhan D, Vanhoucke V, Rabinovich A (2015) Going deeper with convolutions. Comput Vis Pattern Recognit 1–9
12. Simonyan K, Zisserman A (2015) Very deep convolutional networks for large-scale image recognition. In: 3rd International Conference on Learning Representations (ICLR), San Diego, CA, USA
13. Szegedy C, Vanhoucke, Ioffe S, Shlens J, Wojna Z (2016) Rethinking the inception architecture for computer vision. In: Computer vision and pattern recognition, pp 2818–2826
14. He K, Zhang X, Ren S, Sun J (2016) Deep residual learning for image recognition. In: 2016 IEEE conference on Computer Vision and Pattern Recognition (CVPR), Las Vegas, Nevada, USA. https://doi.org/10.1109/CVPR.2016.90
15. Shaban M, Ogur Z, Mahmoud A, Switala A, Shalaby A, Abu Khalifeh H, Ghazal M, Fraiwan L, Giridharan G, Sandhu H, El-Baz AS (2020) A convolutional neural network for the screening and staging of diabetic retinopathy. PLoS ONE 5(6):e0233514. https://doi.org/10.1371/journal.pone.0233514
16. Patel R, Chaware A (2020) Transfer Learning with Fine-Tuned MobileNetV2 for Diabetic Retinopathy. in: International conference for emerging technology (INCET). https://doi.org/10.1109/INCET49848.2020.9154014

17. Gulshan V, Peng L, Coram M, Stumpe MC, Wu D, Narayanaswamy A, Venugopalan S, Widner K, Madams T, Cuadros J, Kim R, Raman R, Nelson PC, Mega JL, Webster DR (2016) Development and validation of a deep learning algorithm for detection of diabetic retinopathy in retinal fundus photographs. JAMA 316:2402–2410
18. Li F, Liu Z, Chen H, Jiang M, Zhang X, Wu Z (2019) Automatic detection of diabetic retinopathy in retinal fundus photographs based on deep learning algorithm. Transl Vis Sci Technol 8:4
19. Sahlsten J, Jaskari J, Kivinen J, Turunen L, Jaanio E, Hietala K, Kaski K (2019) Deep learning fundus image analysis for diabetic retinopathy and macular edema grading. Sci Rep 9:1–11
20. Lam C, Yi D, Guo M, Lindsey T (2018) Automated detection of diabetic retinopathy using deep learning. AMIA J Summits Transl Sci Proc 2017:147–55
21. Singh TM, Bharali P, Bhuyan C (2019) Automated detection of diabetic retinopathy. In: 2019 2nd international conference on advanced computational and communication paradigms, ICACCP 2019. Institute of Electrical and Electronics Engineers Inc. https://doi.org/10.1109/ICACCP.2019.8882914
22. Prabhu P, Selvabharathi S (2019) Deep belief neural network model for prediction of diabetes mellitus. In: 2019 3rd international conference on imaging, signal processing and communication, ICISPC, pp 138–142. Institute of Electrical and Electronics Engineers Inc
23. . Momeni Pour A, Seyedarabi H, Abbasi Jahromi SH, Javadzadeh A (2020) Automatic detection and monitoring of diabetic retinopathy using efficient convolutional neural networks and contrast limited adaptive histogram equalization. IEEE Access 136668–136673 (2020) https://doi.org/10.1109/ACCESS.2020.3005044
24. Tymchenko B, Marchenko P, Spodarets D (2020) Deep learning approach to diabetic retinopathy detection. arXiv:2003.02261
25. Tan M, Quoc VL (2019) Efficientnet: rethinking model scaling for convolutional neural networks. arXiv preprint arXiv:1905.11946
26. Orlando JI, Prokofyeva E, del Fresno M, Blaschko MB (2018) An ensemble deep learning based approach for red lesion detection in fundus images. Comput Methods Programs Biomed 153:115–127
27. Labhsetwar SR, Salvi RS, Kolte PA, Venkatesh VS, Baretto AM (2020) Predictive analysis of diabetic retinopathy with transfer learning. arXiv:2011.04052
28. Chudzik P, Majumdar S, Calivá F, Al-Diri B (2018) Microaneurysm detection using fully convolutional neural networks. Comput Methods Programs Biomed 158:185–192
29. Qiao L, Zhu Y, Zhou H (2020)Diabetic retinopathy detection using prognosis of microaneurysm and early diagnosis system for non-proliferative diabetic retinopathy based on deep learning algorithms. IEEE Access 8:104292—104302. https://doi.org/10.1109/ACCESS.2020.2993937
30. Zago G, Andreão RV, Dorizzi B, Salles EOT (2020) Diabetic retinopathy detection using red lesion localization and convolutional neural networks. Comput Biol Med 116:103537

Wearable Sensor and Machine Learning Model-Based Fall Detection System for Safety of Elders and Movement Disorders

Anju S. Pillai, Sejal Badgujar, and Sujatha Krishnamoorthy

Abstract Due to different working culture of people, elderly people's health gets neglected as they live alone at home. With the rising population, there is a pressing demand for the evolution of fall identification systems. People with age greater than 65 are suffering from highest number of fatal falls. Some of the difficulties and challenges faced by the elders and mobility disordered people can be over passed by implementing algorithms able to anticipate falls. It is possible to have a good to great quality of life for the affected, by providing living assistance through automatic fall detection and alarming systems. We propose implementation of a fall recognition system for real-time tracking of elderly people. The proposed system has wearable sensor unit for detecting falls and alert mechanism to intimate the concerned and the care takers in case of falls by means of messages. The acceleration data are collected by the system using triaxial accelerometer and use machine learning algorithms to detect the falls upon various feature calculations. Extensive computations are carried out to compare the performance of different machine learning algorithms with varying features, and the algorithm giving the highest accuracy with optimal features is identified. The system gains an accuracy up to 99% by using random forest algorithm with tenfold cross validation. Thus, with a secure and reliable fall detection and alarming system, one could reduce the fatal falls, improving social integration, productivity, and quality of life.

Keywords Fall detection · Fall alarming · Activity of daily living · Elderly people assistance · Movement disorders · Machine learning · Wearable sensors · Fall forecast

---

A. S. Pillai (✉) · S. Badgujar
Department of Electrical and Electronics Engineering, Amrita School of Engineering, Amrita Vishwa Vidyapeetham, Coimbatore, India
e-mail: s_anju@cb.amrita.edu

S. Krishnamoorthy
Department of Computer Science, Wenzhou Kean University, Wenzhou, China
e-mail: krishnsu@kean.edu

G. Gupta et al. (eds.), *Proceedings of Academia-Industry Consortium for Data Science*, Advances in Intelligent Systems and Computing 1411,
https://doi.org/10.1007/978-981-16-6887-6_5

1 Introduction

Parkinson's disease (PD) is a chronic progressive neurodegenerative disorder which prominently impacts the older population. Despite this, due to aging, people get lot of difficulties for moving around and around four million people worldwide are living with this condition. The diagnosis is clinical and is complex, while the cause remains largely unknown. While Parkinson's itself is not malignant, disease intricacy can be acute. Under such circumstances, use of a fall detection system becomes a great support. In times of emergency, elderly people require immediate attention in their homes or care centers. In most cases, they are not able to ask for help as they do not have access to technology, many live-in rural areas, and also few may have physical limitations. According to the survey conducted by World Health Organization (WHO), the second major reason for accidental and unintentional injury deaths is falls [1]. To improve quality of life of elderly people, people with movement disorders and to ensure wellbeing of our family member, use of automatic fall detection system is crucial.

This paper proposes a real-time fall detection system to monitor elderly people and people with movement disorders. The system has wearable sensor unit for detecting falls and alerting the concerned person in case of falls by means of messages. It also provides living assistance to elderly people. The system consists of triaxial accelerometer and uses machine learning (ML) algorithms to detect the falls after calculating various features. Six machine learning algorithms are tested for its performance and the most accurate algorithm is identified. The system models are trained by using existing SisFall dataset [2] and tested. It consists of 19 types of daily life activities and 15 types of falls performed by participants. Out of a set of ML algorithms tested, random forest algorithm is found to give best accuracy up to 99%.

This paper is arranged as follows: The detailed literature survey is presented in Sect. 2, and the system overview and methodology are explained in third section. The details of extensive simulations and hardware implementation are illustrated in Sect. 4, and the conclusions and future research directions are summarized in Sect. 5.

2 Related Work

In the literature, there are ample of research works which focus on human activity classification, fall detection, and fall forecast which are a hot research topic [3, 4]. Due to the advancements in the technology, especially, sensor system developments, IoT, and deep learning; the support extended to improve the life of human has increased sufficiently [5]. The literature shows a large variety of methods adopted to identify human falls and efforts to reduce the false alarms and increase the fall identification accuracy. The broad categories of design of fall detection system are using three types of sensors: vision-based devices, ambient devices, and wearable devices [6].

In vision-based fall detection, cameras fitted in the room allow monitoring of the elderly or the affected [7, 8]. Different image processing algorithms are used to detect the fall. Mostly, in this case, simulation dataset is used but on an average 24% of the falls remain unexplored and the presence of false alarms is high. In ambient-based fall detection [9, 10], various sensors such as vibration and passive infrared sensors are incorporated in the fall detection system for identifying the falls, identifying the person, identification of footstep, motion detection, and uncommon inactivity detection. Thus, with the support of ambient sensors, falls are identified [10].

In general, the wearable fall detection systems have buttons incorporated with the fall detection system, so that in case of a not harmful fall, the person can push the button to sending emergency message [6]. Mostly such devices wear like a necklace and the subject needs to be careful during the occurrence of fall to press the button. The other possible versions were a Bluetooth enabled pendant connected to mobile phone. The system enables a communication between the person and the mobile phone; but the phone needs to be within 100 m range of the pendant. The overhead of such device includes activities such as bending toward front are not identified properly due to false reading of falls.

In one of the implemented systems, six wireless sensor units were tightly connected to the subjects' head, chest, waist, right wrist, right thigh, and right ankle. The dataset consists of data acquired from 14 healthy adult volunteers and various features are selected and employed feature reduction using principal component analysis (PCA). In their proposed research, six different ML learning classification algorithms were used for fall identification [11, 12]. The major difficulty faced with the system is presence of large number of sensors to be attached to the elder's body, which is inconvenient [13]. Another method presented is a wearable device attached on human's waist which analyzes the acceleration data to identify the fall [14–16]. Upon fall detection, messages are sent to the care takers by sharing the geographic position. This enables the elderly people to get timely help and reduces the negative influence. The system uses a threshold-based algorithm and thus, accurate identification of threshold is vital in this case. Based on the threshold, the accuracy of the system fluctuates. The literature also supports IoT and big data-based fall detection methods. It has components such as wearable sensor systems, wireless communication network, smart IoT gateway, and cloud services [17–19]. In their proposed system architecture, data get stored for further big data analysis but the installation is costly. Few of the implemented systems, uses a threshold-based mechanism to detect the fall but, generates high false alarms.

3 System Overview and Methodology

We propose to solve this problem with machine learning (ML). The system for identifying the human activities of daily life (ADL) is implemented using SisFall [2] datasets: Here, we propose continuous monitoring of user activities based on Gait analysis through the body sensors attached to the subject via a wearable waist belt.

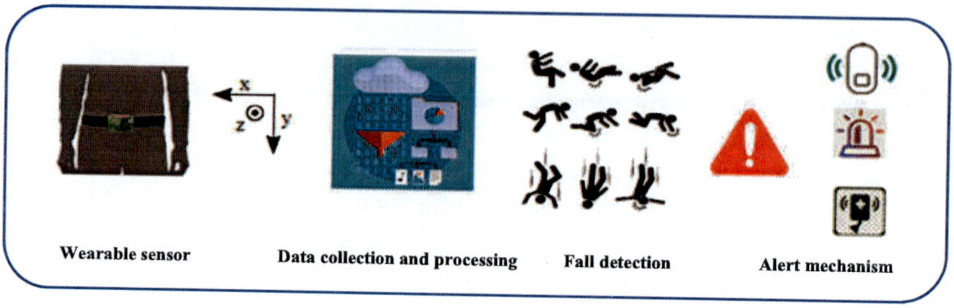

Fig. 1 Proposed fall detection system overview

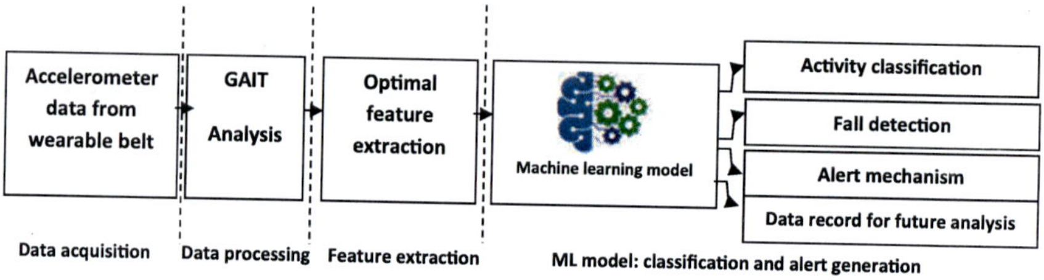

Fig. 2 Implementation of proposed fall detection system

The proposed system consists of wearable waist belt with microcontroller and triaxial accelerometer sensor. The methodology can be explained in five stages as in Fig. 1. The first stage is data acquisition where acceleration data are collected using triaxial accelerometer and in second stage, data preprocessing is carried out. Data processing is done on the collected data to filter it out. In the succeeding stage, features are extracted from the filter data followed and the processed sensor data are fed as input to the ML model which consists of various ML algorithms to classify the activities as fall or not fall. Finally, alert messages or alarm signals are sent to the concerned person or care takers to intimate the fall occurrence. The details of implementation are shown in Fig. 2.

4 Algorithm Implementation

Fall detection system has ML algorithm incorporated which is implemented in two stages: training and testing. The machine learning model is trained using SisFall dataset [2] consisting of 1,546,456 data points and analyzed by using tenfold cross validation. The highest accuracy exhibiting model is taken into account and saved. The saved model is used for testing and the steps of implementation is described in Fig. 3.

Training	
	Read the sensor data from the dataset
	Data pre-processing and filtering
	Feature computation
	Training machine learning model with algorithms
	Save and test the model
	Analyze the model for fall identification

Testing	
	Read the sensor data from the dataset
	Data pre-processing and filtering
	Feature computation
	Test the model
	Save and test the model
	Analyze the model for fall identification

Fig. 3 Steps of algorithm implementation

4.1 Machine Learning Model Training

4.1.1 Sensor Data Procurement

To train the machine learning model, the model is inputted with accelerometer data. The input data ares acquired from the waist band, which is available in SisFall dataset. The triaxial accelerometer used is ADXL345 which gives acceleration values along x, y, and z axis. The accelerometer provides larger span with low energy. The dataset consists of 23 young adults (11 male and 12 female) and 15 elderly people (8 and 7 male and female, respectively). 19 daily life activities and 15 fall activities are carried out by the young adults, whereas, only one participant has performed fall in the elder's group. The data are collected with the sampling rate of 200 Hz. Dataset parameters are acceleration along x-axis, acceleration along y-axis, and acceleration along z-axis detected using ADXL345 accelerometer.

4.1.2 Data Handling

Before the sensor data can be fed as an input to the machine learning model, the unwanted glitches and noise need to be removed. This is accomplished with 4th order Butterworth filter with cutoff frequency of 5 Hz. The filter selection is confirmed as it was capable of generating similar outputs which other sophisticated filters such as FIR and IIR produce and are simple to construct [20]. In [20], analysis off filters at different cutoff frequencies are performed and concluded that 4th order Butterworth filter is most effective. The original sample data and filtered data are plotted as shown below. The data have been smoothened after filtering as seen in Fig. 4.

4.1.3 Feature Extraction

The features are calculated on the entire SisFall dataset. It consists of 19 daily life activities and 15 fall activities. One or more trials are taken for each activity. For every

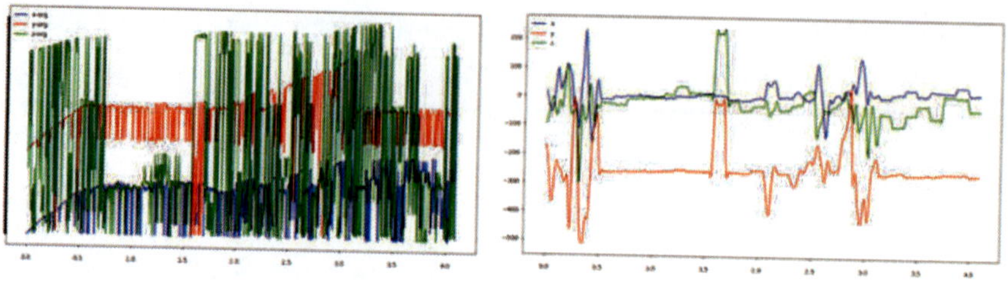

Fig. 4 **a** Acceleration data points before filtering. **b** Acceleration data points after filtering

trial, all the six features along three axes are calculated. Overall, 1789 ADL and 1452 falls exist. Total feature count for ADL class is 1789X 3 axes, and fall feature count is 1452 X 3axes. Overall, 1,546,456 feature points are calculated. Using these features, machine leaning models are trained to detect the fall. The features that are considered with formula are presented in Table 1.

Table 1 Features calculated with its formulae and meaning

Type	Code	Feature	Equation	Meanings
Amplitude	M1	Maximum amplitude	$\max(a[k])$	It is maximum of the amplitude along *x*-axis, *y*-axis, and *z*-axis
	M2	Minimum amplitude	$\min(a[k])$	It is minimum of the amplitude along the three axes
	M3	Mean amplitude	$\mu = \frac{1}{N}\sum a[k]\mu = \frac{1}{N}\sum a[k]$	It is mean of the amplitude along the three axes
Statistical	M4	Variance	$\sigma^2 = \frac{1}{N}\sum (a[k]-\mu)^2\sigma^2 = \frac{1}{N}\sum (a[k]-\mu)^2$	It is variance of the amplitude along the three axes
Skewness	M5	Measure of asymmetry	skewness: $g_1 = m_3/m_2^{3/2}$ $m_3 = \sum (x-\overline{x})^3/n$ and $m_2 = \sum (x-\overline{x})^2/n$	It is measure of asymmetry of distribution curve along the three axes
Kurtosis	M6	Measure of tailedness	kurtosis: $a_4 = m_4/m_2^2 m_4 = \sum (x-\overline{x})^4/n$ and $m_2 = \sum (x-\overline{x})^2/n$	It is measure of tailedness of distribution curve along the three axes

4.1.4 Fall Detection Algorithm

The vital module of the fall detection system is the machine learning model. Various machine learning algorithms are used to test and compare the fall detection accuracy. The algorithms include K-nearest neighbors (KNN), support vector machines (SVM), Naive Bayes, neural network, logistic regression, decision tree, and random forest. Tenfold cross validation is used to detect the accuracy. For the extended testing, the model that exhibits highest accuracy is saved. The major advantage of using machine learning algorithms is it allows detection of different falls with higher accuracy, identify patterns, incompleteness anomalies, and also can detect patterns in the signal [21]. This conforms the superior performance of ML algorithms compared to the threshold-based algorithms. But this demands large data for training of ML algorithms along with high computational complexity.

Dataset is extracted into Python Pandas DataFrame and filtered using Butterworth filter package. Machine learning algorithms are implemented using Python packages. For training, filtered data are divided into 10 parts, 9 parts are used to train the machine learning models using Python sklearn packages. Remaining part of data is used for testing, and accuracy is calculated using sklearn matrices. The procedure is repeated 10 times for tenfold cross validation and average of accuracies is taken as final accuracy. All machine learning algorithms are implemented one by one, and accuracy scores are stored for comparison.

4.1.5 Alert Mechanism

In case of falls, alert messages are sent to the concerned person and care takers as emergency support. Every time when a fall occurs, it is important to alert the concern person in terms of message or call so that immediate actions can be taken to save the life of the person or do any immediate actions to safe guard the person. In the proposed system, SMS is sent by using Twilio as soon as fall occurs. Amazon Web services are used to provide HTTP connectivity and public switched telephone network and formed API called Twilio, which is a Web service to send and receive messages and calls. To protect against wrong outage, it has architectural design principles. Twilio can be used to develop open source software.

4.2 Hardware Implementation

The test algorithm is used in Raspberry Pi3 which is communicated with ADXL345 by serial communication (I2C communication). Sampling rate, i.e., data collection rate from ADXL is 200 Hz, i.e., every sec 200 samples of accelerometer from all three axes X, Y, Z are fetched. The acceleration fetched is in "g" unit which needs to be converted to "m/s^2."

$$\text{Accerleration data} = \frac{\text{Accerleration }[g] * 2^{\text{Resolution}}}{2 * \text{Range}} \quad (1)$$

5 Evaluation Parameters

The performance of different trained ML models is compared on the basis of different parameters. The parameter includes training and prediction time, confusion matrix, accuracy, sensitivity, and specificity. The details of these parameters are given below.

5.1 Sensitivity

Sensitivity is a measure of the actual positive cases that are predicted as positive, i.e., true positive. Sensitivity is also called as recall. Sensitivity gives the proportion of all correctly predicted data points to that of all correct and incorrect predictions (true positive and false negative).

$$\text{Sensitivity(SE)} = \frac{\text{TP}}{\text{TP} + \text{FN}} \quad (2)$$

5.2 Specificity

By definition, specificity is a measure of the actual negative cases that are predicted as negative, i.e., true negative. Specificity gives the proportion of all correctly predicted data points to that of all correct and incorrect predictions (false positive and true negative).

$$\text{Specificity(SP)} = \frac{\text{TN}}{\text{TN} + \text{FP}} \quad (3)$$

5.3 Accuracy

Accuracy is the fraction of predictions, our model got right.

$$\text{Accuracy(AC)} = \frac{\text{SE} + \text{SP}}{2} \quad (4)$$

5.4 Tenfold Cross Validation

While training the model, dataset needs to be divided into training and testing sets and the performance of the model is assessed based on the accuracy of the model. As accuracy varies with various divisions of dataset, this method is not reliable. Therefore, K-fold cross validation is used to divide data into k folds. Every time one-fold acts as test and remaining fold data acts as training set. In the proposed system, tenfold cross validation is used. Total considered data are divided in to 10-folds and every time one of the folds act as test and all other as train.

6 Results

6.1 Implementation of ML Algorithms

The different ML algorithms are fed with the SisFall dataset and used tenfold cross validation to calculate the accuracy. Improved results are obtained and are shown in Table 2 showing a maximum accuracy of 99.18% using random forest algorithm.

The Fig. 5 is a comparison plot of different machine learning algorithms for accuracy with tenfold cross validation. Form the box plot, it is evident that random forest algorithm gives the best accuracy for fall identification when tested with SisFall dataset.

Table 2 Accuracies with training time and prediction time

Algorithms	Accuracy (%)	Training time (s)	Prediction time (s)
KNN ($K = 15$)	91.54	0.031	0.39
SVM	60.09	4.259	3.76
Naïve Bayes	96.89	0.016	0.016
Neural network	75.01	4.742	0.07
Logistic regression	96.36	9.995	0.08
Decision tree	97.00	0.125	0.06
Random forest	99.18	1.154	0.047

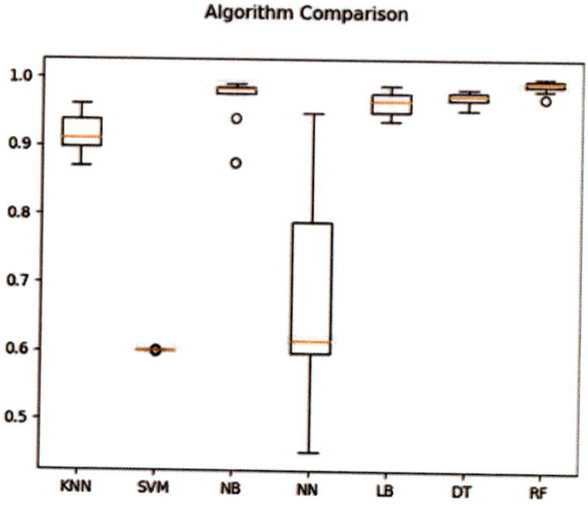

Fig. 5 Accuracy comparison of different machine learning algorithms

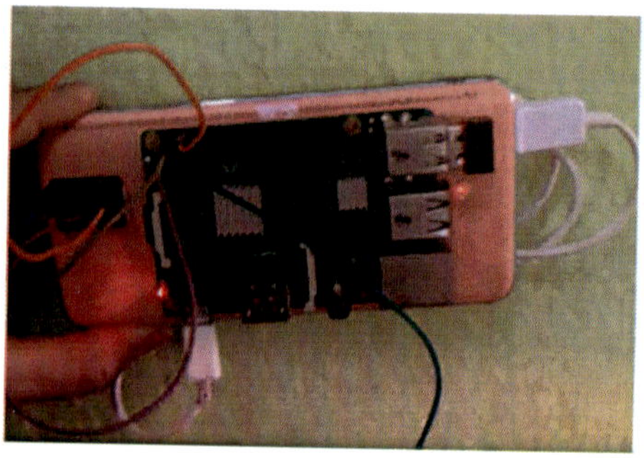

Fig. 6 Hardware prototype

6.2 Hardware Prototype

The hardware prototype is shown in Fig. 6 with positioned sensor and waist belt.

The user will be in upright position when there is no fall, and this state is checked, and readings are taken using VNC server. The result is shown in Fig. 7.

The user will be in bending or lying position when there is a fall. This condition is checked, and readings are taken by using VNC server. Figure 8 shows the respective output.

6.3 Time Analysis

Followed by the testing of the system at real-time, time, and activity analysis is carried out for different sample sizes of 1, 1000 and 2000 for varying time duration of 5 and 10 s, the obtained results are shown in Table 3.

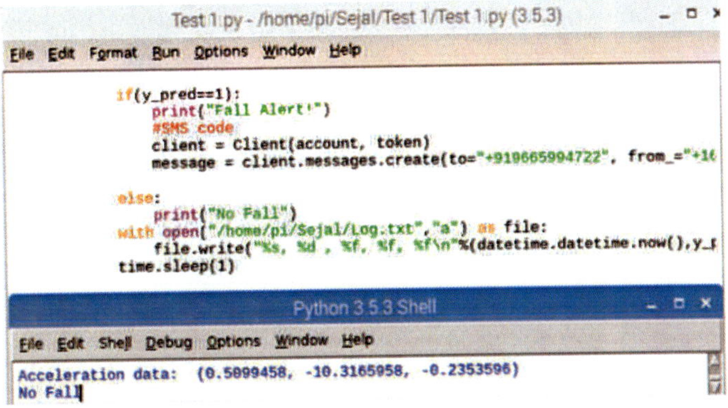

Fig. 7 No fall condition output

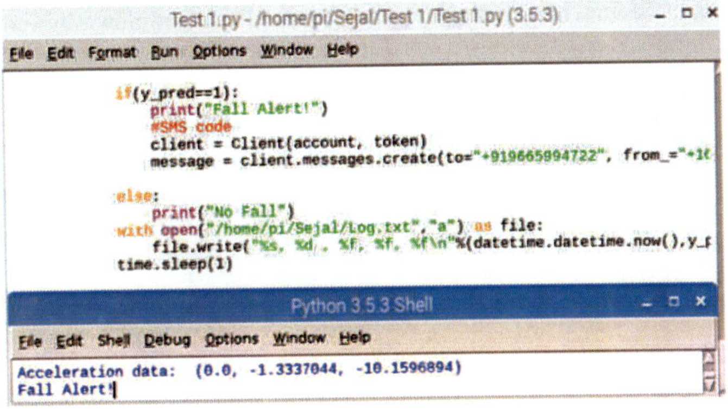

Fig. 8 Fall condition output

Table 3 Time and activity analysis

Steps	For 1 sample	For 1000 sample (for 5 s)	For 2000 sample (for 10 s)
Collection	0.002 s	2 s	4 s
Conversion (g to m/s^2)	0.0 s	0 s	0 s
Saving to data frame	0.017 s	17 s	34 s
Feature calculation		0.143 s	0.146 s
Prediction		0.12 s	0.12 s
Notification		4 s	4 s
Total Time		19–21 s (without fall) 23–25 s (with fall)	35–37 s (without fall) 39–41 s (with fall)

The analysis of the above results shows that, if data are collected for 5/10 s, saving that data to data frame takes more time 17/34 s. So even when the fall is detected on basis of acceleration data of 5/10 s, it takes more time to process and predict it. Overall, to complete one cycle, the system takes about 20 s for 5 s data collection and about 36 s for 10 s data collection. This time can be minimized by using faster microcontrollers.

6.4 Comparison with Literature

In paper [17], an IoT system for fall identification of elderly was implemented. The proposed approach included big data models combined with decision tree machine learning algorithm and achieved 91.67% accuracy. They used the SisFall dataset to the proposed system with different feature extraction methodology. Sliding window-based decision-based algorithms are also used on the SisFall dataset in [2]. They have got different accuracies depending on the set of features used. Our proposed approach is tested with SisFall dataset and used tenfold cross validation to produce 99.18% accuracy using random forest algorithm.

7 Conclusion

In this paper, implementation of recognition and differentiation of fall activities and activities of daily living for the support of elderly and people with movement disorders is carried out. The proposed technique of fall identification includes a wearable sensor-based solution which is more suitable for the elderly, due to the possibility to identify a fall in which environment, the subject is unlike a camera-based or ambient-based fall detection systems, which are confined to a particular room or indoor surrounding. In addition, the wearable sensor-based implementation is less costly when compared to that of camera or PRI sensors.

Falls are appropriately detected using various machine learning algorithms such as KNN, SVM, naïve Bayes, neural network, logistic regression, decision tree, and random forest using SisFall dataset. These ML models are trained, and tenfold cross validation is used to calculate the fall accuracy. The performance of the model is evaluated using significant parameters. Out the different tested ML models, random forest is found to give the highest accuracy of 99.18% for fall detection. But the time overhead associated with the proposed systems is found to be in the order of 20 s for 5 s data collection and in future, we intend to minimize the time using faster microcontrollers and incorporate methods to forecast a fall before its occurrence from the history of stored data. In future, we will try to addresses the complex scenarios of human ADLs to predict a fall with minimum response by determining suitable machine learning algorithms by modeling the human kinematics.

References

1. World Health Organization (2015) World report on ageing and health. World Health Organization, Geneva
2. Sucerquia A, López JD, Vargas-Bonilla JF (2017) SisFall: a fall and movement dataset. Sensors 17:198. https://doi.org/10.3390/s17010198
3. Ranjith R, Pillai AS (2019) Classification of activities of daily living using smart phones. Int J Recent Technol Eng 8(3):2984–2988
4. Badgujar S, Pillai AS (2020) Fall detection for elderly people using machine learning. In: 2020 11th international conference on computing, communication and networking technologies (ICCCNT), Kharagpur, India, pp 1–4. https://doi.org/10.1109/ICCCNT49239.2020.9225494
5. Bhoi SK et al (2018) FallDS-IoT: a fall detection system for elderly healthcare based on IoT data analytics. In 2018 international conference on information technology (ICIT), Bhubaneswar, India, pp 155–160. https://doi.org/10.1109/ICIT.2018.00041
6. Megalingam RK, Pocklassery G, Thulasi AA, Jayakrishnan V, Mourya G, MediSuit: wearable health monitoring system for elders and bed-ridden patients. In: 2016 10th international conference on intelligent systems and control (ISCO), Coimbatore, pp 1–6. https://doi.org/10.1109/ISCO.2016.7727091
7. Debard G, Baldewijns G, Goedem T, Tuytelaars T, Vanrumste B (2015) Camera-based fall detection using a particle filter Glen. IEEE 978-1-4244-9270-1
8. Cucchiara R, Prati A, Vezzani R (2007) A multi-camera vision system for fall detection and alarm generation. Expert Syst 24(5):2007
9. Yazar A, Erden F, Cetin AE (2014) Multi-sensor ambient assisted living system for fall detection. In: Proceedings of the IEEE international conference on acoustics, speech, and signal processing (ICASSP'14), pp 1–3
10. Megalingam RK, Radhakrishnan V, Jacob DC, Unnikrishnan DKM, Sudhakaran AK (2011) Assistive technology for elders: wireless intelligent healthcare gadget. In: 2011 IEEE global humanitarian technology conference, Seattle, WA, pp 296–300. https://doi.org/10.1109/GHTC.2011.94
11. Jian H, Chen H (2015) A portable fall detection and alerting system based on K-NN algorithm and remote medicine. China Commun 12(4):23/31
12. Purushothaman A (2018) Fall detection system using artificial neural network. In: 2nd international conference on inventive communication and computational technologies (ICICCT 2018), Hotel Arcadia, Coimbatore
13. Özdemir Atm, Barshan B (2014) Detecting falls with wearable sensors using machine learning techniques. Sensors 14:10691–10708. https://doi.org/10.3390/s140610691
14. Torres GG, Henriques RVB, Pereira CE, Müller I (2018) An EnOcean wearable device with fall detection algorithm integrated with a smart home system. IFAC-PapersOnLine 51(10):9–14. ISSN 2405-8963. https://doi.org/10.1016/j.ifacol.2018.06.228
15. Delahoz YS, Labrador MA (2014) Survey on fall detection and fall prevention using wearable and external sensors. Sensors 14(10):19806/19842
16. Aslan M, Sengur A, Xiao Y, Wang H, Ince MC, Ma X (2015) Shape feature encoding via sher vector for efficient fall detection in depth-videos. Appl Soft Comput 37:1023/1028
17. Gjoreski H, Lustrek M, Gams M (2011) Accelerometer placement for posture recognition and fall detection. In: Proceedings of 7th international conference on intelligent environments 2011, IE 2011, pp 47–54. https://doi.org/10.1109/IE.2011.11
18. Sucerquia A, López JD, Vargas-Bonilla JF (2018) Real-life/real-time elderly fall detection with a triaxial accelerometer. Preprints. https://doi.org/10.20944/11.0087,v3
19. Diana Y, de Puga JS, Palau C, Esteve M (2018) Fall detection system for elderly people using IoT and big data. In: The 9th international conference on ambient systems, networks and technologies

20. Ozdemir, A.T., Barshan, B. Detecting falls with wearable sensors using machine learning techniques. Sensors 14(6), 10691/10708, 2014.
21. Zhang C, Lai C-F, Lai Y-H, Wu Z-W, Chao H-C (2017) An inferential real-time falling posture reconstruction for Internet of healthcare things. J Netw Comput Appl 89:86–95. ISSN 1084-8045. https://doi.org/10.1016/j.jnca.2017.02.006

Hardware Implementation of Pyramidal Histogram of Oriented Gradients

P. Purushothaman, S. Srihari, Alex Noel Joseph Raj, and M. Bhaskar

Abstract The field of computer vision is gaining more importance because of its varied applications in numerous domains. Object detection and recognition are the most widely used applications. Intelligent and flexible Field Programmable Gate Array—FPGA architectures are replacing the current CPU-based hardware architectures for implementing image processing algorithms. We propose a hardware implementation of a widely used algorithm in object detection—Pyramidal Histogram of Oriented Gradients – PHOG. We have used Vivado High-Level Synthesis—HLS software to implement the PHOG algorithm, through which we can code in the algorithmic level using C/C+ +. In this research work, we explain the PHOG block's implementation and all other underlying blocks. We also discuss the optimizations and modifications introduced in the algorithm and compare the results.

Keywords FPGA · PHOG · High-level synthesis

1 Introduction

Real-time object recognition plays a significant role in the field of computer vision with a myriad of applications. In computer vision, there is an increasing need for compute-intensive algorithms implemented in power-efficient and compact hardware. This can be achieved by developing optimized algorithms and designing FPGA-based implementations. CPUs are optimized for single process systems where code is executed in a sequential manner, they do not possess the ability for parallelization. CPUs are general-purpose chips and do not pose any specialized hardware

P. Purushothaman (✉) · S. Srihari · M. Bhaskar
National Institute of Technology, Tiruchirappalli, Tamil Nadu 620015, India
e-mail: 108117077@nitt.edu

M. Bhaskar
e-mail: bhaskar@nitt.edu

A. N. J. Raj
Shantou Technological University, Guangdong, China
e-mail: jalexnoel@stu.edu.cn

G. Gupta et al. (eds.), *Proceedings of Academia-Industry Consortium for Data Science*, Advances in Intelligent Systems and Computing 1411,
https://doi.org/10.1007/978-981-16-6887-6_6

advancements. FPGA provides with much more flexibility on performing specific high-speed tasks. They are ideal for parallel systems where multiple processes are to be performed simultaneously. FPGAs are programmable chips and their functionality can be modified multiple times.

A generic object detection algorithm's task flow can be expressed in two steps. 1, obtaining a feature description of given input and 2, attributing a class to the descriptor using a feature-based machine learning technique. The PHOG algorithm [1–3] follows a similar scheme. It is based upon the descriptors of the image contour and extracts the sensitive features to scales. The descriptors are then fed into a support vector machine—SVM for classification. The PHOG descriptor encapsulates overall features and suggests that local variance matching can be a powerful method for predicting perceptual resemblances between objects [4, 5].

Implementing object detection algorithms in FPGA have been an active area of investigation [6–9]. Numerous optimization techniques and modifications related to PHOG are being introduced [10–13]. We shall be using the optimizations to improve the PHOG hardware architecture. We used Xilinx's Vivado High-Level Synthesis 2019.2(VHLS) [14, 15] for this research. Our model focuses on implementing PHOG on the Zynq SoC platform considering its cost and features.

This paper is organized as follows: Section 2 illustrates the fundamentals of the PHOG algorithm, Sect. 3 discusses the implementation in hardware and the optimizations introduced, in Sect. 4, we explain various experiments done and their results. Finally, the research is summarized and concluded in Sect. 5.

2 PHOG Algorithm

This algorithm which we are implementing is based upon the work done by Andrew Zisserman [1]. The process comprises of the following steps. Consider a grayscale image stored in an input array "I." The horizontal and vertical directions from top left corner are defined as X and Y, respectively. Gradients of a pixel along X and Y directions are (G_x, G_y). For every pixel in the image, we find the gradient magnitudes (M) and orientation angles (θ) by the following equations.

$$G_x(x, y) = I(x + 1, y) - I(x - 1, y) \tag{1}$$

$$G_y(x, y) = I(x, y + 1) - I(x, y - 1) \tag{2}$$

$$M(x, y) = \sqrt{G_x(x, y)^2 + G_y(x, y)^2} \tag{3}$$

$$\theta(x,\ y) = \arctan\big(G_y(x, y)/G_x(x, y)\big) \tag{4}$$

After computing magnitudes and orientation angles for all the pixels, now, the image is sectioned into multiple equally sized segments. The number of times this splitting happens is called the level of PHOG algorithm. For a 3 level PHOG, the image is split into 1, 4, and 16 segments at 1st, 2nd, and 3rd levels, respectively (1st level signifies the same image). For each image segment in every level, a histogram of oriented gradients is generated. The magnitude of pixels in the corresponding image is added into bins based upon their angle values. The number of bins of the histogram can be varied, however, increasing the bins produce no significant improvement after a limit [16].

Once all the histograms are computed, they are concatenated together to form the PHOG descriptor of the given image [17]. The numerical values of the returned descriptors are not normalized and can range from 0 to very large numbers. In order to normalize the descriptor, the sum of squares of its elements are found and they are divided by the sum. The process can be mathematically presented in Eq. (5). Here, H and H' represent the unnormalized and L2 normalized descriptors, respectively. ε is an insignificant value added to avoid zero division.

$$H'(k) = H(k) \Big/ \left(\sum_{i=0}^{N} H(i)^2 + \varepsilon \right) \quad (5)$$

3 Implementation in Hardware

Since the algorithm from [1] was intended for conventional CPU-based architectures, we introduced few important modifications to the algorithm's implementation in order to make the FPGA execution more efficient. In the following paragraphs, we shall discuss the modules which constitute the PHOG architecture and the optimizations that were introduced.

- Gradient Computation Block: As a first step in the algorithm, the gradients Gx, Gy are to be found for all the pixels (Eqs. 1, 2). The pixel values are sequentially fetched from the RAM and fed into a subtractor for computing corresponding Gx and Gy values. After obtaining these values, we need to compute the magnitude of gradients for each pixel (Eq. 3). To find the orientation angles, we can use inbuilt trigonometric functions available in Vivado HLS. These functions are designed based upon the CORDIC algorithm. However, we utilized a different approach to implement it. The next section describes the block in detail.

- Arctangent Computation and Binning: The purpose of finding angle of a gradient pair is for determining the bin in which the corresponding magnitude gets added to. In this work, all the histograms will contain 8 bins. Each angular bin stores the magnitudes of those pixels in the image, whose phase lies within the range of phase values represented by the bin. The bin angles are chosen in steps of 22.5° from

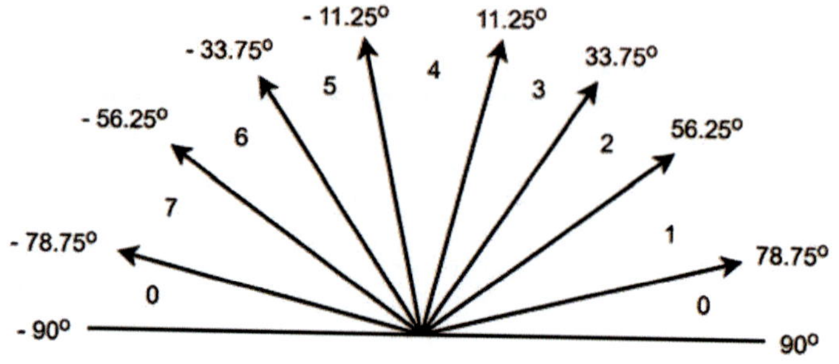

Fig. 1 Angle-bin partitions

− 90° to + 90° as shown in Fig. 1. Implementing the arctangent function using the inbuilt CORDIC function is not an optimal method because of the resource utilization and increased latency. Hence, an approximation method is used [12]. Equation 4 can be modified to Eq. 6 with which, instead of finding arctangent for all the pixels, now we can compare them with fixed tangent constants at the edge cases and assign them a bin index through Eq. 7.

$$G_y(x, y) = G_x(x, y) * tan(\theta(x, y)) \tag{6}$$

$$|G_x(x, y)| * \tan(\theta_1) * 2^n \leq \left|G_y(x, y)\right| * 2^n \leq |G_x(x, y)| * \tan(\theta_2) * 2^n \tag{7}$$

With this modification, instead of finding arctangent for all the pixels, now we can compare them with fixed tangent constants at the edge cases and assign them a bin index. The numerical values of tan(θ(x, y)) in equation are real numbers and are to be stored in floating point look-up tables. Floating-point representations require more resources and increases latency further. Hence, to overcome this issue, we multiply both sides of equation 7 with powers of 2 and round off the tanθ constants. Now, these values can be stored as integers with minimal loss of precision. In binary system, shifting corresponds to multiplying with powers of 2. Figure 2 explains the operations involved in the arctangent approximation block where the tangent constants were multiplied by 256, i.e., left shifted by 8 bits.

- Binarization of Descriptors: The process of comparing a value with a fixed threshold value and assigning a 0/1 is called binarization. The elements of normalized PHOG descriptor obtained using Eq. 5 are of floating-point type. A large-sized bus is needed to transfer it to the machine learning model for further use. Converting the floating-point values into a single bit value by comparing it with a threshold reduces the transmission overhead (Fig. 3).

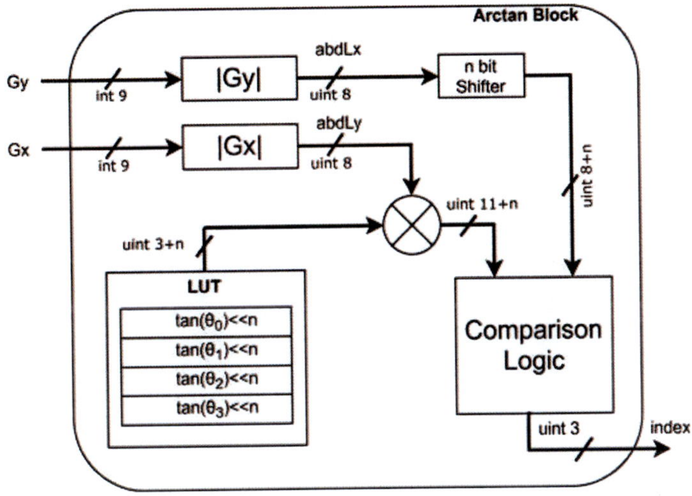

Fig. 2 Architecture of the custom arctangent computation block

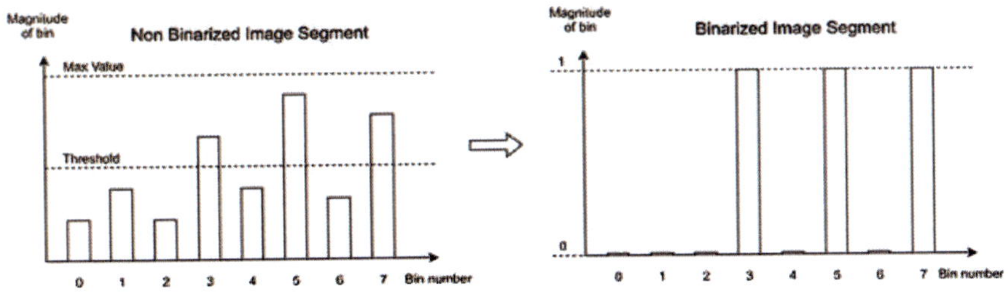

Fig. 3 Binarization of vectors

4 Experiments and Results

4.1 The Bit Width of the Tangent Constants

The tangent values of reference angles, stored in memory, are responsible for the accuracy of finding the bin index of a pixel. We experimented, storing the tangent values by left shifting 8 bits (multiplying by 2^8) in the memory, and left shifting Gx and Gy by 8 bits and use them in Eq. 6 to get Eq. 7. Equation 6 implies that multiplication of decimals and comparing them is more hardware intensive and can increase latency than multiplying and comparing integers. We varied the bit width for left shifting from 4 to 14 bits and the results are given below. For a bit width of 8, the RMSE error with 16 bit reference values and the hardware usage were minimal and latency was also low. The results are shown in Fig. 4.

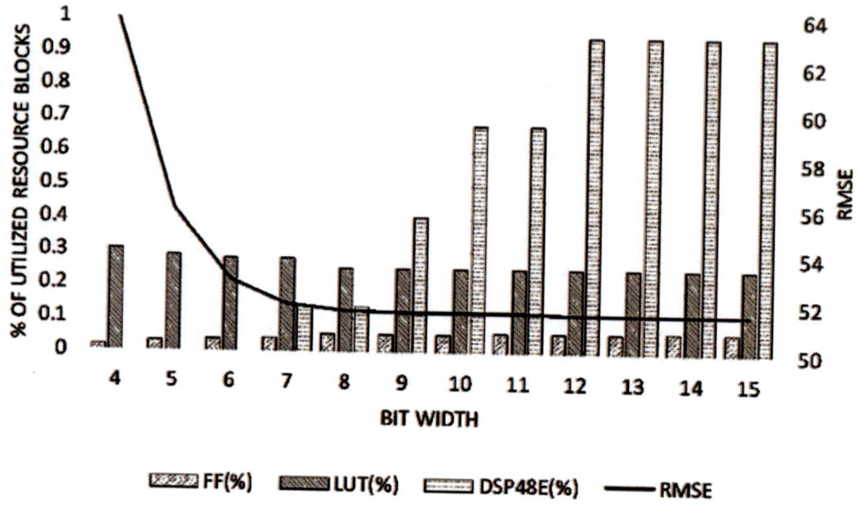

Fig. 4 Utilization versus RMSE for various bit widths

4.2 *Inbuilt Versus Optimized Arctangent Function*

We compared the performance of our optimized arctangent block with the inbuilt HLS function block in finding the bin index, based on hardware usage and latency. The arctan function used in the inbuilt HLS method is resource-intensive and increases latency. Since we require bin index of a pixel and not the orientation angle of the pixel, the approximation method used in Eq. 7 helped to reduce the resource usage as well as the latency. The results of the comparison between both methods are shown in Fig. 5.

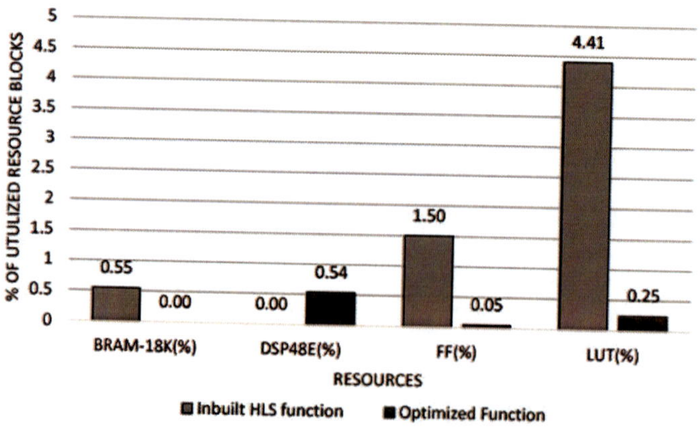

Fig. 5 Utilization comparison between inbuilt and optimized arctangent functions

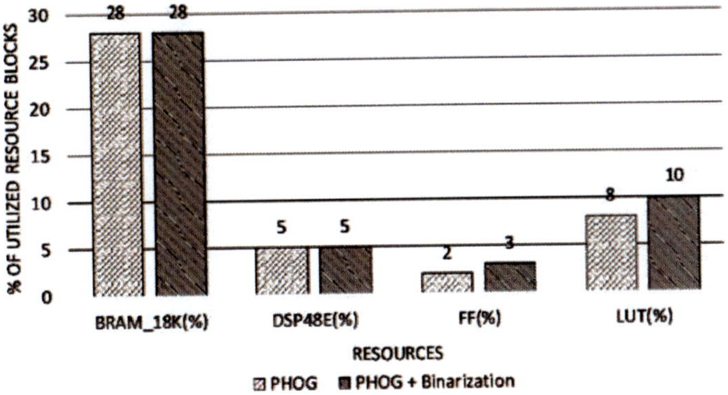

Fig. 6 Utilization comparison of the algorithm with and without binarization

4.3 *Effect of Binarization*

The binarization process reduces the size of a "n"-element PHOG descriptor obtained from Eq. 5. Assuming floating-point values require 2 bytes to represent, the size of "n"-element descriptor consisting of floating values of size "$n \times 16$" bits get reduced to "$n \times 1$" bits—16 times reduction in output size. We found that this can be achieved without a significant increase in hardware resource usage, and the results are given in Fig. 6. The data lost due to binarization may be countered by using stronger machine learning models which is out of scope of this work.

4.4 *Pipelined Versus Non-Pipelined Architecture*

Pipelining is the process of executing blocks with independent inputs simultaneously for efficient hardware utilization. In this experiment, we used pipeline directives in Vivado HLS to pipeline a few blocks in the PHOG algorithm and compared its performance with a non-pipelined version. The latency of pipelined and non-pipelined versions were **8.708** and **11.321 ms**. It was found that the pipelined version can be used if "latency less than 10 ns" is prefered, and the non-pipelined version can be used if minimal resource usage is of more importance. The results are given in Fig. 7.

5 Conclusions

Through this research work, the PHOG algorithm was successfully implemented using Vivado HLS 2019.2 on the Xilinx Zynq Zedboard. The images used for testing were obtained from [18]. The optimization techniques followed for constructing

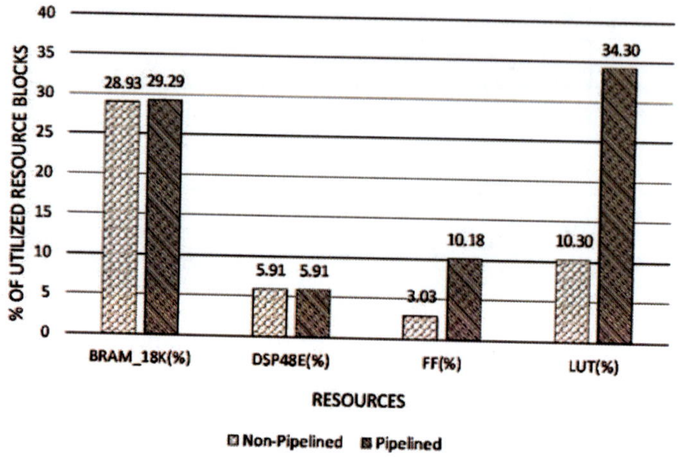

Fig. 7 Utilization comparison between pipelined and non-pipelined architectures

the PHOG descriptor resulted in minimized resource usage. We reduced the arctangent block's utilization by implementing an approximation method and, improved bin index calculation accuracy by comparing bit width vs accuracy. The pipeline vs non-pipeline architectures were contrasted. The non-pipelined and pipelined models achieved a speed of **115.50 MHz** and **110.18 MHz**, respectively. The maximum latency in the non-pipelined and pipelined architectures were **11.321 ms** and **8.708 ms**, respectively.

This research paper introduced a novel FPGA architecture for the PHOG algorithm implemented using High-Level Synthesis flow, discussed the optimization techniques used, and compared the performance of the pipelined and non-pipelined versions of the same.

References

1. Bosch A, Zisserman A, Munoz X (2007) Representing shape with a spatial pyramid kernel. In: Proceedings of the 6th ACM international conference on Image and video retrieval
2. Saïdani A, Echi AK (2014) Pyramid histogram of oriented gradient for machine-printed/handwritten and Arabic/Latin word discrimination. In: 2014 6th international conference of soft computing and pattern recognition (SoCPaR). IEEEs
3. Tan ZR, Tian S, Tan CL (2014), Using pyramid of histogram of oriented gradients on natural scene text recognition. In: 2014 IEEE international conference on image processing (ICIP), Paris, France, pp 2629–2633. https://doi.org/10.1109/ICIP.2014.7025532
4. Bai Y et al (2009) A novel feature extraction method using pyramid histogram of orientation gradients for smile recognition. In: 2009 16th IEEE international conference on image processing (ICIP). IEEE
5. Zhang B et al (2010) Historic Chinese architectures image retrieval by SVM and pyramid histogram of oriented gradients features. Int J Soft Comput 5(2):19–28

6. Lee SE, Min K, Suh T (2013) Accelerating histograms of oriented gradients descriptor extraction for pedestrian recognition. Comput Electr Eng 39(4):1043–1048
7. Mizuno K et al Architectural study of HOG feature extraction processor for real-time object detection. In: 2012 IEEE workshop on signal processing systems. IEEE
8. Hemmati M et al (2014) HOG feature extractor hardware accelerator for real-time pedestrian detection. In: 2014 17th Euromicro conference on digital system design. IEEE
9. Nuno-Maganda MA, Arias-Estrada MO (2005) Real-time FPGA-based architecture for bicubic interpolation: an application for digital image scaling. In: 2005 international conference on reconfigurable computing and FPGAs (ReConFig'05). IEEE
10. Li Y, Su G (2015) Simplified histograms of oriented gradient features extraction algorithm for the hardware implementation. In: 2015 international conference on computers, communications, and systems (ICCCS). IEEE
11. Kadota R et al (2009) Hardware architecture for HOG feature extraction. In: 2009 fifth international conference on intelligent information hiding and multimedia signal processing. IEEE
12. Negi K et al (2011) Deep pipelined one-chip FPGA implementation of a real-time image-based human detection algorithm. In: 2011 international conference on field-programmable technology. IEEE
13. Cao TP, Deng G (2008) Real-time vision-based stop sign detection system on FPGA. In: 2008 digital image computing: techniques and applications. IEEE
14. Georgopoulos K et al (2016) An evaluation of vivado HLS for efficient system design. In: 2016 international symposium ELMAR, Zadar, pp 195–199. https://doi.org/10.1109/ELMAR.2016.7731785
15. Feist T (2012) Vivado design suite. Xilinx Inc., San Jose, CA, USA, White Paper 5, 2012, p 30
16. Dalal N, Triggs B (2005) Histograms of oriented gradients for human detection. IEEE computer society conference on computer vision and pattern recognition (CVPR'05), vol 1, pp 886–893
17. Lazebnik, Svetlana, Cordelia Schmid, and Jean Ponce. "Beyond bags of features: Spatial pyramid matching for recognizing natural scene categories." 2006 IEEE Computer Society Conference on Computer Vision and Pattern Recognition (CVPR'06). Vol. 2. IEEE, 2006
18. Burkardt J (2020) TIFF Files, Accessed on: Aug. 21, 2020. [Online]. Available http://people.math.sc.edu/Burkardt/data/tif/tif.html

Manoeuvre of Machine Learning Algorithms in Healthcare Sector with Application to Polycystic Ovarian Syndrome Diagnosis

Piyush Bhardwaj and Parul Tiwari

Abstract Artificial intelligence (AI) is the technology that is transforming every step of life and especially valuable when issue is of healthy lives. It consists wide ranging tools that enable a person to rethink how one can integrate knowledge, use data and interpret the cognizance to improve decision-making. The present work focuses on the importance of machine learning algorithms in healthcare sector. This study uses the clinical data to diagnose polycystic ovarian syndrome in patients based on different features. A detail descriptive analysis is performed before applying any AI technique to understand the nature of the data, and any discrepancy observed is resolved by pre-processing of the data. Features were selected using Pearson correlation, and several advanced classification algorithms such as support vector machine, random forest classifier, XG boost and multi-layer perceptron are applied, analysed and compared to get the overall accuracy for correct prediction. Train–test split approach is used to overcome the problem of overfitting. Overall performance of the model is discussed and validated based on precision, true positive rate, true negative rate, F-1 and ROC-AUC score. According to the results, support vector machine with RBF kernel and multi-layer perceptron model achieved highest accuracy of 93% with true positive rate of 0.89. Moreover, decision tree classifier was implemented in order to calculate the feature importance using Gini index in which average left follicle size, hair growth, number of left follicles, prolactin and average right follicle size were ranked highest. Data visualization and results are also supplied with recently developed machine learning tools. It is worth noting that obtained results are in good agreement with the existing literature.

Keywords Polycystic ovarian syndrome · Machine learning · Multi-layer perceptron

P. Bhardwaj · P. Tiwari (✉)
Centre for Advanced Computational Solutions (C-fACS), Lincoln University, Lincoln, New Zealand

G. Gupta et al. (eds.), *Proceedings of Academia-Industry Consortium for Data Science*, Advances in Intelligent Systems and Computing 1411,
https://doi.org/10.1007/978-981-16-6887-6_7

1 Introduction

"The key to artificial intelligence has always been the representation".—Jeff Hawkins. There is no such formal definition of artificial intelligence (AI) in literature and is defined in several ways by experts and scientists. It is a subfield of computer science, and machine learning (ML) is one of the ways to achieve AI. Several ML algorithms are being used for analysis and prediction of disease progression. These algorithms together with human and machine capabilities not only provide evidence-based decision-making, coordinate care and support in patients, but also reduce cost and time involved in clinical tests. Though AI provides a broad-spectrum in almost all healthcare sectors, its success relies on keeping some key concepts in mind: to recognize what kind of AI technology is to be used for a particular task; to set out the context/circumstances of its proper use; inaugurating performance standard; to make sure the safety, security and privacy of the patients and following the medical ethics. It is really crucial and important to pre-diagnose and prognosis of critical diseases such as cancer, Alzheimer, dementia, diabetes mellitus and others. Literature shows the importance of ML algorithms in prediction of these diseases with high degree of accuracy. Several classification algorithms have been broadly applied in prediction of cardiovascular diseases and improved the development and understanding of clinical practices [1, 2]. A number of review papers are available in literature for predicting diabetes mellitus at early onset [3–5]. Several aspects of cancer based on gene expression signatures are discussed in [6, 7].

All machine learning algorithms are developed on strong mathematical background, and their exposition is clear and understandable. Pure statistical methods face challenges while using multi-omics data for cancer prognosis; however, ML techniques have shown good results in this direction [8, 9]. Machine learning techniques are designed to keep in mind that real data may be noisy and biased. Renowned organizations and universities are setting up advanced research centres which are responsible for developing new AI techniques useful for healthcare. For example: Apple's health App, Mayo clinic's text analyser and Google's DeepMind laboratory and many more. However, not only in medical science, but machine learning is also providing effective solutions in several other fields including retail banking, manufacturing, E-Commerce, Logistics and Supply Chain Management. Thus, there is no harm in saying that machine learning is the process that powers many of the services we use today and is deriving the lives.

PCOS is a complex endocrine disorder related to elevated level of certain hormones affecting 5–10% woman in reproductive age. Mainly, increased level of two hormones (insulin and testosterone) is responsible for PCOS. As per the World Health Organization reports, percentage of PCOS patients is highly fluctuating ranging from 2.2% to 26% globally. A thorough study to manage the dysregulated functions of PCOS, its prediction and most responsible features using machine learning techniques is performed in [10, 11]. The objective of current study is to understand the true potential of AI in accurately predicting the possibility of a disease in a patient with reference to polycystic ovarian syndrome [PCOS] and to gain insights about how one can predict the onset of this disease based on available features and parameters.

2 Material and Methods

Data analysis throughout this study was conducted using various libraries such as Scikit learn, Pandas and Matplotlib in Python environment (Version 3.7).

2.1 *Data Visualization*

The PCOS dataset were downloaded from Kaggle which contains two type of data (https://www.kaggle.com/prasoonkottarathil/polycystic-ovary-syndrome-pcos), PCOS without infertility and with infertility. For machine learning analysis, we chose data without fertility which contains the medical profile of 541 female patients (Table 1). Dataset contains PCOS related medical information about all patients and has in total 41 independent variable and one dependent variable. Data visualization is done for few features (Fig. 1) with different labels. There are no missing values in the data. Figure 1 shows some violin plots for selected features on vertical axis with respect to two different labels.

2.2 *Data Pre-Processing and Dimensionality Reduction*

Before analysis, patient file number and serial number were removed from dataset. After that, correlation coefficient is calculated with the help of Pearson correlation method. Only one feature among highly correlated group were selected for further analysis. Features were considered highly correlated with correlation score higher than 0.8 and above. Descriptive analysis of the data is presented in Table 2. Next step after feature selection is the scaling of the features. For current study, standard scaling method is applied and ensured that values of all the features are centred around zero mean and unit standard deviation.

2.3 *Model Building and Evaluation*

Train, test and split approach is used for machine learning analysis where models are trained with 70% (378 samples) of the total data and remaining 30% (163 samples) are used to test the model. Main aim of doing this is to avoid machine learning models being over fitting and under fitting. Several classification algorithms are used in this

Table 1 Data information

Name	Abbreviation	value	Name	Abbreviation	value
Serial number	SI. No	1–541	Waist–hip ratio	WHR	0.75–0.979
Patient file number	Patient file No	1–541	TSH (Miu/L)	TSH	0.04–65
Polycystic ovarian syndrome	Label	1(yes) and 0(no)	AMH (ng/mL)	AMH	0.1–66
Age (years)	Age	20–48	PRL (ng/mL)	PRL	0.4–128.4
Weight (kg)	Weight	31–108	PRG (ng/Ml)	PRG	0.25–0.35
Height (cm)	Height	137–180	Vitamin D3 (ng/mL)	Vit D3	6.07–90
Pulse rate (BPM)	PR	70–82	Weight gain	WG	1-yes, 0-no
RR (breathe/min)	RR	16–28	Hair growth	HG	1-yes, 0-no
Haemoglobin (g/dl)	HB	8.5–14.8	Blood pressure diastolic (mmHg)	Diastolic	60–100
Cycle (R/I)	Cycle	2-May	Blood pressure systolic (mmHg)	Systolic	120–140
Cycle length (days)	CL	0–12	Body mass index	BMI	
Marriage status (years)	Marriage	0–30	Follicle number (right)	Right follicle	0–22
Pregnant	Pregnant	1-yes, 0-no	Follicle number (left)	Left follicle	0–20
Number of abortions	Abortions	0–5	Average follicle size (left)(mm)	Average left follicle size (ALFS)	0–24
I BetaHCG (mIU/ml)	Beta I	1.3–32,460.97	Average follicle size (right)(mm)	Average right follicle size (ALRS)	0–24
II BetaHCG (mIU/ml)	Beta II	0.99–250,001.99	Endometrium (mm)	Endometrium	0–18
FSH (mIU/mL)	FSH	0–65.4	Fast food	Fast food	1(yes) 0 (no)
LH (mIU/mL)	LH	0.02–20.18	Exercise	Exercise	1(yes) 0 (no)

(continued)

Table 1 (continued)

Name	Abbreviation	value	Name	Abbreviation	value
FSH/LH	FLR	0.21456–327	Skin darkening	Skin darkening	1(yes) 0 (no)
Hip(inch)	Hip	26–48	Hair loss	HL	1(yes) 0 (no)
Waist (inch)	Waist	24–47	Pimples	Pimples	1(yes) 0 (no)
Blood group	BG	$A^{+}= 11$, $A^{-}=$ 12, B += 13, B −= 14, O += 15, O −= 16, AB += 17, AB −= 18	RBS (mg/dl)	RBS	60–350

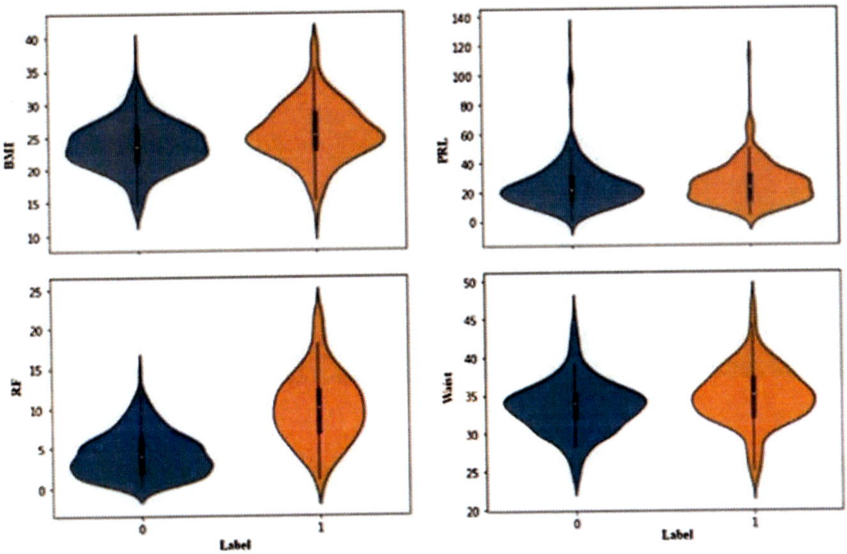

Fig. 1 Pictorial representation (Violin Plots) of some features with different labels

study such as random forest, K-nearest neighbour, logistic regression, multi-layer perceptron, support vector machines (SVM) with and without kernel trick and XG boost classifier. In the end, we selected best two classifiers namely MLP and SVM with RBF kernel based on their performance to predict PCOS. MLP classifier is the feed forward neural network defined by the combination of different layers with some number of identical units.

Equation (1) demonstrates the neuron-like processing unit where a is the unit activation, x_j are the inputs to the unit, b is the bias, w_j are the weights and ϕ is the nonlinear activation function [12]. Every layer is fully connected with next

Table 2 Descriptive analysis of numerical data

Feature	Mean	Standard deviation	Min. value	Max. value	Feature	Mean	Standard deviation	Minimum value	Maximum value
BMI	24.31	4.056	12.4	38.90	AMH	5.61	5.8811	0	66.00
Diastolic	76.93	5.574	8	100.00	FSH	14.60	217.022	0.21	5052.00
RF	6.13	4.229	0	22.00	LH	6.47	86.673	0.02	2018.00
Left Follicle	6.64	4.436	0	20.00	Marriage	7.67	4.8055	0	30.00
ALFS	15.02	3.566	0	24.00	Pregnant	0.38	0.486	0	1.00
ALRS	15.48	3.253	0	24.00	Abortion	0.29	0.6925	0	5.00
Endometrium	8.49	2.134	0	18.00	HB	11.16	0.8669	8.5	14.80
Fast food	0.51	0.5	0	1.00	Cycle	2.56	0.9019	2	5.00
Exercise	0.25	0.432	0	1.00	CL	4.94	1.492	0	12.00
Systolic	114.66	7.384	12	140.00	BG	13.80	1.8408	11	18.00
Skin darkening	0.31	0.461	0 ara>	1.00	PR	73.25	4.4302	13	82.00
HL	0.45	0.498	0	1.00	RR	19.24	1.6886	16	28.00
Pimples	0.49	0.5	0	1.00	Height	156.48	6.0335	137	180.00
RBS	99.84	18.559	60	350.00	Age	31.43	5.411	20	48.00
WG	0.38	0.485	0	1.00	Beta I	664.55	3348.92	1.3	32,460.97
HG	0.27	0.446	0	1.00	Beta II	238.23	1603.82	0.99	25,000.00
PRL	24.32	14.97	0.4	128.24	WHR	0.89	0.0463	0.75	0.98
Vit D3	49.92	346.206	0	6014.66	Hip	37.99	3.9679	26	48.00
PRG	0.61	3.8088	0.05	85.00	Weight	59.64	11.0282	31	108.00
Waist	33.84	3.59689	24	47.00	FLR	6.90	60.6918	0.002	1372.83
TSH	2.98	3.7555	0.04	65.00	Label	0.33	0.46961	0	1.00

layer together with a bias neuron. Input layer is the first layer taking inputs from the features followed by one or more hidden layer and an output layer. Hidden layers contain input units, output units and hidden units, and its mathematical computation is demonstrated in Eq. (2). For binary classification problem, output layer contains single unit for each value, on the other hand, it contains N units for N-class classification.

$$a = \phi(\sum_j w_j x_j + b) \tag{1}$$

$$h_i^{(1)} = \phi^{(1)}\left(\sum_{ij} w_{ij}^{(1)} x_j + b_i^{(1)}\right) \tag{2}$$

Another best classifier concluded in this study is SVM with RBF kernel. It performs classification by finding the best possible hyperplane to separate classes [13, 14]. The results of the SVM algorithm rely on the support vector points because they are the points near to the hyperplane. The goal of SVM is to find a hyperplane that has maximum margin from the closest points. In other words, the goal is to maximize the minimum distance for perfect separation. SVM algorithm has a special feature called kernel techniques with the objective of providing accurate results. Kernels are the set of mathematical functions, and linear, RBF, polynomial and sigmoid are some of the widely used kernels as part of SVM. Equation (3) explains how kernels work where K is the kernel function, x and y are n-dimensional inputs, f defines the map from n dimensional space to m dimensional space (generally n is smaller than m) and $\langle x, y\rangle$ signifies the dot product.

$$K(x, y) = \langle f(x), f(y)\rangle \tag{3}$$

All the machine learning models in this study are evaluated based on the classification report that consists of accuracy, precision, sensitivity, specificity and F1 score. Classification report helps in comparative study of various classification algorithms and guides in choosing the best model for future prediction. Accuracy is defined as the percentage of correct predictions out of all predictions. Moreover, classification models are evaluated based on ROC-AUC score.

2.4 Feature Importance

Feature importance techniques play critical role in finding the important input variables within the data by assigning score to the features. This can improve the efficiency and effectiveness of machine learning models. In this study, we implemented decision tree classifier for the feature importance.

3 Results

3.1 *Dimensionality Reduction Using Pearson Correlation Method*

Pearson correlation method is used to determine the correlation among independent variables and for dimensionality reduction. According to the analysis, BMI is highly correlated with weight, and waist is highly correlated with hip. In addition to that, FSH is found to be in correlation with FLR. Hence, BMI, FSH and waist are chosen together with other features for machine learning analysis.

3.2 *Model Evaluation*

Main aim of this study is to predict PCOS (Class 1), so class 1 is treated as positive chances and class 2 as negative chances. Eight machine learning classifiers are implemented and evaluated based on the accuracy, ROC-AUC, precision, sensitivity, specificity and F1 score. All the models are trained on the 70% (378 samples) and tested on remaining 30% (163 samples). In terms of accuracy, all the classifiers predicted true positives and true negatives with overall accuracy from 87 to 93% (Fig. 2) given as Eq. (4). Among all the classifiers, MLP and SVM with RBF kernel performed exceptionally well by classifying 93% samples correctly.

$$\text{Accuracy} = \frac{\text{Number of correct predictions}}{\text{Total number of predictions}} \tag{4}$$

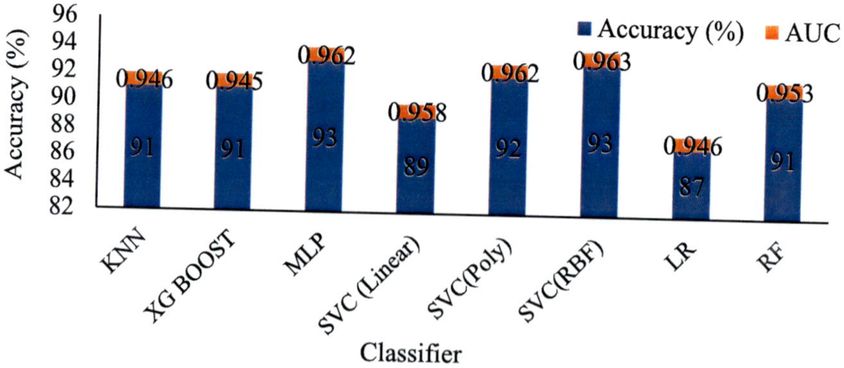

Fig. 2 Comparison between overall accuracy (blue colour) achieved by different classifiers together with ROC- AUC score (orange colour) (SVC is support vector classifier for SVM)

It is advisable to evaluate machine learning models other than accuracy; hence, we measured ROC-AUC score by plotting true positive rates versus false positive rates. ROC curve works on the notion of the decision variable, where curve near to the upper left-hand corner by any classifier considered as best in differentiating between classes.

In addition, ROC-AUC score area covered by the classifiers were calculated. ROC-AUC score achieved by all classifiers was higher than 0.945. MLP and SVM (RBF) attained highest score of 0.962 and 0.963, respectively (Fig. 3).

Almost all the classifiers achieved an acceptable precision score for class 0. However, MLP and SVM with RBF and polynomial kernel predict class 0 with highest precision of 0.96. On the other hand, KNN classifier outperformed all the classifiers in terms of precision for class 1 by attaining 0.90 score. Precision for

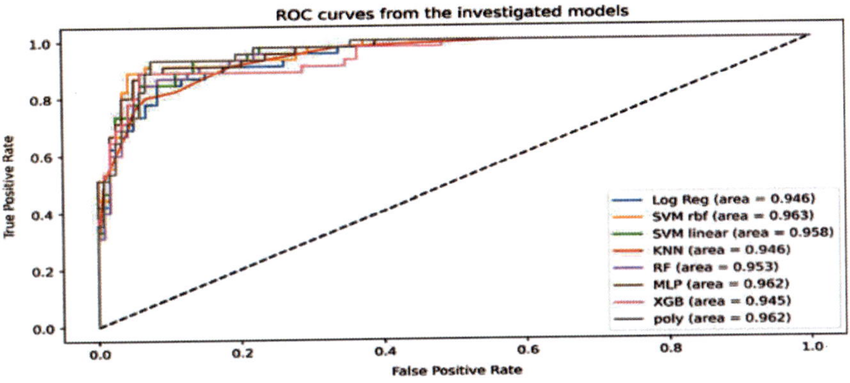

Fig. 3 ROC curves (true positive rate against true negative rate) achieved by different classifiers

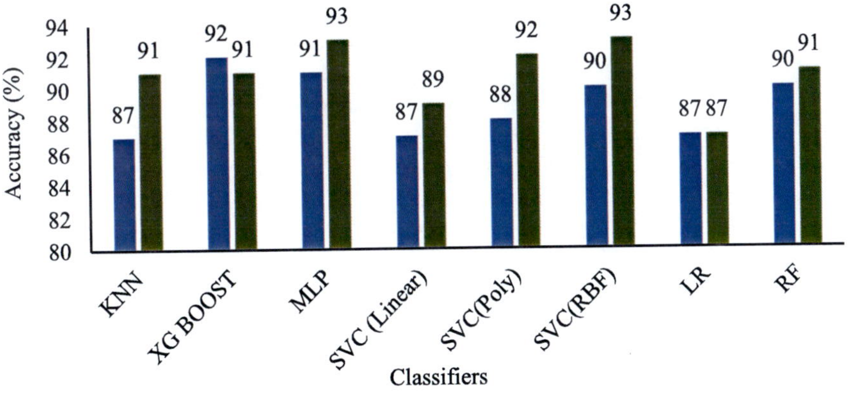

Fig. 4 Comparison between overall accuracy achieved by different classifier before (blue colour) and after dimensionality reduction using Pearson correlation (green colour) (SVC is support vector classifier for SVM)

Table 3 Model performance of different classifiers

Classifier	Label	Precision	Sensitivity/TPR	F1 score	Specificity/TNR
KNN	0 1	0.92 0.90	0.78	0.89	0.96
LR	0 1	0.93 0.84	0.87	0.85	0.87
SVM (RBF)	0 1	0.96 0.85	0.89	0.91	0.94
SVM (linear)	0 1	0.94 0.78	0.84	0.87	0.91
SVM (polynomial)	0 1	0.96 0.82	0.85	0.90	0.92
Random forest	0 1	0.93 0.84	0.82	0.89	0.94
MLP	0 1	0.96 0.85	0.89	0.91	0.94
XG boost	0 1	0.93 0.86	0.80	0.88	0.95

positive class is defined in Eq. (5) and for negative class is explained by Eq. (6).

$$\text{Precision} = \frac{\text{True Positives}}{\text{True Positives} + \text{False Positives}} \tag{5}$$

$$\text{Precision} = \frac{\text{True Negatives}}{\text{True Negatives} + \text{False Negatives}} \tag{6}$$

Similarly, classifier's performance was evaluated based on their ability to predict net true positive rate or sensitivity given by Eq. (7) and true negative rate or specificity given by Eq. (8). Sensitivity and specificity scores achieved by classifiers (Table 3) show that all the classifiers predict true negatives higher than the true positive.

As per the results, MLP and SVM with RBF kernel produced highest sensitivity score of 0.89 each. On the other hand, KNN and XG boost classifier scored highest specificity of 0.96 and 0.95, respectively. As far as the F1 score is concerned, MLP and SVM (RBF) achieved high values which are calculated by using Eq. (9). Overall, all the classifiers performed better for classifying true negatives than true positives during testing. This could be because of the lesser sample of class 1 present in the dataset and because of that models come across more with negative elements (class 0) during training.

$$\text{Sensitivity} = \frac{\text{True Positives}}{\text{True Positives} + \text{False Negatives}} \tag{7}$$

$$\text{Specificity} = \frac{\text{True Negatives}}{\text{True Negatives} + \text{False Positives}} \tag{8}$$

$$\text{F1} - \text{Score} = 2*\frac{\text{Precision*Recall}}{\text{Precision} + \text{Recall}} \tag{9}$$

4 Discussion

Present study provides evidence how machine learning can play critical role in health sector by predicting medical complications with higher accuracy. In order to provide evidence, we conducted a machine learning investigation to predict PCOS. According to the results, we found two classification algorithms (MLP and SVM-RBF) which can predict PCOS with overall accuracy of 93% each, true positive rate of 0.89 each and ROC-AUC score of 0.962 and 0.963, respectively.

By using Pearson correlation method, we reduced the dimension of the data with the aim to improve the accuracy of machine learning classifiers. Data with high dimension increases the distance, and it becomes harder to predict correct results [15]. Objective of dimensionality reduction is the removal of irrelevant and highly correlated features for faster and stable models with higher accuracy. Irrelevant and unwanted variable can affect the ability of the algorithms to differentiate between classes and can increase the modelling error. For example, KNN algorithm do not operate with stability in the presence of irrelevant variables because their presence increase the sample complexity exponentially [16]. Reduction of variables reduces the noise within the data and helps the models to perform better. As shown in Fig. 4, all the classifiers achieved high accuracy (except XG Boost and LR) after removing three variables from the dataset with the help of Pearson correlation method.

KNN algorithm predict correct outcomes with 91% accuracy after variable reduction, previously which was 87% before variable reduction. On the other hand, MLP and SVM (RBF) classifiers got accuracy of 93% from 91 and 90%, respectively, after variable reduction. Hence, it is quite important to implement variable reduction methods in order to predict correct true positives and true negatives. MLP and SVM with RBF kernel performed exceptionally well in both scenarios (with and without variable reduction) by predicted true positives (class 1) and true negatives (class 0) with higher accuracy as compared to other algorithms. As our focus is on the positive class, both the algorithms predict true positive with rate of 0.89 each higher than any other classifiers implemented in this study. According to the scores generated by decision tree classifier, 16 features are considered less important due to their ranking score (0).

Other 25 features as mentioned in Table 4 achieved score in range of 0.0034–0.43. Top five variables important in determining PCOS are average left follicle size (ALFS), hair growth (HG), left follicle, Prolactin (PRL) and average right follicle size (ALRS).

Decision tree works like a tree where branch indicate the decision rule, nodes demonstrate the nodes, and outcome is represented by leaf node. This algorithm identifies important feature using attribute selection measures such as information

Table 4 Feature ranking using decision tree classifier

Features	Score	Features	Score	Features	Score
ALFS	0.43131	AMH	0.01937	Vit D3	0.0077
HG	0.08613	Hip	0.01851	Skin darkening	0.007
Left follicle	0.07702	WG	0.01717	Exercise	0.0042
PRL	0.07166	Beta I	0.01655	TSH	0.0042
ALRS	0.03401	CL	0.012	RR	0.0034
BG	0.02204	Pregnant	0.01027	LH	0.01651
Beta II	0.01988	PRG	0.01008	Endometrium	0.01554
FSH	0.01964	Fast food	0.0098		
Age	0.01958	Weight	0.00947		

gain, Gini index and gain ratio. In this study, we predicted important feature by calculating Gini index or Gini impurity [17]. Equation (10) indicates the Gini impurity index in which P_i is the probability of an instance being classified as distinct class.

$$\text{Gini Index} = 1 - \sum_{i=1}^{n} P_i^2 \qquad (10)$$

Several studies on different types of PCOS data are available in literature. PCOS with 89.02% accuracy through RF classifier was done in [18]. Clinical and metabolic parameters are used to study PCOS in [19, 20], whereas in [20], ultrasonic images were used. RNA sequence and microarray datasets are used in [11] to diagnose PCOS patients and resulted in better performance than already used marker genes. In [10], 12 important features are suggested based on several classification algorithms applied on clinical PCOS data. Though, there is no such confirmed theory for important features in literature for diagnosing PCOS, but based on decision tree's Gini index, ranking of individual features is presented in Table (4). Present model predicted PCOS with overall accuracy of 93% (SVM-RBF and MLP) together with true positive rate of 0.89. As a summary, present study found two suitable classifier to predict PCOS in patients with overall accuracy of 93% and true positive rate of 0.89.

5 Conclusion

In this paper, we tried to explain the importance and usage of machine learning in healthcare sector. The machine learning algorithms are being implemented in medical science to diagnose several critical diseases at their onset. This paper contains the application of these algorithms to visualize the PCOS dataset, and predictions are made based on classification report. Current study suggests that multi-layer perceptron and support vector machines with radial basis function (SVM-RBF) have the same level of accuracy for the used set of data. Though the performance of machine learning methods is better as compared to traditional statistical methods, the accuracy level can be improved by using large datasets. It is our belief that with the advancement of technology and their easy application will ensure that machine learning will be used more frequently in hospitals, laboratories and clinical settings in the near future.

References

1. Marimuthu M, Abinaya M, Madhankumar KSK, Pavithra V (2018) A review on heart disease prediction using machine learning and data analytics approach. Int J Comput Appl 181(18):20–25.: https://doi.org/10.5120/ijca2018917863
2. Mathur P, Srivastava S, Xu X, Mehta JL (2020) Artificial intelligence, machine learning, and cardiovascular disease. Clin Med Insights Cardiol 14. (SAGE Publications Ltd). https://doi.org/10.1177/1179546820927404
3. Xiong Z et al (2018) A machine learning aided systematic review and meta-analysis of the relative risk of atrial fibrillation in patients with diabetes mellitus. Front Physiol 9:835. https://doi.org/10.3389/fphys.2018.00835
4. Swapna G, Soman KP, Vinayakumar R (2018) Automated detection of diabetes using CNN and CNN-LSTM network and heart rate signals. Procedia Comput Sci 132:1253–1262. https://doi.org/10.1016/j.procs.2018.05.041
5. Cai L, Wu H, Li D, Zhou K, Zou F (2015) Type 2 diabetes biomarkers of human gut microbiota selected via iterative sure independent screening method. PLOS ONE 10(10):e0140827. https://doi.org/10.1371/journal.pone.0140827
6. Zhu W, Xie L, Han J, Guo X (2020) The application of deep learning in cancer prognosis prediction. Cancers 12(3). (MDPI AG) https://doi.org/10.3390/cancers12030603
7. Cruz JA, Wishart DS (2006) Applications of machine learning in cancer prediction and prognosis. Cancer Inf 2:59–77. (Libertas Academica Ltd.) https://doi.org/10.1177/117693510600200030
8. Zhang G et al (2019) OSblca: a web server for investigating prognostic biomarkers of bladder cancer patients. Front Oncol 9:466. https://doi.org/10.3389/fonc.2019.00466
9. van IJzendoorn DGP, Szuhai K, Briaire-De Bruijn IH, Kostine M, Kuijjer IH, Bovée , JVMG (2018) Machine learning analysis of gene expression data reveals novel diagnostic and prognostic biomarkers and identifies therapeutic targets for soft tissue sarcomas. PLoS Comput Biol 15(2). https://doi.org/10.1371/journal.pcbi.1006826
10. Aggarwal S, Pandey K (2020) An analysis of PCOS disease prediction model using machine learning classification algorithms. Recent Patents Eng 15. https://doi.org/10.2174/1872212115999201224130204
11. Xie NN, Wang FF, Zhou J, Liu C, Qu F (2020) Establishment and analysis of a combined diagnostic model of polycystic ovary syndrome with random forest and artificial neural network. BioMed Res Int 2020. https://doi.org/10.1155/2020/2613091
12. Chen C, He W, Zhou H, Xue Y, Zhu M (2020) A comparative study among machine learning and numerical models for simulating groundwater dynamics in the Heihe River Basin, northwestern China. Sci Rep 10(1):1–13. https://doi.org/10.1038/s41598-020-60698-9
13. Doran G, Ray S (2014) A theoretical and empirical analysis of support vector machine methods for multiple-instance classification. Mach Learn 97(1–2):79–102. https://doi.org/10.1007/s10994-013-5429-5
14. Huang S, Nianguang CAI, Pacheco PP, Narandes S, Wang Y, Wayne XU (2018) Applications of support vector machine (SVM) learning in cancer genomics. In: Cancer genomics and

proteomics, vol 15, issue no 1. International Institute of Anticancer Research, pp 41–51. https://doi.org/10.21873/cgp.20063
15. Keogh E, Mueen A (2017) Curse of dimensionality. In: Encyclopedia of machine learning and data mining. Springer US, pp 314–315
16. Albattah W, Khan RU, Khan K (2020) Attributes reduction in big data. Appl Sci 10(14):4901. https://doi.org/10.3390/app10144901
17. Menze BH et al (2009) A comparison of random forest and its Gini importance with standard chemometric methods for the feature selection and classification of spectral data. BMC Bioinform 10(1):213. https://doi.org/10.1186/1471-2105-10-213
18. Denny A, Raj A, Ashok C, Ram M, George R (2019, October) I-HOPE: detection and prediction system for polycystic ovary syndrome (PCOS) using machine learning techniques. In: IEEE region 10 annual international conference, proceedings/TENCON, vol 2019, pp 673–678. https://doi.org/10.1109/TENCON.2019.8929674
19. Mehrotra P, Chatterjee J, Chakraborty C, Ghoshdastidar B, Ghoshdastidar S (2011) Automated screening of polycystic ovary syndrome using machine learning techniques. https://doi.org/10.1109/INDCON.2011.6139331
20. Cheng JJ, Mahalingaiah S (2019) Data mining polycystic ovary morphology in electronic medical record ultrasound reports. Ferti Res Prac 5(1):1–7. https://doi.org/10.1186/s40738-019-0067

Medical Image Authenticity Using Lifting Wavelet Transform

Anuj Bhardwaj

Abstract In medical science, imaging plays vital importance in the correct diagnosis of illness. Therefore, to take the opinion of medical experts available globally, imagery data is shared through the Internet over a secured communication environment. In this sharing process, there are chances to deform or steal the important information of a patient. Hence, there is a need to secure these medical images during the process of telemedicine. In the present work, a wavelet-based watermarking scheme is presented to secure the secret information, and LWT and SVD are used for this purpose. To show the effectiveness of the proposed scheme, simulation results are evaluated over a set of database taken from a local center and discussed. The PSNR and NCC are used as evaluation parameters.

Keywords Peak signal-to-noise ratio · Lifting wavelet transform · Normalized correlation coefficient · Singular value decomposition

1 Introduction

It has been widely observed that not only the telemedicine but also others like telediagnosis and teleophthalmology, etc., have been popular in almost every field of medical science [1, 2]. This process of sharing medical information always has the challenge to provide the security, authenticity, and ownership [3, 4]. The watermarking methods provide the solution to these problems by embedding some secret information called a watermark in the cover image. There are many ways of implementing watermarking in spatial and transform domains as [5–11]. Recently, it has been observed in the research community a great interest in digital watermarking for protecting the medical information [12].

Rastegar et al. [13] presented a scheme using radon transform and SVD. A watermarking scheme using RDWT and SVD is proposed by Nasrin et al. [14]. The scheme shows a good imperceptibility with a large capacity of embedding. Boonyapalanant et al. [15] proposed a watermarking scheme which is based on the concept of least

A. Bhardwaj (✉)
Jaypee Institute of Information Technology, Noida 201309, India

G. Gupta et al. (eds.), *Proceedings of Academia-Industry Consortium for Data Science*, Advances in Intelligent Systems and Computing 1411,
https://doi.org/10.1007/978-981-16-6887-6_8

significant bit for medical images. In this work, a QR code is considered as a watermark having the patient's details. Araghi et al. [16] proposed a hybrid watermarking scheme using DWT and SVD. In this scheme, encryption is used for improving the security. Authors have shown that its performance is better than the scheme proposed by Bao and Zhou [17]. Dai et al. [18] presented a hybrid scheme using discrete SWT and DCT. For improving the authenticity, an encryption is applied using chaotic map. Singh et al. [19] proposed a secure scheme for medical image watermarking using the hybrid scheme of DWT and SVD. For the improvement of imperceptibility, the watermark is embedded in a RONI.

Recently, Sabbane and Tairi [20] suggested a scheme using polynomial decomposition for medical images. This scheme is region-based and integrates some numerical information as a watermark in the host image. The novelty in the scheme is that they use polynomial transform to decompose the original image into two parts as structure and texture. Hemdan [21] presented a hybrid scheme for the security to medical images using WF, SVD, and MDWT. In this scheme, fusion is used to combine two watermarks into single watermark results enhancing the payload. For increasing security, the fused watermark is scrambled using chaotic and Arnold algorithms. Finally, the resultant watermark is embedded using the SVD and DWT. For the security of medical images, Anand and Singh [22] presented a watermarking scheme using the SVD and DWT with hamming code. The authors have shown high imperceptibility and good robustness.

It has been observed in the literature that there is a research gap for providing the security of medical images using the lifting scheme. Here, an effort is made to accomplish the security of medical images using a hybrid scheme of LWT and SVD. The use of LWT reduces the computation time as compared to conventional wavelet transforms, and SVD preserves the image stability. Remaining part of this paper is arranged as follows: Sect. 2 deals preliminaries including LWT and SVD, Sect. 3 shows proposed methodology including embedding and extraction scheme, results and discussion are explained in Sect. 4, and conclusion is covered in last section. The abbreviations used in the text of this paper are shown in Table 1.

Table1 List of abbreviations

Details	Abbreviation
Discrete cosine transform	DCT
Discrete wavelet transform	DWT
Multi-level discrete wavelet transform	MDWT
Normalized correlation	NC
Peak signal-to-noise ratio	PSNR
Region of non-interest	RONI
Robust discrete wavelet transform	RDWT
Singular value decomposition	SVD
Stationary wavelet transform	SWT
Wavelet fusion	WF

2 Preliminaries

A brief introduction of LWT and SVD is presented in this section.

2.1 LWT

The lifting scheme was developed by Sweldens [23, 24]. The lifting scheme takes half of the computations than the discrete wavelet transform. It is simple in calculations and easy to implement. Split, predict, and update are the three main steps that are used for the lifting scheme. These steps are given as.

Split: Here, the input signal $x(k)$ is split into even and odd sub-signals as $x(2k)$ and $x(2k+1)$, respectively.

Predict: In this high pass filtering step, odd sub-signal is used to predict the value of even sub-signal and the difference is recorded as

$$\text{diff}(k-1) = x(2k+1) - P[x(2k)] \tag{1}$$

where P represents the operator predict.

Update: In this step, the low-frequency component $h(k)$ gives a coarse approximation to $x(k)$, the original signal, and using the operation $U[.]$, it is defined as

$$h(k-1) = x(2k) + U[\text{diff}(k-1)] \tag{2}$$

The decomposition steps of an input signal are elaborated in Fig. 1.

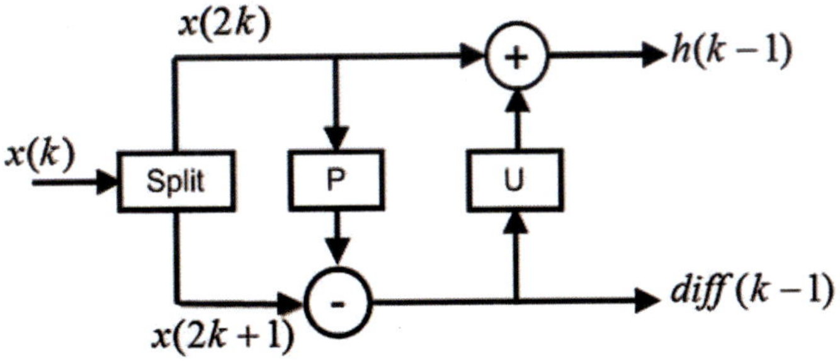

Fig. 1 Decomposition of signal using LWT

2.2 SVD

Initially SVD is considered as a numerical analysis tool for matrix analysis, but recently, SVD has been observed in combination of other methods for various image processing applications as information hiding, watermarking, compression, and noise removal. Since SVD takes high-end computations when applied alone [25]; therefore, it is preferred with other methods in a hybrid way. When we use the singular values obtained by SVD, then these singular values do not change with a small change in pixel value of an image and hence provides a good stability. With the help of SVD, a matrix is decomposed into three components as U, S, and V. Consider a matrix of size $m \times m$

$$\left[U_{m\times m}\ S_{m\times m}\ V_{m\times m}\right] = \mathrm{SVD}(A_{m\times m}) = U_{m\times m} S_{m\times m} V^{T}_{m\times m} \quad (3)$$

where U, V, and S represent three matrices in which first two are orthogonal and the last one is a singular matrix.

3 Proposed Methodology

Here, we discuss the methodology used for providing the security to medical images. For this, we embed secret information into the medical image, and then after this, watermarked image is transferred to receiver end, and then, its authenticity is judged by extracting the watermark. The steps of embedding and extraction are explained below.

3.1 Embedding Scheme

To embed the secret data to a host image, the following steps are used as

(i) Take a medical image A of size $m \times m$ as host image.
(ii) Apply two-level LWT to the cover image to obtain HL_2 and LH_2 sub-bands as shown in Fig. 2.

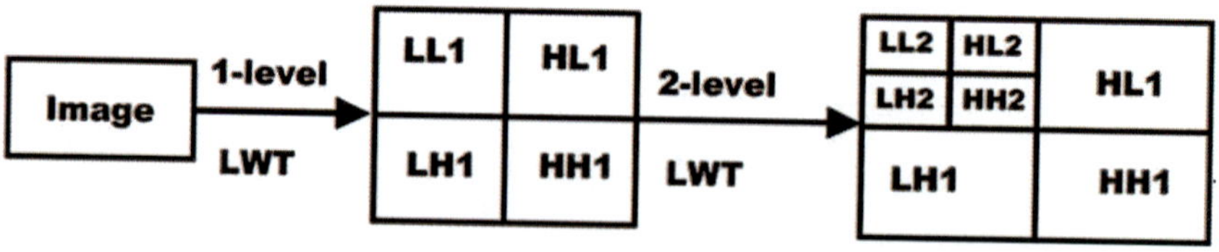

Fig. 2 Two-level LWT decomposition of image

(iii) Apply SVD on each sub-bands HL_2 and LH_2 separately as

$$\left[u_{HL_2}\, s_{HL_2}\, v_{HL_2}\right] = SVD(HL_2) \tag{4}$$

$$\left[u_{LH_2}\, s_{LH_2}\, v_{LH_2}\right] = SVD(LH_2) \tag{5}$$

(iv) Consider a watermark image and convert it into binary as

$$w = \text{binary}(\text{"}Math''). \tag{6}$$

(v) Divide the binary watermark into two equal non-overlapped partitions w_1 and w_2.
(vi) Update the singular values obtained in step iii as

$$s^*_{HL_2} = s_{HL_2} + \beta w_1 \text{ and } s^*_{LH_2} = s_{LH_2} + \beta w_2, \tag{7}$$

where β is a strength parameter.

(vii) Apply inverse SVD to get updated coefficients HL_2^* and LH_2^* of HL_2 and LH_2, respectively, as

$$HL_2^* = u_{HL_2} s^*_{HL_2} v^T_{HL_2} \text{ and } LH_2^* = u_{LH_2} s^*_{LH_2} v^T_{LH_2} \tag{8}$$

(viii) Apply inverse two-level LWT to get the watermarked image, W_A.

3.2 Extraction Scheme

To get the secret information from the watermarked image, the steps are followed as

(i) Consider the watermarked image and apply two-level LWT to obtain HL_2 and LH_2 sub-bands.
(ii) Apply SVD on each sub-bands HL_2 and LH_2 separately to find $s^*_{HL_2}$ and $s^*_{LH_2}$ as

$$\left[u_{HL_2}\, s^*_{HL_2}\, v_{HL_2}\right] = SVD(HL_2) \tag{9}$$

$$\left[u_{LH_2}\, s^*_{LH_2}\, v_{LH_2}\right] = SVD(LH_2) \tag{10}$$

(iii) Use Eq. (7) to extract the embedded secret information e_1^* and e_2^*. Construct the extracted watermark e^* image using extracted data e_1^* and e_2^*.

4 Results and Discussion

To understand the efficiency of this scheme, the simulation is carried out over medical images which are shown in Fig. 3. These images are obtained from a local diagnostic center and are in DICOM format of size 512 × 512. The original watermark for the experiment is also shown in Fig. 3. NC and PSNR are used as evaluation parameters for this experiment.

The PSNR can be evaluated as

$$\text{PSNR} = 10\log_{10}\left|\frac{255^2}{\text{MSE}}\right|,$$

where $\text{MSE} = \frac{1}{mn}\sum_1^m\sum_1^n[I(x,y) - O(x,y)]^2$.

The NC is given as

$$NC = \frac{\sum_1^m\sum_1^n e(x,y)e^*(x,y)}{\sum\limits_1^m\sum\limits_1^n[e(x,y)]^2}$$

where e and e^* represent embedded and extracted watermarks.

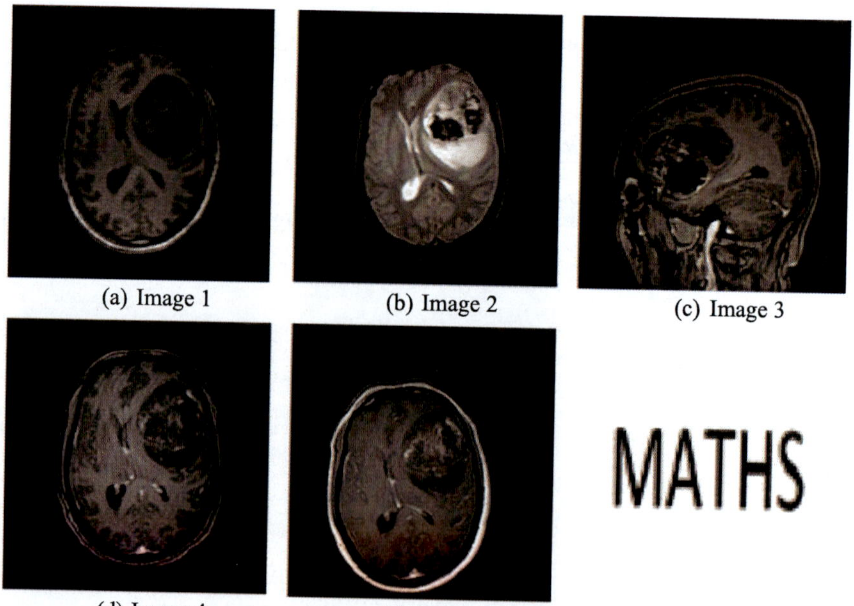

(a) Image 1 (b) Image 2 (c) Image 3

(d) Image 4 (e) Image 5 (f) Original watermark

Fig. 3 Original host and watermark images

Table 2 PSNR values between input and output images

Image name	PSNR (dB)
Image 1	60.9824
Image 2	61.7586
Image 3	60.3376
Image 4	61.2016
Image 5	61.0950

Table 3 Different attacks with abbreviation

Attack name	Abbreviation
Scaling	Scl
Salt-and-pepper noise	Sp
Speckle noise	Sn
Gaussian filter	GF
Median filter	MF
Average filter	AF
Histogram equalization	HE
JPEG compression (quality factor)	JPEG
Amplification	Ap

The PSNR values between the input and the output images of embedding process are presented in Table 2. The high value of PSNR supports good imperceptibility, while the robustness is measured by NC. A good value of NC shows that the proposed scheme has high robustness. The experimental results for NC values are presented in Table 4. The different geometric and image processing attacks used in Table 4 are explained in Table 3. The proposed scheme is rich in robustness against the attacks GF, MF, AF, JPEG, and in Scl. The scheme underperforms in case of Sp and Sn.

Table 4 NC values under different attack conditions

Attack name	NC
Without attack	1
Scl 0.5	0.9234
Sp (0.001)	0.6865
Sn	0.6012
GF 3 × 3	0.9421
MF 3 × 3	0.9610
AF 3 × 3	0.9756
HE	0.7045
JPEG (60)	0.9820
JPEG (80)	0.9981
Ap 1.5	0.9387

5 Conclusion

The proposed scheme presents a hybrid watermarking scheme to secure the medical images. In this scheme, LWT followed by SVD is used to secure the watermark. Experimental results demonstrate that this scheme has good imperceptibility and it confirms with a very high value of PSNR. The NC values for the attacks GF, MF, AF, JPEG are reasonably good which shows its robustness against these attacks. However, it underperforms in case of Sp and Sn.

References

1. Bennett K, Bennett AJ, Griffiths KM (2010) Security considerations for E-mental health interventions. J Med Internet Res 12:e61. https://doi.org/10.2196/jmir.1468
2. Singh A, Dutta MK (2017) Imperceptible watermarking for security of fundus images in teleophthalmology applications and computer-aided diagnosis of retina diseases. Int J Med Inform 108:110–124. https://doi.org/10.1016/j.ijmedinf.2017.10.010
3. Sang J, Alam MS (2008) Fragility and robustness of binary-phase-only-filter-based fragile/semi-fragile digital image watermarking. IEEE Trans Instrum Meas 57:595–606
4. Wu HT, Cheung YM (2010) Reversible watermarking by modulation and security enhancement. IEEE Trans Instrum Meas 59(1):221–228
5. Liu R, Tan T (2002) An SVD-based watermarking scheme for protecting rightful ownership. IEEE Trans Multimedia 4(1):121–128
6. Luo L, Chen Z, Chen M, Zeng X, Xiong Z (2010) Reversible image watermarking using interpolation technique. IEEE Trans Inf Forensics Security 5(1):187–193
7. Das C, Panigrahi S, Sharma VK, Mahapatra K (2014) A novel blind robust image watermarking in DCT domain using inter-block coefficient correlation. Int J Electron Comm 68(3):244–253
8. Mishra A, Agarwal C, Sharma A, Bedi P (2014) Optimized gray-scale image watermarking using DWT-SVD and firefly algorithm. Expert Syst Appl 41(17):7858–7867
9. Ali M, Ahn CW, Pant M, Siarry P (2015) An image watermarking scheme in wavelet domain with optimized compensation of singular value decomposition via artificial bee colony. Inform Sci 301:44–60
10. Bhardwaj A, Verma VS, Jha RK (2018) Robust video watermarking using significant frame selection based on coefficient difference of lifting wavelet transform. Multimedia Tools Appl 77(15):19659–19678
11. Verma VS, Bhardwaj A, Jha RK (2019) A new scheme for watermark extraction using combined noise-induced resonance and support vector machine with PCA based feature reduction. Multimedia Tools and Applications 78(16):23203–23224
12. Kumar C, Singh AK, Kumar P (2017) A recent survey on image watermarking techniques and its application in e-governance. Multimedia Tools Appl 77(3):3597–3622
13. Rastegar S, Namazi F, Yaghmaie K, Aliabadian A (2011) Hybrid watermarking algorithm based on singular value decomposition and radon transform. AEU Int J Electron Commun 65(7):658–663
14. Makbol NM, Khoo BE (2013) Robust blind image watermarking scheme based on redundant discrete wavelet transform and singular value decomposition. AEU Int J Electron Commun 67(2):102–112
15. Boonyapalanant A, Ketcham M, Piyaneeranart M (2019) Hiding patient injury information in medical images with QR code. Adv. Intell Syst Comput 258–267
16. Araghi TK, Manaf AA (2019) An enhanced hybrid image watermarking scheme for security of medical and non-medical images based on DWT and 2-D SVD. Future Gener Comput Syst 101:1223–1246

17. Bao L, Zhou Y (2015) Image encryption: generating visually meaningful encrypted images. Inform Sci 324:197–207
18. Dai Q, Li J, Bhatti UA, Chen YW, Liu J (2019) SWT-DCT-based robust watermarking for medical image. Smart Innov Syst Technol 93–103
19. Singh HV, Rai A (2018) Medical image watermarking in transform domain. Adv Intell Syst Comput 48:5–493
20. Sabbane F, Tairi H (2019) Medical image watermarking technique based on polynomial decomposition. Multimedia Tools Appl 78(23):34129–34155
21. Hemdan EED (2020) An efficient and robust watermarking approach based on single value decompression, multi-level DWT, and wavelet fusion with scrambled medical images. Multimedia Tools and Applications 1–29
22. Anand A, Singh AK (2020) An improved DWT-SVD domain watermarking for medical information security. Comput Commun 152:72–80
23. Sweldens W (1996) The lifting scheme: a custom-design construction of biorthogonal wavelets. Appl Comput Harmon Anal 3(2):186–200
24. Sweldens W (1998) The lifting scheme: A construction of second generation wavelets. SIAM J Math Anal 29(2):511–546
25. Lai CC, Tsai CC (2010) Digital image watermarking using discrete wavelet transform and singular value decomposition. IEEE Trans Instrum Meas 59(11):3060–3063

Data Security Through Fog Computing Paradigm Using IoT

Jayant Kumar Singh and Amit Kumar Goel

Abstract At first, change happened painfully. The Internet has grown in forms we never might have expected. Fog calculation is a new computing paradigm which extends cloud computing to the edges of the network. Nowadays, privacy and security is one of very challenges for the Internet of Things (IoT). Tiny devices are facing such a challenge due to lack of awareness in user for improper devices updates from time to time, lack of knowledge and not following proper security channel. I would like to explain the IoT systems background and its different kind of security parameter which is having different kind of security parameter and privacy challenges, we will explain here our various scheme and demonstrate for network security so that we can determine network security about the fog-computing model and it's used to boost the reliability and efficacy of the delivery of IoT certificates. The basic concept of the bloom philtre is seen in Fig. 2. While a false positive match in the bloom philtres is probable, a false negative is not probable. I will describe the fog networking security summary how to solve this problem in our method and the overall it will see in result graph also.

Keywords Cloud computing · Fog computing · Privacy and security issues and · IoT · Etc.

1 Introduction

Nowadays, the Internet has become the very important part of our daily lives. Innovation and connectivity are taking place at an excellent pace today. The Internet of Things (IoT) is one of the technologies that will offer our world infinite benefits. In the recent years [1], both business and academics have generated considerable interest

J. K. Singh (✉)
School of Computing Science and Engineering, Galgotias University, Greater Noida, India
e-mail: jayant.singh_phd17@galgotiasuniversity.edu.in

A. K. Goel
School of Computing Science and Engineering, Galgotias University, Greater Noida, Uttar Pradesh, India
e-mail: amit.goel@galgotiasuniversity.edu.in

G. Gupta et al. (eds.), *Proceedings of Academia-Industry Consortium for Data Science*, Advances in Intelligent Systems and Computing 1411,
https://doi.org/10.1007/978-981-16-6887-6_9

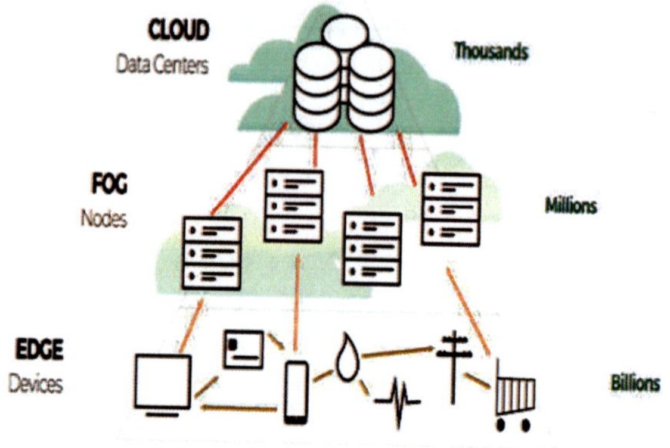

Fig. 1 Cloud computing architecture

from the Internet of Things, cloud computing, edge computing and fog computing. However, it is difficult to locate in literature a straightforward and tidy description of these programming paradigms and their association. Fog computing framework contrasts these disturbances with robust cloud-oriented complementary capabilities focused on micro-cloud implementation (fog knots) at the edge of data sources. This computational model is not, however, an alternative to cloud storage, as it poses several protection and privacy problems [2].

Fog computing may be called an intermediary between consumers and the cloud provider, which effectively links end-users and eliminates data latency. The popularity of IoT has become famous day by day and increases its demand promptly, and this kind of technologies is used for different purposes like end-to-end communication, various transportation, all education segment and the various business development. One decade ago IoT introduces as a hyper connectivity concept. Where large organizations and individuals can talk simultaneous with each other from user's remote locations without any network hurdle [3] (Fig. 1).

IoT incorporation with cloud infrastructure provides numerous benefits for multiple IoT applications [4]. "However, because several IoT devices with heterogeneous architectures exist, it is challenging to create new IoT applications. While we use IoT application, then during this inter transferring process the data transferring generates huge data from various sources. This data evaluated to settle on separate behaviour. All of these data has to be transmitted to the cloud via high network bandwidth. Fog computing comes into play to overcome these problems.

1.1 Fog Computing

Fog computing is defined as a storage system that utilizes cloud computing concepts and technology to handle complex services at the network boundary, i.e. edge of the

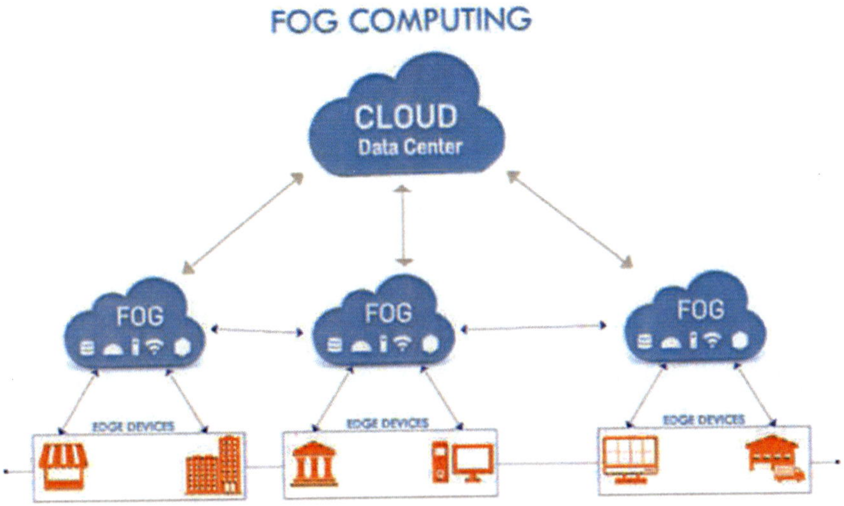

Fig. 2 Fog computing in terms of IoT

nodes [5]. Fog computing facilitates latency dependent systems with lower energy usage and decreased data flow, as compared to conventional cloud computing. In lieu of those living in the cloud, fog computing model uses local computational tools to process the data obtained from sensors connected to physical devices on an IoT network [6]. Fog computation is used interchangeably since the two performances and storage speeds are intermediate. However, the biggest distinct between the two is the device position. In fog networking, LAN which defines local area network is a portal where machines like programmable automation controllers (PACs) communicate with smart devices and edge environments."

1.2 There is Below Mention Attacks Name on Fog Computing

- Virtual machine-based attacks
- DoS attacks
- Session hijacking attack: In this attack, a user catches by another user in same session.
- Side-channel attack: When device's cryptography is reversed by engineering, then such an attack can be fall.

1.3 Fog Computing Benefits

- Place data close to the end-user.

- Globally distributed network helps minimal downtime.
- **No problems with bandwidth**—Data and metadata information can be saved for each session.
- Create dense geographical distribution.
- Load balancing
- Adopt numerous verticals.
- **Better Security**: Fog computing gives better security to fog nodes as fog nodes can be protected using the same controls, procedures and policy you use in other areas of IT environment.
- Maximize network bandwidth utilization.
- **Privacy**: The level of privacy may be regulated by fog computing. Every important consumer data may be processed locally rather than transferred to a centralized cloud system. This allows the IT team to monitor and manage the respective computer. In addition, if some subset of data requires review, it can be submitted to the cloud.
- **Greater Business Agility**: With the help of good tools, software engineer can develop various IoT applications like RFID, biometrics some example of it [7].

1.4 Fog Computing Disadvantages

- **Limited scalability**—Fog is not as scalable as the cloud.
- Sometimes developer is not focusing on encryption algorithms and security policies which they are generating, so any small mistakes in security algorithms lead to exposure of data to the hackers [8].
- **Power Consumption**: In a fog area, the amount of fog nodes is directly proportional to their energy consumption. This indicates that those fog nodes need high energy to operate. There is often more resource demand in a fog grid. Some businesses also utilize these fog nodes to reduce their prices.
- **Additional expenses**—Each company must need to buy edge devices like routers, hubs, gateways, biometric, etc. [9].

2 Fog Computing Under IoT

Fog computing is such a terminology that combines cloud and network edge technologies. "The features of fogging are its proximity to the end-users. Fog infrastructure redistributes data and computer, so that much is done right at your fingertips on your edge devices [8]. The application resources are situated on the edges of the network. The data from the point of concern where sensors are connected is transmitted to the network's edge (a fog network). The data is distributed from the fog network to remote Internet providers for storage, handling and processing of data (cloud network).

2.1 Review of Literature

Chiang et al. (2017) furnish a fog computing demonstration. The distinctions between fog computing, edge computing and cloud computing are addressed at a rather high level [10]. They also address and explore the benefits of fog computing. Moreover, numerous studies of fog computing in general and particular application fields have also been released, including radio access networks (RAN), ad hoc vehicles (VANETs) and the Internet of Things.

Dastjerdi et al. (2016) 'As with cloud networks, a fog environment involves IaaS, PaaS, SaaS, and network storage infrastructure, framework and applications in storage' [9] and the fog platform's technological design. The Cisco IOx API, which incorporates a Linux and CISCO IOS network framework, produces the fog IaaS interface. Fog Node with processing, computing and end to end network access can be a example of IoT such as manage/unmanaged switches and layer 2 routers, firewall, servers or even cameras also.

3 Research Methodology

We are going to propose here very important and new certificate revocation schemes towards fog computing in terms of IoT. With the help of algorithms, we will analyse and find out the various results for our scheme. We will also give brief summary of this certificate revocation schemes and their output in terms of IoT.

We are going to define here two certificate revocation schemes as mentioned below.

1. Certificate Revocation List, i.e. (CRL).
2. OCSP: OSCP is nothing but it defines about Online Certification Status in terms of protocol parameter.

4 Proposed Scheme

Cloud manages multiple CA's revocation list and sends to intended fog node. Fog node stores the list and generates the bloom filter and distributes to IoT devices. IoT device stores the bloom filter for further usage.

Due to the nature of bloom filter, false positive can occur. When IoT device finds a certificate identity in the bloom filter, it will send packet containing the certificate to the fog, and the fog will check this certificate status and reply.

5 Results and Discussions

I am explaining here scheme algorithm and demonstrating of the fog computing model and efficiency of the IoT certificates. "Proposed scheme is an up-to-date approach which removes the risk of accepting revoked certificate."

Bloom filter reduces the computation and storage overhead with little price to pay (false positive).

- Storage:

bloom filter: $m = \frac{-b.\ln(p)}{(\ln(2)^2)}$, where m is showing number of bits, p defines the false positive probability(0.01), and b is average number of revoked certificate each day.

CRL: $b * 20 + 700$ bytes, where 20 bytes is serial number length, and 700 bytes is a CA's signature

OCSP: Since it is online method, it does not have storage overhead.

We use a bloom philtres in our scheme to build and developed a short list that can easily minimize the number size of the revocation list with appropriate overheads. A flower philtre is a spatial efficiency of data structure used to keep a set of items in a bit-vector as given (0, ... $m..m - 1$) to decide if an object is in the category. We have taken A null bloom philtres vector is 0 in initially mode also store certificate cancellation records for maintaining the records, we use the serial number of the certificate, that is called its special CA identification. The serial number should be hacked to a collection of bit positions which should be set to 1 with k separate hash functions. The basic concept of the bloom philtre is seen in Fig. 2. While a false positive match in the bloom philtres is probable, a false negative is not probable. We describe in the following summary how to solve this problem in our method (Fig. 3).

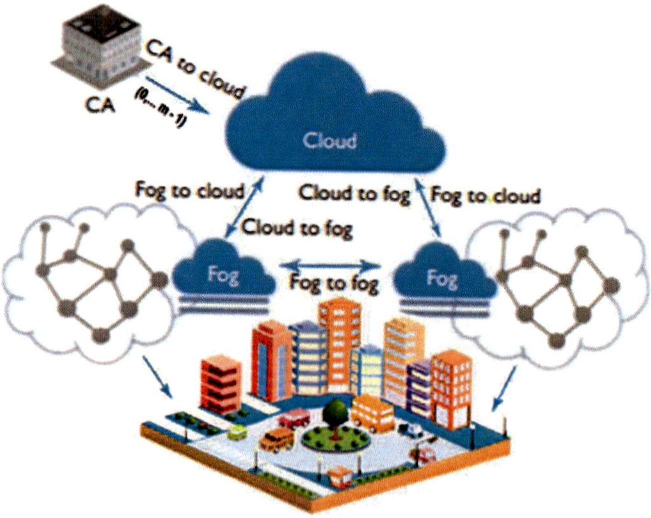

Fig. 3 Scheme overview

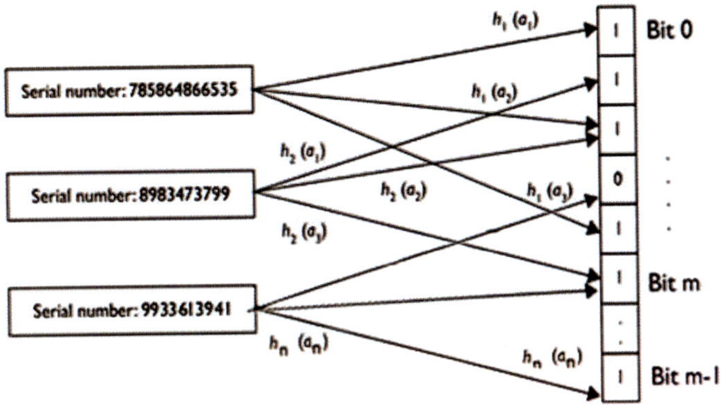

Fig. 4 Revocation certificate graph

Figure 4 In this figure we are trying to show bloom filter from initial level bit (0, … *m*… *m* − 1) i.e. 0 to bit *m* and reverse on the same mode Bit $M-1$ with n no. of time interval, which shows the different signature serial.

5.1 *Results*

In the current system, the certificate revocation knowledge is transmitted effectively. Our system is an up-to-date solution, in which all withdrawn certificates are transferred to cloud and then fog computing devices directly from the CAs. The upgrading of the cancellation records automatically guarantees the authentication and reduces the possibility of receiving the revoked certificate. Another protection aspect is that the fog signatures per flower philtre to show its durability. Since the signature of the fog cannot be changed or made, it provides an absolute evidence that the variable was generated from the fog. The benefit of our scheme is that the bloom philtre will reduce the overhead computing on IoT devices.

"We must request an appraisal of our system quantitatively and equate it to the global revocation certificate regimes. Our primary concern is the use of services on IoT computers, such as data and overhead connectivity."

In order to measure the overhead connectivity, we must assume the typical transmission level, which is expected to be 10 Mbps. For each revocation strategy, we may quantify the bandwidth usage as Fig. 5 indicates. Our device requires less bandwidth consistent with the packet size (Fig. 6).

Less bandwidth shows good compatibility with the packet size that is why we have taken low bandwidth.

Storage: The overall flower philtre size can be determined as follows in our proposed scheme:

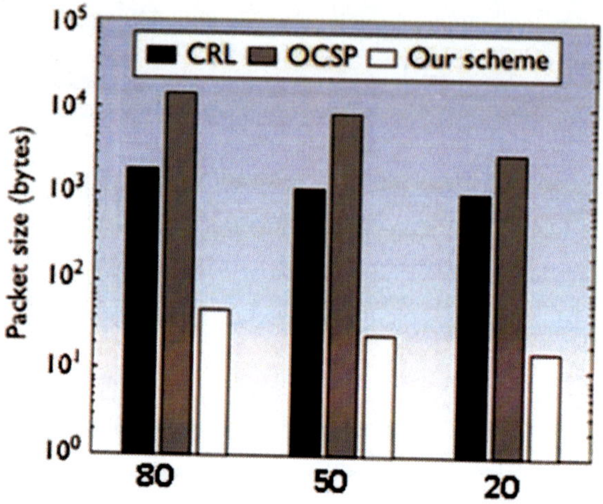

Fig. 5 Comparison shows the different packet size, when value $b = 20, 50, 80$

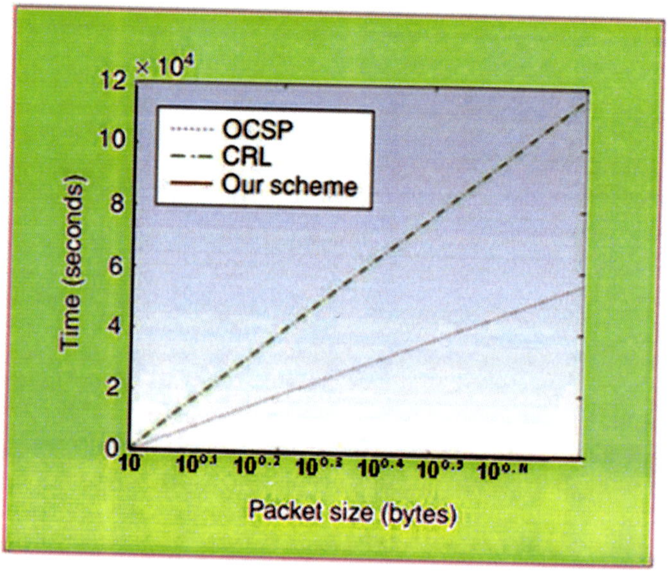

Fig. 6 Comparison of the communication overhead in different certificate revocation schemes

$$m = \frac{-b.\ln(p)}{\left(ln(2)^{2...n}\right)}$$

where m indicates the amount of bits required for the floral philtre, and p indicates a specified false positive likelihood, which in our experiment is 0.001 in order to get the CRL file size, we have taken serial number length of the certificate and the signature duration of the CA. The serial number duration of a standard certificate varies from 15 to 20 bytes, and the signature duration of a CA is approximately 700 bytes. The average value of the CRL file is $b \cdot 80 + 1700$ bytes. While for OCSP, the procedure does not require any overhead storage since the solution is online.

6 Conclusion

Fog computing provides cloud storage to manage the greater regular IoT results. It also increases exposure and reaction to incidents by taking out n number of data for cloud review. It helps overcome explosive data level, range and volume problems. We researched and addressed protection and privacy issues in IoT defined by implementing fog computing. It is important to concentrate further time on how to address and determine these difficulties. In this paper, we investigated about security and privacy challenges in terms of IoT with fog computing framework and also proposed a case study of certificate revocation distribution to show the ability of fog.

References

1. Hassan WH (2019) Current research on internet of things (IoT) security: a survey. Comput Netw 148:283–294
2. Vaquero LM, Rodero-Merino L (2014) Finding your way in the fog: towards a comprehensive definition of fog computing ACM SIGCOMM. Comput Commun Rev 44:27–32
3. Yi S, Li C, Li Q (2015) A survey of fog computing: concepts, applications and issues. In: Proceedings of the 2015 workshop on mobile big data, New York, pp 37–42
4. Dastjerdi AV, Gupta H, Calheiros RN, Ghosh SK, Buyya R (2016) Fog computing: principals, architectures, and applications. arXiv preprint arXiv:1601.02752
5. Yannuzzi M, Milito R, Serral-Gracia R, Montero D, Nemirovsky M (2014) Key ingredients in an IoT recipe: fog computing, cloud computing, and more fog computing. In: 2014 IEEE 19th international workshop on computer aided modeling and design of communication links and networks (CAMAD), pp 325–329
6. Siby S, Maiti RR, Tippenhauer NO (2017) Iotscanner: detecting privacy threats in IoT neighborhoods. In: Proceedings of the 3rd ACM international workshop on IoT privacy, trust, and security, Abu DhabiUnited Arab Emirates, pp 23–30
7. Chiang M, Ha S, Risso CLIF, Zhang T (2017) Clarifying fog computing and networking: 10 questions and answers. IEEE Commun Mag 55(4):18–20
8. Roman R, Lopez J, Mambo M (2016) Mobile edge computing, fog et al.: a survey and analysis of security threats and challenges. Future Gener Comput Syst
9. Loke SW (2015) The internet of flying-things: opportunities and challenges with airborne fog computing and mobile cloud in the clouds. arXiv preprint arXiv:1507.04492
10. Conti M, Dehghantanha A, Franke K, Watson S (2018) Internet of things security and forensics: challenges and opportunities. Future Gener Comput Syst 78:544–546 [CrossRef]

One-Way Cryptographic Hash Function Securing Networks

Vijay Anant Athavale, Shakti Arora, Anagha Athavale, and Ruchika Yadav

Abstract Due to the rapid growth of Internet technologies, voluminous data is generated every minute from multifarious sources. It is impossible for conventional tools to analyse and process such an immense volume of data. Cryptographic hashing is one of the best approaches to efficiently deal with accuracy and consistency of the data on the cloud. It also plays a crucial role in protecting message creditability during transmission. A new one-way cryptographic hashing algorithm has been suggested in this paper, which produces a digest of 128 bits for any message length (short, medium and long). With the basic properties of the hashing algorithm, the suggested algorithm is implemented and compared with existing ones. Outcomes shows that proposed algorithm is stronger than the current ones in terms of execution and protection.

Keywords MD5 · SHA · Attacks · Hash · Collision · Network · Cloud

1 Introduction

Every use whether professional or home maker is accessing the Internet and deploying the maximum data on the cloud, such as uploading and downloading file, for connecting to other users, to provide the online services and many more. Because of this, a hacker makes use of vulnerable websites to hack the system and misuse or corrupt the network data. Cryptographic algorithms and hash functions are to deal with certain kinds of security and privacy problems [1–4]. For authentication purposes, cryptographic hash functions are auxiliary functions. The standard and are Message Digest-5 (MD5), Secure Hash Algorithm 1 (SHA-1) and Hash-based Message Authentication Code (HMAC) are the basic and most common cryptographic hash functions used for signatures, message authentication code, key

V. A. Athavale (✉) · S. Arora
Panipat Institute of Engineering and Technology, Panipat, Haryana, India

A. Athavale
Bhabha Atomic, Research Centre, Mumbai, Maharashtra, India

R. Yadav
SNRL Jairam Girls College, Lohar, Kurukshetra, Haryana, India

G. Gupta et al. (eds.), *Proceedings of Academia-Industry Consortium for Data Science*, Advances in Intelligent Systems and Computing 1411,
https://doi.org/10.1007/978-981-16-6887-6_10

derivations, random number generators and so forth. In addition, hashing plays an important role in blockchain technology which is the bitcoin fundamental process [5–7] and SHA-256 is the predominant hashing algorithm in blockchain. Shakti et al. [4, 8] proposed an integrity algorithm which provides authentication and consistency to the data, over the cloud and applied checksum while data is being transmitted on the cloud Internet. Wang et al. [9] used hash maps for the arbitrary string of data to convert into fixed length message digest as output. There are many characteristics of hash functions, but one-way, collision resistance, second pre-image resistance and avalanche effects are essential one.

The research by Bellare et al. [10] and Gauravram et al. [11] shows that the hash functions of MD and SHA family cryptography are susceptible to brute force attacks and cryptanalytical attacks, leading to the creation of secure hashing function. Therefore, to deal with such types of attacks, this paper introduces a more efficient hashing algorithm.

The organization of the paper is as follow: literature survey is discussed in Sect. 2. The proposed hashing algorithm with a working of flowchart is discussed in Sect. 3. The proposed work is exemplified by Sects. 4 and 5. This paper is eventually concluded in Sect. 6.

2 Literature Survey

Kim et al. [12] provided a new hash function family called LSH. For superior software performance, they used parallel processing instructions, for example SSE, AVX2, and EON. Their findings showed that software output of the proposed work was faster by more than four times as that of the SHA-3 and it was 1.5–2.3 times faster as compared to other SHA-3 family. They also showed that their hashing function was immune to all critical hash function attacks.

Arya et al. [13] proposed a novel hashing scheme that implemented key use so that without the help of key attacker is not able to crack the hash code Their algorithm created digest of 128 bits for an arbitrary message length. They have splitted it in two stages: pre-processing and hash calculation. Similar work was done by researchers Shakya et al. [7]; they enhanced MD5 by introducing a key into it and produced 128 bits of digest for arbitrary input length. Their algorithm satisfied the criteria of confidentiality, authentication and integrity for a message while in transmitting due to key integration. They showed that the number of operations, i.e. 2^{64} to 2^{192}, increased the time complexity by 15–20% for data range of 50–100 KB.

Chowdhury et al. [14] developed a light-weight and one-way cryptography hash algorithm, which they named LOCHA. The algorithm was primarily developed in such a way that it consumes low energy in transmission. It is useful for energy starved wireless networks. Their algorithm produced a digest of 96 bits for any input string length and satisfied all the key of hashing properties, that is one-way; second is pre-image resistance and collision resistance. Mondal et al. [15] discussed the timestamp defined hash function called TDHA for the safe transmission among vehicles. In this technique, sender vehicle transmitted deformed message and incomplete message digest and receiver vehicle produced digest from the intermediate digest and distorted message type. Their algorithm was simulated and compared with MD5, SHA-1 and LOCHA using comparison metrics such as overhead (communication and computation) and overhead storage. Their design outperform the other techniques both qualitatively and quantitatively.

Abutaha et al. [16] designed a one-way hash algorithm that used two steps; that is, they have arranged the input data in a non-invertible matrix by using conversions and generated the initial hash initially. Secondly, they used the previous step's performance and applied additional value to it, which was sent to the receiver. They have compared their proposed work with MD5, SHA-1 and SHA-512.

Tiwari et al. [17] developed a dedicated cryptography hash technique called MNF-256 that was based on the NewFORK-256 concept. Their design used three branch parallel structure, and each branch consisted of eight operations. The results and thorough analysis claimed that the proposed work was robust against cryptanalytic attacks and faster than NewFORK-256. Manel et al. [18] provided a Pizer hash function based on two techniques, elliptic curve and expander graph. After computing the hash with the help of Pizer function, they signed a message by ECDSA. They used MATLAB to simulate their results and proved that their work was collision resistant.

Mirvaziri et al. [19] introduced a hybrid cryptography hash function that was a combination of SHA-1 and MD5. The compression function used in the proposed hash function was made from the four rounds of encryption function, and each and every round consisted of 20 transformation steps. Researchers Rubayya et al. [15] designed a new technique that was based on HMAC/SHA-2. The results were analysed and compared using various design goals and strategies like balanced and area reduction. They proved that their proposed work utilized less area and consumed less power.

Moe et al. [20] developed a new honeyword generation technique that stored the users' password as honeywords to decrease the typo-safety problem, storage overhead and some other drawbacks which were existing in older honeyword generation methods. They have used special hashing technique to store the passwords and honeywords in database which reduced the time complexity of the algorithm.

Chen et al. [21] developed an enhanced algorithm that followed the fundamentals of store hash and rehash for context-triggered piecewise hashing technique (CTPH) also known as FKSUM. They have compared it with spamsum, and results proved that the performance and ability of the proposed method are better. The new design was valuable for forensics practice.

Rasjid et al. [22] surveyed several types of attacks on hash algorithms. They reviewed existing and present methods which are used in digital forensic tools to create an attacks. They have compared the common features of MD series and SHA series in terms of output size, number of round, collision found and performance (MiB/s). Sharma et al. [8] discussed various intrusion detection mechanism to maintain the confidentiality and other security challenges, and comparative analysis of various security methods has been discussed.

3 Proposed Work

Hash function is composed of two functions; firstly, one is compression function, which applies mapping function to translate a large input string into a small output, and secondly, the mechanism by which the compression function is being repeatedly called to process a variable-length message.

Working of Proposed Hashing:

Step 1: *Padding*—Perform padding in such a way that the length of message after padding is congruent to 448 modulo 512. In padding, first bit is 1 and rest of the bits are 0.

Step 2: *Appending*—After padding, append the length of original message (represented in binary form) to the output of the step 1. After doing this step, the length of message is now in multiples of 512.

Step 3: *Message into Blocks*—The message obtained from the previous step is divided into n blocks of size 512 bits. That is: BL0, BL1, BL2, BL3, BL4, BL5, BL6, BL7,... BLn.

Step 4: *Blocks into Sub-Blocks*—Each block which is obtained in previous block is of bigger size. In this step, we have divided the blocks into sub-blocks of size 128 bits., that is:

(BL01, BL02, BL03, BL04), (BL11, BL12, BL13, BL14), (BL21, BL22, BL23, BL24), ... (BLn1, BLn2, BLn3, BLn4).

Step 5: *Apply XOR operation*—Next step is to perform XOR among each part of the block and override the block's value with its output. As the size of sub-block is 128 bits, after performing XOR the overridden block's size is 128 bits, that is:

BL0 = BL01 XOR BL02 XOR BL03 XOR BL04.

BL1 = BL11 XOR BL12 XOR BL13 XOR BL14.

BL2 = BL21 XOR BL22 XOR BL23 XOR BL24.

... and so on.

Step 6: *Overall XOR*—The size of blocks which is produced from the previous step is 128 bits. In this final step, apply XOR among all the blocks. This will result into message digest of size 128 bits.

output = BL0 XOR BL1 XOR BL2 XOR BL3.

3.1 Flowchart

In Fig. 1, the proposed algorithm is depicted in a flowchart.

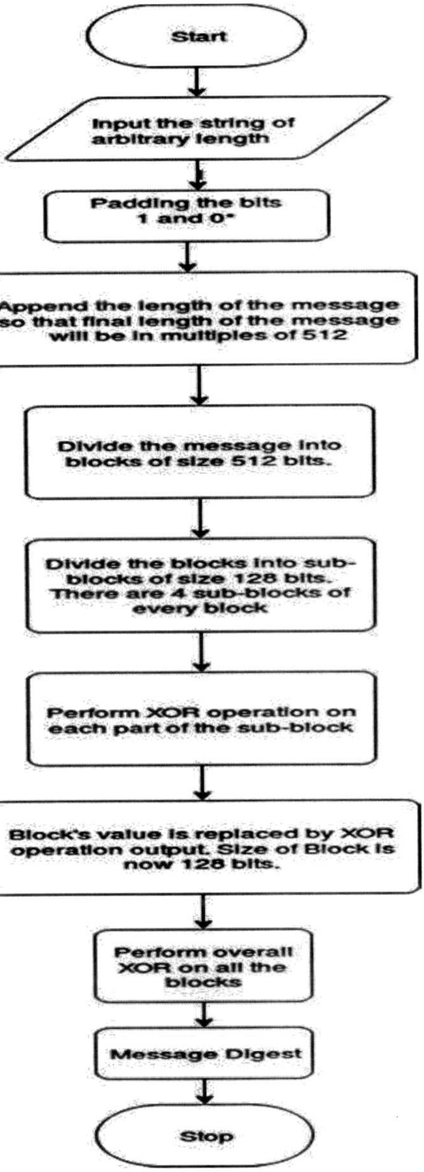

Fig. 1 Flowchart of proposed hashing algorithm

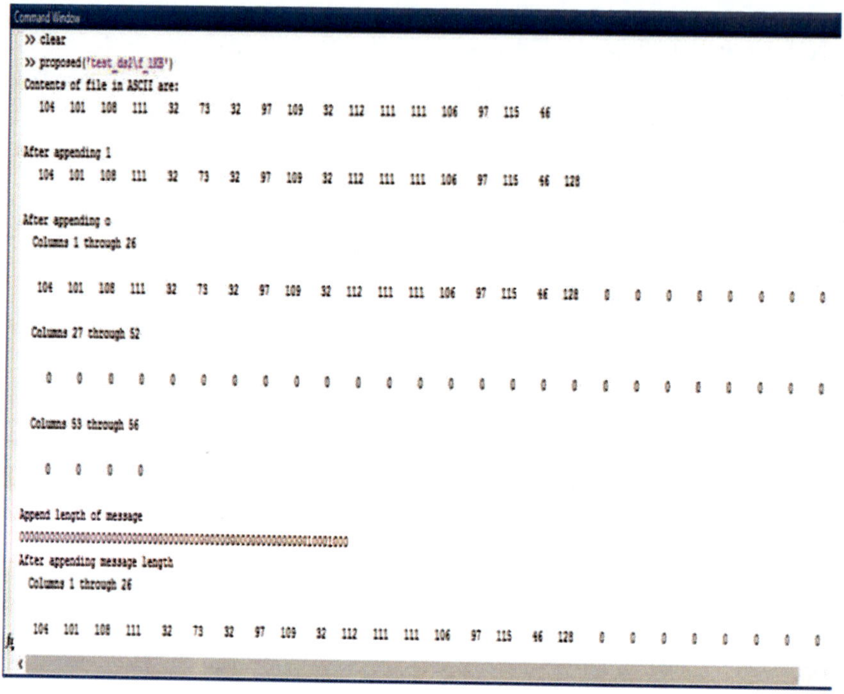

Fig. 2 Demonstration of proposed algorithm

4 Simulation

4.1 Execution of the Proposed Hashing Algorithm

The proposed hashing algorithm is implemented on MATLAB, and the following figure shows the step by step demonstration of proposed algorithm.

The file of 1KB has been considered for the simulation which contains the string "I am shakti." In Fig. 2, steps 1 and 2, i.e. ASCII code of the string, padding, appending and adding length to the string, have been displayed.

4.2 Comparison

Compared with current standard techniques, the suggested hashing technique was compared. To determine their execution time, the aforementioned hashing techniques have been implemented on MATLAB. In order to evaluate the time consisting of various text files with size varying from 10 to 80 KB, a dataset was considered. Each hashing techniques have been simulated 5 times. In terms of average execution time

Table 1 Comparison of proposed hashing algorithm with standard hashing techniques

Parameters	Average execution time (s)	Hash size (bits)
SHA-1	0.1882	160
MD5	0.1922	127
SHA-384	0.1887	256
SHA-256	0.1888	384
SHA-512	0.1895	512
Proposed algorithm	0.119	128

and hash size, the suggested hashing algorithm has been compared with standard hashing algorithms, i.e. MD5, SHA-1 and SHA-256, SHA-384 and SHA-512.

4.3 Analysis

The proposed hashing algorithm has been compared with standard hashing techniques, i.e. MD5, SHA-1, SHA-256, SHA-384 and SHA-512, in terms of average execution time and hash size. In Table 1, average execution time and hash size for the file size 10 KB have been presented.

From Table 1, it has been observed that average execution time of proposed hashing algorithm is less as compared to other algorithms and the hash size of MD5 and proposed is same, i.e. 128 bits, which is smaller than other hashing algorithms.

5 Implementation

One-way: If a hash function is given, $h(x) = y$, where x is the message and y is the calculated hash value, then one-way property tells the infeasibility of computing x for the given value of y. Here, x is the pre-image of y. From Figs. 2, 3 and 4, it is clear that for the given hash value, original message cannot be retrieved. *Second is pre-image resistance*: As per this property, it should be infeasible to find x' for given x and $x' \neq x$ such that $h(x) = h(x')$. *Collision resistance*: According to this property, it should be infeasible to find x and x' such that $x \neq x'$ and $h(x) = h(x')$. *Avalanche effect*: This effect says that if there is even a single bit of modification in a message then the generated hash value of new message will be drastically different. The input string and hash value of three hash algorithms, i.e. MD5, SHA-256 and proposed algorithm, have been implemented. It was analysed that hash values of MD5 and SHA-256 are extremely different even a single character of modification in a message. In proposed hash algorithm, hash value is different for every random message, but it is not strikingly different as in MD5 and SHA-256. This is because the proposed algorithm uses only XOR operations and other hashing techniques use circular shift operations, XOR, AND, OR, NOT operations which change the output hugely on single bit of modification.

6 Conclusion

The latest digest generating is vulnerable to brute force attacks. Therefore, a new hashing technique that uses blocks subgrouping and XOR operation in a specific way has been proposed in this paper which makes difficult for an attacker to break an algorithm. Hence, the proposed technique is more secure than the existing ones.

The proposed technique is very simple and applies basic mathematical operation which increases the speed of computation of hash. Therefore, proposed hashing is simpler than the existing algorithms. In addition, fundamental properties of hash function are fulfilled.

This work can be extended further by designing an algorithm more complex so that results of avalanche effect would be better, and it would be kept in mind that execution time will not be compromised.

References

1. Alizadeh M, Zamani M, Baharun S, Abdul MA, Sakurai K, Anada H (2015) Cryptanalysis and improvement of a secure password authentication mechanism for seamless handover in proxy mobile IPv6 networks. PLoS ONE 10(11):1–21
2. Tanwar S, Thakkar K, Thakor R, Singh PK (2018a) M-Tesla-based security assessment in wireless sensor network. In: 2018 international conference on computational intelligence and data science. Procedia computer science, 132, pp 1154–1162
3. Tanwar R, Singh K, Zamani M, Verma A, Kumar P (2019) An optimized approach for secure data transmission using spread spectrum audio steganography, chaos theory, and social impact theory optimizer. Hindawi J Comput Netw Commun 2019:1–10
4. Arora S, Dalal S (2019) Trust evaluation factors in cloud computing with Openstack. J Comput Theor Nanosci 16(6):5073–5077
5. Feng Q, He D, Zeadally S, Khan MK, Kumar N (2018) A survey on privacy protection in Blockchain system. J Netw Comput Appl, pp 45–58
6. Makhdoom I, Abolhasan M, Abbas H, Ni W (2018) Blockchain's adoption in IoT: the challenges, and a way forward. J Netw Comput Appl 125:251–279
7. Aggarwal S, Chaudhary R, Aujla GS, Kumar N, Choo KKR, Zomaya AY (2019) Blockchain for smart communities: applications, challenges and opportunities. J Netw Comput Appl 144:13–48
8. Sharma R, Athavale V (2019) Survey of intrusion detection techniques and architecture in wireless sensor networks. Int J Adv Netw Appl 10(4):3925–3937
9. Wang L, Shen X, Li J, Shao J, Yang Y (2019) Cryptographic primitives in blockchains. J Netw Comput Appl 127:43–58
10. Bellare M, Kohno T (2003) Hash function balance and its impact on birthday attacks. In: 2003 23rd annual eurocrypt conference, lecture notes in computer science, 3027, pp 1–32

11. Gauravram P (2003) Cryptographic hash functions: cryptanalysis, design and applications. Ph.D. thesis, Faculty of Information Technology, Queensland University of Technology, Brisbane, Australia
12. Kim DC, Hong D, Lee JK, Kim WH, Kwon D (2015) LSH: a new fast secure hash function family. In: 2014 ICISC. Springer International Publishing Switzerland, LNCS, vol 8949, pp 286–313
13. Arya RP, Mishra U, Bansal A (2013) Design and analysis of a new hash algorithm with key integration. Int J Comput Appl 81(1):33–38
14. Chowdhury AR, Chatterjee T, DasBit S (2014) LOCHA: a light-weight one-way cryptographic hash algorithm for wireless sensor network. In: 2014 5th international conference on ambient systems, networks and technologies, Procedia Computer Science, India, 126, pp 497–504
15. Monadal A, Mitra S (2016) TDHA: a timestamp defined hash algorithm for secure data dissemination in VANET. In: 2016 international conference on computational modelling and security, procedia computer science, India, pp 190–197
16. Abutaha M, Hamamreh R (2013) New one way hash algorithm using non-invertible matrix. In: 2013 international conference on computer medical applications (ICCMA), Sousse, Tunisia, pp 1–5
17. Tiwari H, Asawa K (2012) A secure and efficient cryptographic hash function based on NewFORK-256. Egypt Inf J 13(3):199–208
18. Manel D, Raouf O, Ramzi H, Mtibaa A (2013) Hash function and digital signature based on elliptic curve. In: 2013 international conference on sciences and techniques of automatic control computer engineering, Sousse, Tunisia, pp 388–392
19. Mirvaziri H, Jumari K, Ismail M, Hanapi ZM (2007) A new hash function based on combination of existing digest algorithms. In: 2007 5th student conference on research and development, Malaysia, pp 1–6
20. Moe KSM, Win T (2017) Improved hashing and honey-based stronger password prevention against brute force attack. In: 2017 international symposium on electronics and smart devices (ISESD), Yogyakarta, Indonesia, pp 1–5
21. Chen L, Wang G (2008) An efficient piecewise hashing method for computer forensics. In: 2008 first international workshop on knowledge discovery and data mining, Adelaide, Australia, pp 635–638
22. Rasjid ZE, Soewito B, Witjaksono G, Abdurachman E (2017) A review of collisions in cryptographic hash function used in digital forensic tools. Procedia Comput Sci 116:381–392
23. Rubayya RS, Resmi R (2014) Memory optimization of HMAC/SHA-2 encryption. In: 2014 1st international conference on computational systems and communications (ICCSC) Trivandrum, India, 282–287
24. Shakya A, Karna N (2019) Enhancing MD5 hash algorithm using symmetric key encryption. In: 2019 international conference on cryptography, security and privacy. ACM, pp 18–22
25. Chowhan R, Tanwar R (2019) Password-less authentication: methods for user verification and identification to login securely over remote sites. In: Book: machine learning and cognitive science applications in cyber security, pp 190–212
26. Arora S, Dalal S (2019) An optimized cloud architecture for integrity verification. J Comput Theor Nanosci 16(6):5067–5072

A Deep Learning Based Convolution Neural Network-DCNN Approach to Detect Brain Tumor

Hewan Shrestha, Chandramohan Dhasarathan, Manish Kumar, R. Nidhya, Achyut Shankar, and Manoj Kumar

Abstract Medical image analysis is one of the important and high critical applications of deep learning. In complex body organs like brain, it may be crucial to detect the brain tumor which may lead to some negative results. So, it is very important to safely detect the tumor in the brain which is not that easy to find out. In this research, the focus is on Deep Learning (DL) based Convolution Neural Networks (CNNs) approach as DCNN. For optimizing the weights, we applied Adam optimization technique. Having a huge dataset, it gives good accuracy which can be used for proper detection of tumor during the analysis and can go for diagnosis further. Different algorithms are implemented to test the proposed method, and their precision, recall performance parameters are estimated. This approach has given an accuracy of 98% during the prediction.

Keywords Medical image analysis · Brain tumor · Segmentation · Deep learning (DL) · Convolution neural networks (CNNs)

1 Introduction

Deep Learning (DL) is a division of machine learning which is absolutely based on artificial neural networks. As neural networks are going to impersonator of human brain in depth to identify the damaged spot using deep learning approach. It is also a kind of mimic of human brain. DL is used for analyzing the medical images with an explicit design as a customized system. DL is scrupulous machine learning

H. Shrestha · C. Dhasarathan · R. Nidhya
Department of Computer Science & Engineering, Madanapalle Institute of Technology & Science, Madanapalle, Andhra Pradesh, India

M. Kumar
School of Computer Science and Engineering, VIT, Chennai, Tamil Nadu, India

A. Shankar
Department of Computer Science and Engineering, Amity University, Noida, Uttar Pradesh, India

M. Kumar (✉)
School of Computer Science, University of Petroleum and Energy Studies, Dehardun, India

G. Gupta et al. (eds.), *Proceedings of Academia-Industry Consortium for Data Science*, Advances in Intelligent Systems and Computing 1411,
https://doi.org/10.1007/978-981-16-6887-6_11

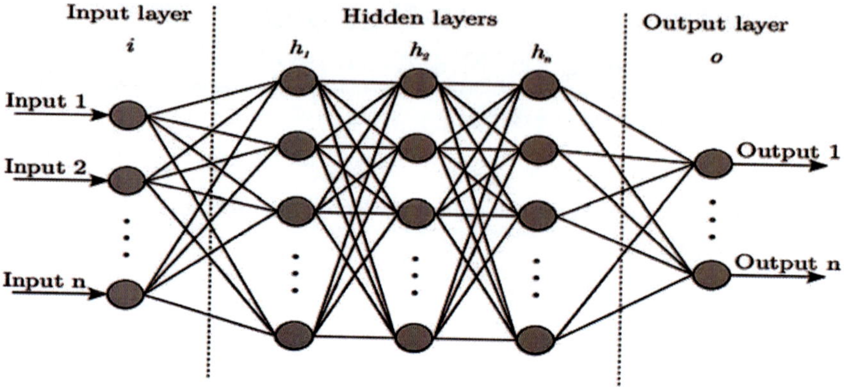

Fig. 1 Neural network architecture

that accomplish enormous power and litheness by learning to signify the world as a nested ladder of notion [1–5], with apiece impression defined in next of kin to simpler perception, and more intangible representation compute in requisites of less abstract ones [6–15]. Deep Learning Algorithms are impressive to identify a neural network and be links between a set of inputs and outputs. A neural network is unruffled of input, hidden, and output layers all are tranquil of 'nodes.' Input layers seize in a numerical illustration of data, output layers provide output predictions, while hidden layers are unified with most of the computation. Today, computers can not only categorize photos, but also portray assorted rudiments present in pictures [16–23].

As shown in Fig. 1, Deep Learning Network called Convolution Neural Networks (CNN), which actually learns patterns that in nature occur in photos. For improved accuracy, image classification using CNN is not effective. First and foremost, it is a need to set of images (Fig. 2).

Deep learning is having a wide range of applications including self-motivating Cars, news bulletin Aggregation, hoax information recognition, Natural Language Processing, Virtual subordinate, Visual detection, Image Analysis, Healthcare, Automatic engine conversion, Automatic Text Summarization, and others as well [9, 24, 25]. Among all the applications of Deep Learning, healthcare is solitary of the majority imperative appliance in contemporary world. Having a huge impact in healthcare, deep learning can be used in many subfields such as analyzing blood samples, tracking glucose levels in diabetic patients, detecting heart problems [26–28] using image analysis to detect tumors, detecting cancerous cells and diagnosing cancer and many more. Image classification is one of the problems of image processing, computer vision and machine learning areas. Classification is a systematic arrangement in groups and categories depending on its features. Image classification is achieved by distinguishing the image into appropriate category anchored in the substance of the representation [29, 30].

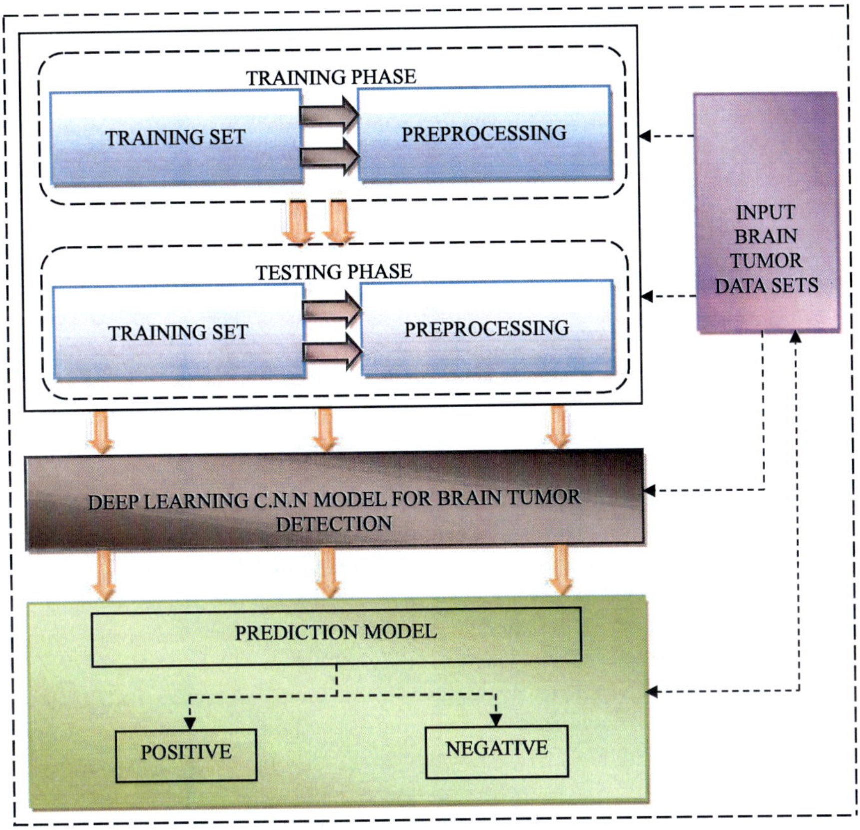

Fig. 2 Brain tumor detection using deep learning convolution neural networks-DCNN

2 Related Work

Shahamat et al. (2019), concerning medical image analysis, convolution neural networks (CNNs) models have presented impressive performance. Research community have, a three-dimensional convolution neural network (3D-CNN) has been engaged to categorize brain MRI scans into two previously distinct groups. Furthermore, a Genetic Algorithm Based Brain Masking (GABM) technique has been anticipated to visualize the new insights in the function of 3D-CNN. Pomponio et al. [2], harmonizing of image data diagonally various cronies with changing etiquette have been difficult due to the increasing needs of big data analytics in the modern information era. The authors designed, a large-scale dataset of 10,477 structural brain MRI scans from people exclusive of a notorious neurological disorder has been taken and analyzed. Dong et al. [3], Traumatic brain injury (TBI) direct to high transience. TBI is a community vigor problem worldwide having no valuable healing. The researchers discussed, the neuro-protective role of cur cumin in mouse

TBI is dependent on the NF-E2-related factor (Nrf2) pathway. Hussein [4], after the discovery of radioactivity, natural radiation has found various imaging application in various fields. In this paper, overview of many applications in which radiation is used to image structures is provided. Khademi et al. [5], segmentation of brain from MRI is a beginning step in many neuroimaging processing tasks. $T1$ and $T2$-weighted MRI are some of the available technologies however, a brain extracted method for Fluid-Attenuated Inversion Recovery (FLAIR) MRI has to be recognized. FLAIR MRI is fetching important for the scrutiny of neuro-related syndrome. Yin et al. (2020), the irregular intensification of cells in the brain which can be banned for the next jeopardy juncture by proper opinion is a critical brain tumor. The authors designed, a new meta-heuristic line of attack has been proposed for early verdict of brain tumor to thwart from auxiliary escalation of the tumor. Elahi et al. (2020), in the field of computer vision and image processing, gradient-based consistency study is fashionable and having many applications in medical image analysis. Research community approach of rising a texture feature mining method to distinguish Amyotrophic Lateral Sclerosis (ALS) patients is proposed. Maharjan et al. [9], for the multi-class classification of brain tumor, use of sigmoid function is badly used in deep learning due to its boundary of dual taxonomy. Authors highlighted, an approach to increase the categorization truth has been planned. The proposed model consists of a convolution neural network with customized softmax loss gathering and regularization. Cheng et al. (2020), there are unique advantages in segmentation of brain hankie in distribution of MRI image liberty. The author states, a novel segmentation scheme using direction-averaged diffusion weighted imaging (DWI) signal has been proposed. Kociołek et al. (2020), in medical image analysis, image consistency is a very important section. Due to noise and artifacts, medical images are quite corrupted and affected. Markousis-Mavrogenis et al. [13], there are a number of imaging modalities in PsA, among which magnetic resonance imaging (MRI) holds an important role, as it is suitable for assessment of brain association. Although there are not any official recommendations, it could be considered for evaluation with respect to cognitive dysfunction. Chatterjee et al. (2019), Hearing the word ‘brain tumor,’ it sounds threat among children and adults all over the world. If we can detect and diagnose the tumor pretty early then, we can improve the ability of endurance among the folks. Although segmentation and categorization are good based on their applicability, manual segmentation is time-consuming and also results in inaccurate results. Knier et al. [15], brain meta-sets is one of the leading causes of mortality these days. Studying about brain sickness is also exigent for the reason that of the imperfect investigational methods available. Soleman et al. [17], for the diagnosis and analysis of brain tumors, technical advancements and availability of neuroimaging is playing a crucial role. Proper management and risk factors of these tumors are not well defined which may lead to further growth of the tumor. Chandra et al. [18], in the early stage of tumor, it is very hard to detect and is quite challenging task due to the surrounding of non-cancerous cells. Using image segmentation as a contrivance for detecting the tumor makes it easy to segment the tumorous section. Özyurt et al. (2019), when we have an image with high resolution which we call as super-resolution these days, it becomes easy to view the image. If we increase the resolution of an image with Magnetic Resonance Imaging (MRI), it makes the image more visible and clearer which provides the borders of tumor cells to be found more easily.

Toğaçar et al. [20], a brain tumor is nothing but a mass of muscle which grows erratically in the brain and influence human life. This mass transpires unexpectedly due to the tissues adjacent roughly the skull. Though surgical process is commonly favored for brain tumor treatment, deep learning practice is used to the diagnosis and treatment of brain tumor these days. Çinar et al. [22], when it comes to brain tumor, premature and perfect verdict is important for the healing process. It is always the most important step to distinguish brain tumor using computer classification. These provide professional to detect the tumor more simply and perfectly. Hashemzehi et al. [23], evil brain tumor is one of the most outrageous type of cancer with undeviating concern into cognition and meager brain implementation. Chandamohan et al. [27], designed a data dissemination approach for evaluating unstructured irrelevant data to contribute for the research community with novel process for identification and classification.

Making use of Magnetic Resonance Imaging (MRI) for analyzing the brain tumors, formulate easy for former recognition of tumors. Maharjan et al. [9], although CT scan images are swamped using breaking point segmentation and morphological maneuver to separate the brain tissues and the tumor, it fails to disconnect the tumor region perfectly. Hence, using neural networks and segmenting, the Magnetic Resonance Imaging (MRI) images shall be painstaking for the suitable exposure of the tumor. Saba et al. [25], one of the decisive diseases in the world currently is brain tumor. Having tumor in the brain reimbursement vigorous tissues, it might foundation wrecked of the brain and eventually to death as the augmentation of tumor augment. So, early diagnosis is necessary to spot the tumor perfectly. Sharif et al. [26], it is one of the motivations for escalating death rate among children and adults. Han et al. [28], CNN can be used to accomplish outstanding opinion with proper amount of instruction data. However, medical imaging datasets are moderately small and disjointed. Generative Adversarial Networks (GAN) can be used to produce dissimilar supplementary training images to fill the records require. Li et al. [29], as brain tumor grows, more assorted it transpire for healing. Brain tumors can undergo diagnosis only after proper analysis of brain tumor images. Using Magnetic Resonance Imaging (MRI) images with CNN can help with advance process. Noreen et al. [30], classifying whether it is tumor or not, is one of the exigent chore authorities have to go through. MRI have been sustaining to scrutinize the brain tumor. These days, a scheme of attribute origin and aspect mixture can be used for untimely opinion of brain tumor. Some techniques can be referred from [31–34].

3 Deep Learning Based Convolution Neural Network (DCNN)

Machine Learning (ML) is making everybody's life easier with automotive works. DL, which is one of the subset of ML, can be used for high end image analysis and automotive tasks. The article describes a novel DL technique for image analysis of brain using DCNN. Brain, being one of the most crucial part of human body, is responsible for all human activities. If there is any injury in brain may lead to poor health condition and even death. This paper talks about the image analysis for brain tumor detection using CNN model. The proposed image processing and classification using CNN model are highlighted to identify the difference between normal and abnormal study of both 2-dimensional (2-D) and 3-dimensional (3-D) structure of a health organ is imperative. CNNs are appropriate to carry out image-recognition tasks. Not only image-recognition, Fig. 3 CNNs have been used in image categorization, localization, recognition, segmentation and muster. CNNs are one of the most popular ML algorithm used for image taxonomy errands. CNNs can take both 2-D images and 3-D images as inputs with few modifications and process them. CNN is one of the examples of direct ML algorithm, and it involves considerable amounts of guidance data. The units of intricacy deposit '1' deduct their commencement $A_j^{(l)}$ based only on subset of component in attribute maps $A_i^{(k-1)}$ of the foregoing layer '$l-1$' by convolving the kernels $k_{ij}^{(l)}$ as given below in Eq. (1).

$$A_j^{(l)} = f(x) = \sum_{i=1}^{M(l-1)} A_i^{(l-1)} * k_{ij}^{(l)} + b_j^{(l)} \tag{1}$$

where, M $(l-1)$ denote magnitude of power atlas in layer $l-1$, '*' denotes a convolution operator, $b_j^{(l)}$ is a bias parameter.

Pooling layer: Pooling layer condense the quantity of constraint to be taught and sum of computation. Max-pooling layer is most commonly used pooling layer. Max-pooling acquires the ceiling constituent from a province of attribute map sheltered by the filter.

Rectified Linear Unit (RELU) layer: It is a convention to apply a non-linear layer (activation layer) immediately after each convolution layer. The RELU deposit is an establishment occupation that situates pessimistic participation values to zero.

Mathematically, it is distinct as shown in Eq. 2,

$$f(x) = \max(0, x) \tag{2}$$

Fully connected layers: There are copious associated input layers and abundant associated output layers as well. In Figure 4, the input layers take the output of preceding layers, squash them and spin to a sole vector which can be input for subsequently peak. But, the output layer gives the final likelihood for each label.

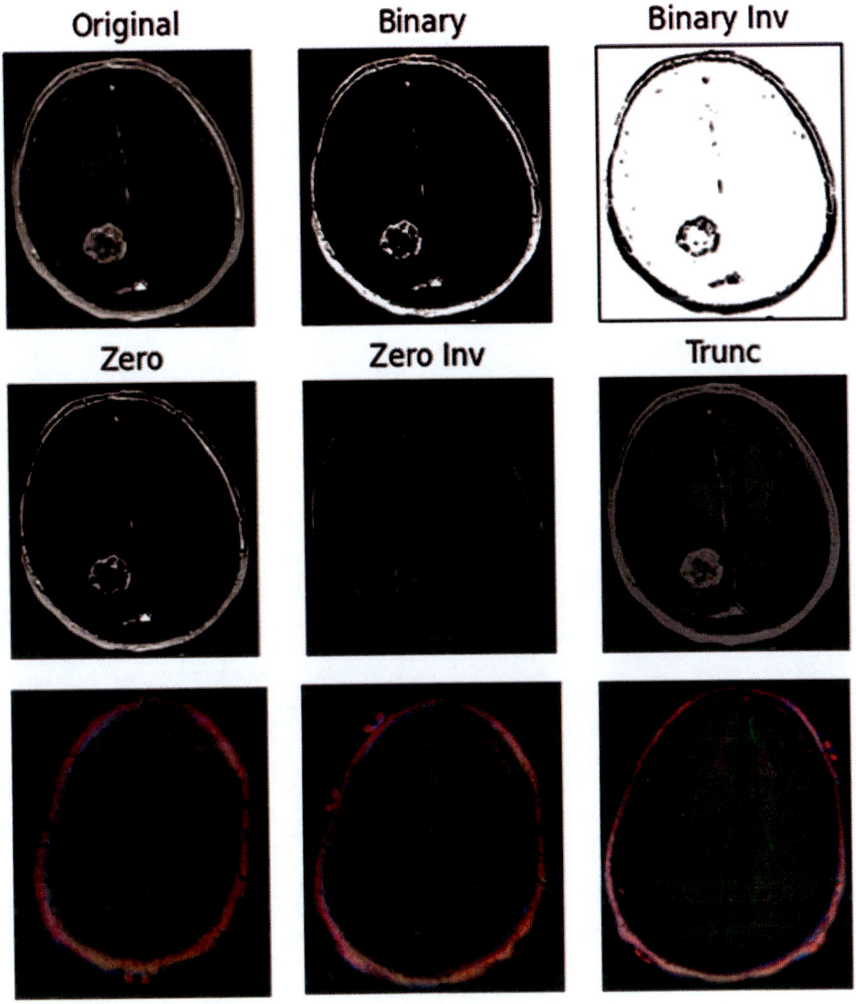

Fig. 3 Segmentation of brain tumor detected from MRI image

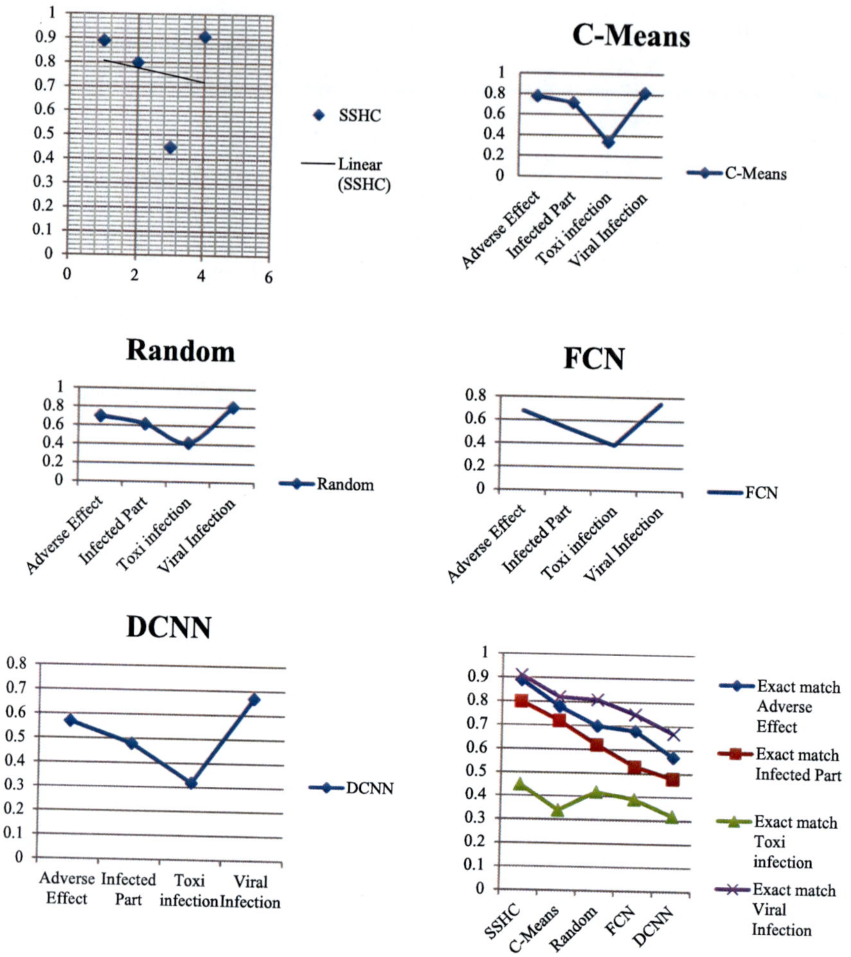

Fig. 4 Exact brain tumor match detection using deep learning convolution neural networks-DCCN

PSEUDOCODE FOR MODEL PREPARATION of DCNN
 	
Begin
Initialize CNN.
model = Sequential()
for each CNN_Sequence do
if initial stage Add 2 CL then
md(C2D(ft=32,ks=(3,3),isp=(64,64,3),atv='rul'))mdx(C2D(ft=32,ks=(3,3),atv='rul'))
else if next stage reached to max then
Add a max-pool layer
mdx(MP2D(pool_size=(2,2)))
end if
else if it reached convolution stage then
Add two more CL
mdx(C2D(ft=32,ks=(3,3),atv='rul'))mdx(C2D(ft=32, ks=(3,3), atv='rul'))
end if
else if one more max then
Add one more max-pooling layer
mdx(MP2D(pool_size=(2,2)))
end if
else if another two more max reached then
Add two more CL mdx(C2D(ft=32, ks=(3,3), atv='rul')) mdx(C2D(ft=32, ks=(3,3), atv='rul'))
end if
else if extra max polling included then
	Add one more max-pooling layer
	mdx(MP2D(pool_size=(2,2)))
	end if
	else if average layer is added then
	Add global average pooling layer
	mdx(GlobalAveragePooling2D())
	end if
	else if all available connections are added then
	Add full connection mdx(Dense(units=2, atv='softmax'))
	end if
	end if
	else no more model present then
	Compile the model
	model.compile(optimizer='adam', loss='binary_crossentropy', metrics=['accuracy'])
	end for CNN
	for each DCNN_Sequence do
	if initialize CL then
	mdx(C2D(ft=32,ks=(3,3),isp=(64,64,3), atv='BrainTumer'))mdx(C2D(ft=32,ks=(3,3),atv='Tumer'))
	else if next stage reached to max then
	Add a max-pooling layer
	mdx(MP2D(pool_size=(2,2)))
	end if
	else if it reached DCNN stage then
	Add two more CL
	mdx(C2D(ft=32,ks=(3,3),atv='BrainTumer'))mdx(C2D(ft=32, ks=(3,3), atv='Tumer'))
	end if
	else if one more max then
	end for DCNN end

Training Phase: In the training phase, we have taken the training dataset as input and preprocessed the dataset. For preprocessing the dataset, we rescaled the dataset to 1/255 and gave a batch size of 32.

Testing Phase: In the testing phase, we have taken the testing dataset as input and preprocessed the dataset as we did in the training phase. After completing the training phase and the testing phase, we created our CNN model and saved it as 'model.h5.'

Predicting Phase: In the predicting phase, we loaded the CNN model 'model.h5' and used that model for the input image. We converted the image to numpy array and predicted the image classification whether the image is having tumor or not.

4 Experimentation and Result Analysis

Setup: For the entire process, we have used Keras library and TensorFlow library as backend for the association preparation. The arrangement weights were learnt using 'adam' optimizer. In Fig. 5, different algorithms are tested with the number of epochs and its precision, recall is illustrated in Tables 1 and 2 with a pre-training of $40 + 2e$. Dataset: Experiments were conducted on dataset with MRI brain tumor detection.

However, Table 3 illustrates the tartan correctness and defeat for different number of epochs with exact brain tumor detection and evaluation. In Fig. 3, reflects the accuracy and its corresponding loss for the model of $40 + 2e$ epochs and got accuracy and loss of 0.9997 and 0.0021, respectively. It is also observed in Fig. 4. We have the segmentation of input image into different forms by using the OpenCV library.

Finding the tumor in inner regions of the brain is a very difficult task. If there are any internal injuries in the brain, it may be difficult to find out where the tumor is. For proper analysis of image, our proposed approach as illustrated in Table 3 has an accuracy of 98% and also shown in Fig. 5. The deep learning CNN model has been implemented to classify the images with tumor from those which are not having tumor. Currently, DCNN is tested with healthcare dataset tested the performance in all scenarios. Because of having moderate sized dataset, there is a disparity breach between guidance and taxing concert. For more complex augmentation, some other models like GAN, would be preferably espouse for expectations vocation. Also, proposed approach cannot be trained properly because of computer's GPU memory.

Table 1 Different number of epochs noted for precision and recall using DCNN

Algorithm	Epochs	Precision	Recall
SSHC	10	0.70	0.75
C-means	12	0.21	0.34
Random	15	0.42	0.56
FCN	17	0.65	0.69
DCNN	10	0.73	0.82

Table 2 Accuracy and loss distribution for different number of epochs using DCNN

Epochs	Accuracy	Loss
10	0.9874	0.0448
12	0.9929	0.0256
15	0.9949	0.0222
17	0.9992	0.0080

Table 3 Exact brain tumor match detection and evaluation

Techniques	Exact match			
	Adverse effect	Infected part	Toxi infection	Viral infection
SSHC	0.89	0.8	0.45	0.91
C-means	0.78	0.72	0.34	0.82
Random	0.7	0.62	0.42	0.81
FCN	0.68	0.53	0.39	0.75
DCNN	0.57	0.48	0.32	0.67

It discriminates and conveys various dilapidation on the whole performance of the DCNN model.

5 Conclusion

The article articulates the importance of DCNN with a dynamic approach to analyze brain MRI images having tumor using Deep learning methodology. Nowadays, accidents causing brain tumor is increasing more and more, while the traditional brain tumor detection method may not be effective for all cases. Today, most researchers are focused toward extracting information from MRI scan of brain images using latest technologies available which includes deep learning as well. However, this article is providing a practical approach to classify brain tumor images for individual's healthcare researches willing to contribute for the enhancement of image segmentation and classification related works.

References

1. Shahamat H, Abadeh MS (2020) Brain MRI analysis using a deep learning based evolutionary approach. Neural Netw. https://doi.org/10.1016/j.neunet.2020.03.017
2. Pomponio R, Erus G (2020) Habes, "Harmonization of large MRI datasets for the analysis of brain imaging patterns throughout the lifespan." Neuroimage. https://doi.org/10.1016/j.neuroimage.2019.116450
3. Dong W, Yang B, Wang L, Li B, Guo X, Zhang M, Jiang Z, Fu JJ, Pi D, Zhao GR (2018) Curcumin plays neuroprotective roles against traumatic brain injury partly via Nrf2 signaling. Toxicol Appl Pharmacol https://doi.org/10.1016/j.taap.2018.03.020
4. Hussein EMA (2019) Imaging with naturally occurring radiation. Appl Radiat Isotopes 145. https://doi.org/10.1016/j.apradiso.2018.12.006
5. Khademi A et al (2019) Whole volume brain extraction for multi-centre, multi-disease FLAIR MRI datasets. Magn Reson Imaging. https://doi.org/10.1016/j.mri.2019.08.022
6. Dhasarathan D, Dayalan T, Ponnurangam (2015) A multi-agent approach: to preserve user information privacy for a pervasive and ubiquitous environment. Egypt Inf J 16(1):151–166. ISSN 1110-8665. https://doi.org/10.1016/j.eij.2015.02.002

7. Bo Y, Chao W, Francis A (2020) New brain tumor classification method based on an improved version of whale optimization algorithm. Biomed Signal Process Control 56. https://doi.org/10.1016/j.bspc.2019.101728
8. Mashrur GM, Elahi E, Kalra S, Zinman L, Genge A, Korngut L, Yang Y-H (2020) Texture classification of MR images of the brain in ALS using M-CoHOG: a multi-center study. Comput Med Imaging Graph 79. https://doi.org/10.1016/j.compmedimag.2019.101659
9. Maharjan S, Alsadoon A, Prasad PWC, Salam M, Alsadoon OH (2019) A novel enhanced softmax loss function for brain tumour detection using deep learning. J Neurosci Methods. https://doi.org/10.1016/j.jneumeth.2019.108520
10. Hu C, Sharlene N, Eleftherios G, Maryam A, Shreyas SF (2020) Segmentation of the brain using direction-averaged signal of DWI images. Magnet Reson Imaging 69. https://doi.org/10.1016/j.mri.2020.02.010
11. Marcin K, Michał S, Rafał O (2020) Does image normalization and intensity resolution impact texture classification? Comput Med Imaging Graph. https://doi.org/10.1016/j.compmedimag.2020.101716
12. Chandramohan D, Vengattaraman T, Dhavachelvan P (2017) A secure data privacy preservation for on-demand cloud service. J King Saud Univ Eng Sci 29(2): 144–150. https://doi.org/10.1016/j.jksues.2015.12.002. ISSN: 1018-3639
13. Markousis-Mavrogenis G, Nurmohamed MT, Koutsogeorgopoulou L, Dimitroulas T, Katsifis G, Vartela V, Mitsikostas D, Kolovou G, Tektonidou M, Voulgari P, Sfikakis PP, Kitas GD, Mavrogeni SI (2020) Current understanding and future perspectives of brain–heart–kidney axis in psoriatic arthritis. Rheumatol Int. https://doi.org/10.1007/s00296-020-04633-1
14. Subhashis C, Ananya D (2019) A novel systematic approach to diagnose brain tumor using integratedtype-II fuzzy logic and ANFIS (adaptive neuro-fuzzy inference system) model. Soft Comput. https://doi.org/10.1007/s00500-019-04635-7
15. Knier NN, Hamilton AM, Foster PJ (2020) Comparing the fate of brain metastatic breast cancer cells in different immune compromised mice with cellular magnetic resonance imaging. Clin Exp Metas. https://doi.org/10.1007/s10585-020-10044-0
16. Dumka DC, Dilipkumar V, Loganathan J (2020) Data dissemination for green-VANETs communication: an opportunistic optimization approach. Int J Pervasive Comput Commun https://doi.org/10.1108/IJPCC-04-2020-0030
17. Soleman J, Kozyrev DA, Ben-Sira L, Constantini S, Roth J (2020) Management of incidental brain tumors in children: a systematic review. Child's Nervous Syst. https://doi.org/10.1007/s00381-020-04658-8
18. Chandra SK, Bajpai MK, Fractional Crank-Nicolson finite difference method for benign brain tumor detection and segmentation. Biomed Sig Process Control. https://doi.org/10.1016/j.bspc.2020.102002
19. Özyurt F, Sert E, Avcı D (2019) An expert system for brain tumor detection: fuzzy C-means with super resolution and convolutional neural network with extreme learning machine. Med Hypotheses. https://doi.org/10.1016/j.mehy.2019.109433
20. Toğaçar M, Ergen B, Cömert Z (2019) BrainMRNet: brain tumor detection using magnetic resonance images with a novel convolutional neural network model. Med Hypotheses. https://doi.org/10.1016/j.mehy.2019.109531
21. Chandramohan D, Vengattaraman T, Rajaguru D, Dhavachelvan P (2016) A new privacy preserving technique for cloud service user endorsement using multi-agents. J King Saud Univ Comput Inf Sci 28(1):37–54. ISSN 1319-1578. https://doi.org/10.1016/j.jksuci.2014.06.018
22. Çinar A, Yildirim M (2020) Detection of tumors on brain MRI images using the hybrid convolutional neural network architecture. Med Hypotheses. https://doi.org/10.1016/j.mehy.2020.109684
23. Hashemzehi R, Mahdavi SJS, Kheirabadi M, Kamel SR (2020) Detection of brain tumors from MRI images base on deep learning using hybrid model CNN and NADE. Biocybern Biomed Eng. https://doi.org/10.1016/j.bbe.2020.06.001
24. Shrestha H, Chandramohan D, Munisamy S et al (2020) Natural language processing based sentimental analysis of Hindi (SAH) script an optimization approach. Springer Int J Speech Technol 23:757–766. https://doi.org/10.1007/s10772-020-09730-x

25. Saba T, Mohamed SA, El-Affendi M, Amin J, Sharif M (2019) Brain tumor detection using fusion of hand crafted and deep learning features. Cogn Syst Res 59. https://doi.org/10.1016/j.cogsys.2019.09.007
26. Sharif M, Amin J, Wasif Nisar M, Almas Anjum M, Ahmad N, Ali shad S (2019) A unified patch based method for brain tumor detection using features fusion. Cogn Syst Res. https://doi.org/10.1016/j.cogsys.2019.10.001
27. Chandramohan D, Dumka A, Jayakumar L (2020) 2M2C-R2ED: multi-metric cooperative clustering based routing for energy efficient data dissemination in green-VANETs. Technol Econ Smart Grids Sustain Energy 5:15. https://doi.org/10.1007/s40866-020-00086-4
28. Han C et al, Combining noise-to-image and image-to-image GANs: brain MR image augmentation for tumor detection. https://doi.org/10.1109/ACCESS.2019.2947606
29. Li M et al, Brain tumor detection based on multimodal information fusion and convolutional neural network. https://doi.org/10.1109/ACCESS.2019.2958370
30. Noreen N et al, Deep learning model based on concatenation approach for the diagnosis of brain tumor. https://doi.org/10.1109/ACCESS.2020.2978629
31. Aggarwal A, Kumar M (2020) Image surface texture analysis and classification using deep learning. Multimedia Tools Appl (MTAP). https://doi.org/10.1007/s11042-020-09520-2
32. Kumar M, Srivastava S, Image authentication by assessing manipulations using illumination. Multimedia Tools Appl 78(9): 12451–1246. https://doi.org/10.1007/s11042-018-6775-x
33. Kumar M, Srivastava S, Uddin N (2017) Forgery detection using multiple light sources for synthetic images. Aust J Forensic Sci 51(3):243–250
34. Kumar M, Alshehri M, AlGhamdi R, Sharma P, Deep V (2020) A DE-ANN inspired skin cancer detection approach using fuzzy C-means clustering. Mobile Netw Appl 25:1319–1329. https://doi.org /https://doi.org/10.1007/s11036-020-01550-2

Procuring and Analysing Vital Signs Datasets and Diagnosing the State of Comatose Patients

Shinjini Biswas, Pragya Choudhary, Lipika Rustagi, and Nidhi Goel

Abstract Continuously monitoring coma patients to understand their health state is an essential and critical step. It becomes extremely important to accurately determine the state of the coma patient to provide further treatment. There have been researches in this domain based on image processing techniques; however, these techniques have slower computational speed. The paper proposes a sensor-based technique which aims to rectify this drawback as sensors provide numeric data and are thus computationally faster than image processing techniques. The research work is broadly divided into two parts. The first part focuses on monitoring of the coma patients, whereas the second part discusses a diagnostic test app developed for measuring Glasgow Coma Scale (GCS) which is a standardized test for determining the state of coma. The data procured from the patients is continuously updated on both a real-time database and an app. The paper also aims to calculate the error in the measured vital signs through our system as compared to those obtained from analog devices.

Keywords Internet of things · Comatose · Glasgow Coma scale · Real-time database · App

1 Introduction

Coma is that state of a person where he or she cannot respond to the internal and external stimulus. Comatose state is a state of extended unconsciousness that can be caused due to a variety of problems like brain tumor, traumatic brain injury, drug or alcohol intoxication, or illnesses such as diabetes or any other infection [1].

It becomes a tedious job for the medical staff to continuously monitor each patient for 24 h since the proportion of the medical staff to patient is very low. Hence, it is difficult to monitor each patient regularly. In the ICUs, continuous monitoring of patients is done. But the paper focuses on the multiple coma patients in wards, who are sometimes neglected due to the low doctor-patient ratio.

S. Biswas (✉) · P. Choudhary · L. Rustagi · N. Goel
IGDTUW, New Delhi, Delhi, India

G. Gupta et al. (eds.), *Proceedings of Academia-Industry Consortium for Data Science*, Advances in Intelligent Systems and Computing 1411,
https://doi.org/10.1007/978-981-16-6887-6_12

Thus, it is necessary to regularly monitor the patients to determine further treatment for the patient. Also, a continuous real-time update to the doctors/medical personnel of the level of vital signs as well as seizure detection is very important.

The existing research papers [2–4] focus on only the monitoring of vital signs of patients using either image processing techniques or sensor. The image processing techniques have slower computational speed as compared to sensor-based techniques. Moreover, the existing sensor-based techniques only monitor the vital signs of the patients which can be easily done nowadays.

The paper proposes a sensor-based technique using IoT which aims to rectify this drawback as data from the sensors is numeric data and is thus computationally faster than image processing techniques. Also, it proposes to develop a standardized diagnostic test for determining the state of the comatose patient.

This research work is broadly divided into two parts. The first part focuses on the monitoring of the vital signs of the coma patients using a wearable motion sensor system and other vital signs sensors. Various parameters are being monitored such as body temperature, pulse/heart rate, limb movement and urine level. Monitoring all these vital signs becomes extremely necessary as even a slight change in any of these vital signs requires immediate medical attention.

In the second part, an app-based diagnostic test has been developed for measuring the GCS, which is a standardized test for determining the state of coma. This test determines the state of the coma patient based on two kinds of responses namely verbal response and motor response [5].

The next section of the paper discusses the existing systems in the state-of-the-art. Section 3 then discusses the proposed solution. Section 4 then presents the result analysis. Further, the last section presents the conclusion and future scope.

2 Literature Survey

Some related developments have been done to monitor the coma patient's health. The associated developments have been discussed below.

Lokesh et al. [6] proposed a system that monitors and alerts the caretaker about the transition of the patient from vegetative state to minimally conscious state. Continuous monitoring of the vital signs namely heartbeat, pulse rate, physiological signals along with the muscle movements is performed. Any change in the behavioral approach is recorded with the help of different sensors integrated to Node MCU, and the alerts are sent to the doctor.

Subha et al. [7] developed a method for coma patients using IoT as a dynamic service non-dependency verification that is based on equipment interface. The method creates an online application with a web server in a verified way for the comatose patients by interfacing multiple sensors with the controller board and the data is made available over the cloud. Jie et al. [8] proposed a wearable IoT cloud-based health monitoring system that uses body area sensor network (BASN) for real-time health monitoring. BASN is a collection of three vital sensors that uses a

Wi-Fi module to send the data over the cloud. Data is displayed on LCD instead of smartphone which is generally used for such cases.

Several other papers propose the different types of health care systems based on sensor technique. Arduino Uno board has been used in some research work for measuring parameters like body temperature, oxygen saturation percentage, and eye movement using appropriate sensors [9]. Also, the sensor parameter data in digital form can be displayed on the smartphone using Bluetooth communication between Arduino UNO and smartphone [10]. Sensor data could be updated over the cloud using a Wi-Fi module and ethernet shield and accessed to monitor real-time health statistics [11]. Even a smartphone can also be used as a real-time health parameter monitoring device, where some of the vital parameters can be measured using the sensors embedded in the smartphone and displayed on the same [12, 13]. The basic monitoring could be done with the sensors integrated to the Raspberry Pi board and the data transmitted into the IoT network [14].

Banerjee et al. [15] proposed a system based on pulse rate detection that uses non-invasive technique using plethysmography process; the method had better result as compared to invasive techniques. The GCS test is traditionally performed manually wherein a trained professional assesses the response based on their perception [16].

3 Proposed Solution

The present paper proposes a 2-tier solution based on Internet of Things (IoT). First, continuous recording of the vital signs of the coma patient is performed. Initially, the patient is monitored for any changes in the physical movement or the vegetative state of the patient, followed by sending an alert to the doctor or paramedical staff when attention is needed. Second, an automated diagnostic test that determines the state of the coma patient is performed. Block diagram of the proposed solution is shown in Fig. 1.

3.1 Monitoring System

The monitoring system collects the vital parameters of the patient continuously using the different sensors, all the retrieved data is updated on the cloud on a real-time database and reflected parallelly on an app. Using both of these methods, the doctors and the paramedical staff can have a check on the patient's condition and appropriate

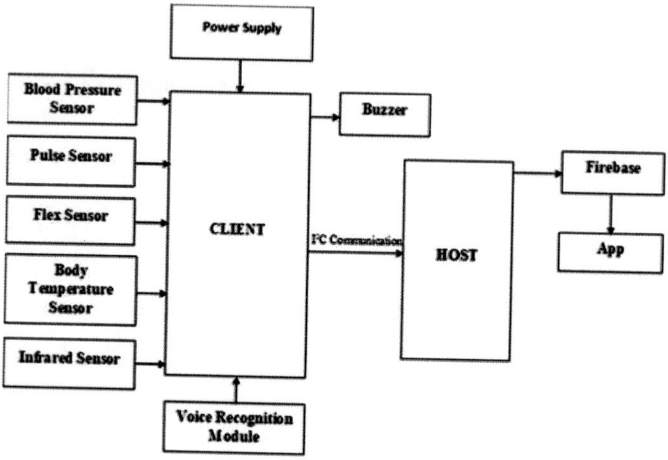

Fig. 1 Block diagram of the proposed solution

attention could be provided to the patient. It has two parts described in the sections below.

Continuous Recording and Analysis of Vital Signals of the Patient

The coma patient is continuously monitored for the different vital parameters that are body temperature, pulse rate, urine level and limb movement using body temperature sensor, pulse rate sensor, infrared sensor, and flex sensor, respectively.

The infrared sensor for urine level measurement is mounted onto a urine bag and continuously records the urine level. The flex sensor detects slightest of the movement of fingers as whenever the flex sensor bends, and there is a change in its internal resistance which leads to detection of movement. These parameters are continuously recorded using the sensor modules.

From the block diagram, the data retrieved from the sensors is then processed using the microcontroller board which acts as the client. Using the I2C communication protocol, the data is transferred from the client to the host. The host further sends the data to the real-time database over the cloud. Google Firebase is used as a real-time database for storing and displaying results. The data from firebase is also displayed on an app, alarming the doctor for any parameter that goes out of range and any physical movement that requires any attention such that suitable action can be taken on time.

Wearable Motion Sensor System (Glove)

In case of a coma patient, it is essential that even the smallest limb movement is recorded. The wearable motion sensor system in form of a glove is used to monitor the limb movements such as twitching to detect any seizure or the conscious state of the patient. The limb movement is recorded using the flex sensor as shown in Fig. 1, in case of any movement, the data is sent to the microcontroller board which processes the input and sends the response to the host via I2C communication. Hence, the movement is reflected on both the real-time database and the app that immediately alerts the doctor or the paramedical staff.

3.2 *Diagnostic Test for Glasgow Coma Scale (GCS)*

The GCS test is a practical method for the assessment of impairment of conscious level of a person in response to a defined stimulus [17]. This is a diagnostic test that helps to determine the stages of coma. There are three levels of disordered consciousness underlined in the label of coma. The disorders that affect an individual's ability to be awake are combined under the name of Disorders of Consciousness (DOC). DOC includes Coma, Vegetative State (VS), and Minimally Conscious State (MCS) [18].

The app-based test gives the score based upon two responses namely verbal response and motor response. The verbal response is recorded using the voice recognition module, and the motor response is recorded using the wearable motion sensor glove that uses flex sensor as shown in Fig. 1. The data is processed and sent to the doctor via real-time database and app.

For the measure of GCS, a score is given for the verbal and motor response. Tables 1 and 2 summarize the scores and ratings assigned for a particular category of verbal response and motor response, respectively. The scores from the above two tables are combined for the total score. Calculated total score determines the state of coma patient. The total score range 1–3 would indicate a Permanent Coma State, 4–7 would indicate a Vegetative Coma State and 8–10 would indicate a Minimal Conscious State.

Table 1 Criterion for verbal response score

Criterion	Rating	Score
Correct name and place	Oriented	5
Disoriented communication	Confused	4
Intelligible single words	Words	3
Only moans/groans	Sounds	2
No audible	None	1

Table 2 Criterion for motor response score

Criterion	Rating	Score
Full extension	Obeys command	5
Near to full extension	Localizing	4
Half extension	Normal flexion	3
Little movement	Extension	2
No movement	None	1

4 Result Analysis

In this section, the result of the proposed system has been shown which includes the implementation of the system and display of the required data on the cloud and app.

4.1 Monitoring System

In the first part, the vital parameters such as pulse rate, body temperature, limb movement and urine level are recorded. Figure 2a, b shows the data of the vital signs of a particular patient on the real-time database and app, respectively. The data from the sensors is recorded every 10 s making the monitoring real-time. Body temperature is measured in degree Celsius, and pulse rate in beats per minute (bpm). For limb-movement, there are three conditions to be analyzed that are when there is no twitch or a slight twitch or a severe twitch. For urine level, urine bag is monitored and when it becomes full, an alert is sent to the hospital staff so that the container can be emptied. There are two states for analysis, i.e., whether the urine bag is full. A range has been set for all the parameters and if any of them goes out of range, it immediately alerts the doctor and the hospital staff. From Fig. 2, it can be seen that the recorded body temperature for the patient was 37 °C, pulse rate was 72 bpm. No twitch was detected, and the urine bag was not full at the time of recording the result. The sensor data retrieved is transferred from client to host using I2C communication and is updated on a real-time database, firebase as shown in Fig. 2a and displayed on an App as shown in Fig. 2b.

Comparison has been done between the actual and observed data. The sensors used for measuring the pulse rate and body temperature were tested to

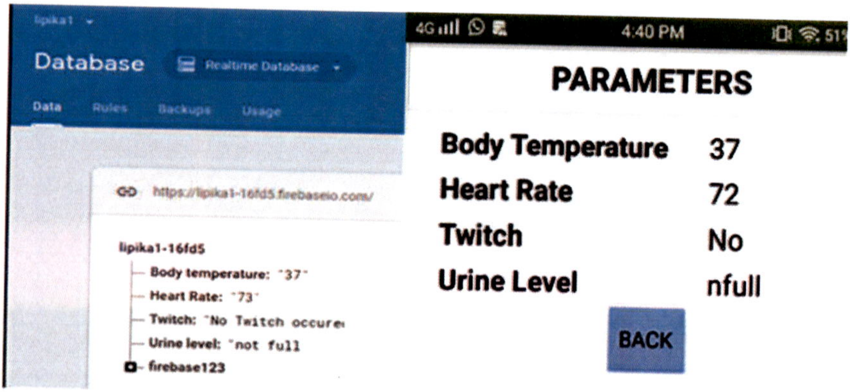

(a) (b)

Fig. 2 Real-time uploading of vital signs data on **a** firebase **b** App ensure whether they are giving accurate data. The actual values of pulse rate and body temperature were determined manually using actual devices and the observed values from the developed system.

Tables 3 and 4 summarize the deviation between the actual values and observed values of pulse rate (in bpm) and body temperature (in degree Celsius), respectively. Deviation has been studied by calculating the error percentage between the actual and observed data. As shown in Tables 3 and 4, the data has been recorded for seven patients, and the error percentages for each patient is recorded in the table. The highest and lowest error rates in sensing the pulse rate are 2.89% and 1.35%, respectively. The highest and lowest error rates for the body temperature measurement are 0.87% and 0.36%, respectively. For the proposed paper, as seen from Tables 3 and 4, the average error percentages for pulse rate and body temperature are 1.79 and 0.56, respectively.

The deviation between the actual and the observed values is depicted in a graphical manner in Figs. 3 and 4 for pulse rate and body temperature, respectively. For the graph in Fig. 3, *Y*-axis represents the pulse rate values in bpm and the *X*-axis represents the patients. For the graph in Fig. 4, *Y*-axis represents the body temperature values in degree Celsius and the *X*-axis represents the patients. The deviation as seen from the two graphs is not very large. The average error rates of <2% and <1% for pulse rate and body temperature, respectively, strengthen the work for the proposed paper.

Table 3 Pulse rate data recorded by analog machine (actual data) and proposed system (observed data)

Patients	Actual data (bpm)	Observed data (bpm)	Error (%)
1	73	74	1.36
2	72	73	1.38
3	74	73	1.35
4	69	71	2.89
5	72	71	1.38
6	70	71	1.42
7	71	73	2.81

Table 4 Body temperature data recorded by analog machine (actual data) and proposed system (observed data)

Patients	Actual data (bpm)	Observed data (bpm)	Error (%)
1	36.27	36.55	0.77
2	36.31	36.44	0.36
3	36.43	36.75	0.87
4	36.56	36.47	0.57
5	36.59	36.42	0.54
6	36.66	36.49	0.46
7	36.74	36.60	0.38

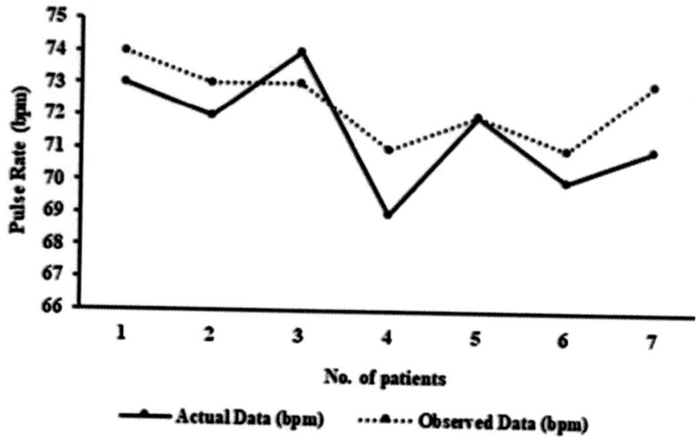

Fig. 3 Test results of the actual and observed value of pulse rate

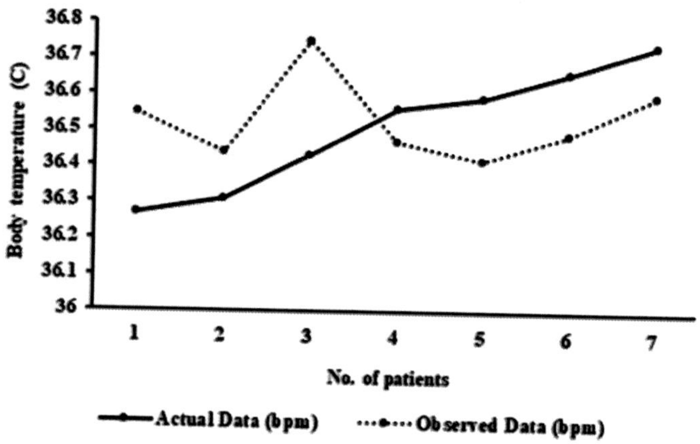

Fig. 4 Test results of the actual and observed value of body temperature

There have been papers where monitoring of normal patients has been studied. But, no paper has been found with a monitoring system for Comatose patients along with detection of state of coma using GCS. Hence, no comparison with the state of the artwork has been presented.

The monitoring system has been implemented by integrating all the sensors used to monitor the vital signs, handling and analyzing the data of the parameters obtained as depicted in Fig. 5.

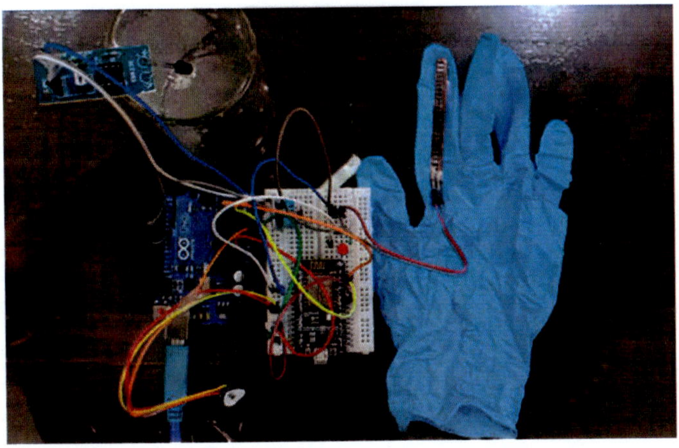

Fig. 5 Prototype of the monitoring system

4.2 *Diagnostic Test for Measuring Glasgow Coma Scale (GCS)*

The scores of motor and verbal responses are calculated and then added to determine the stage of the coma a patient is in using the prototype shown in Fig. 6. The scores are allocated as per Tables 1 and 2. Figure 7a shows the uploading of the score of individual responses namely verbal response as score A and motor response as Score B and the total score as well as the state of coma in Firebase and Fig. 7 b shows all these three scores and the state of coma on the app.

For Fig. 7, the verbal response score is 1 and the motor response score is 1, making the total score for GCS depicted in Fig. 7a, b as 2, i.e., sum of scores depending upon the category verbal response and motor response to a defined stimulus.

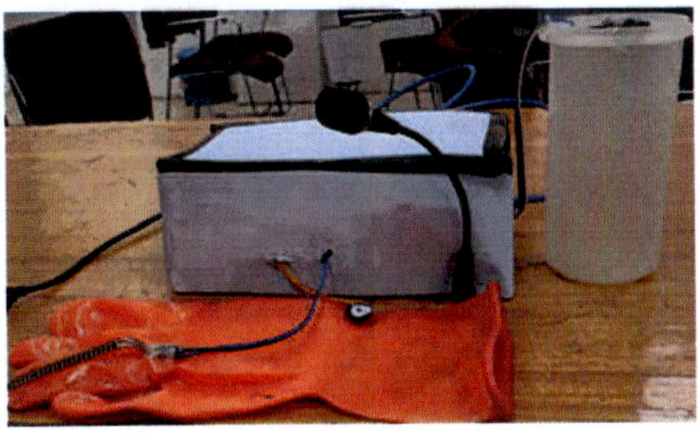

Fig. 6 Prototype for the diagnostic test for measuring GCS

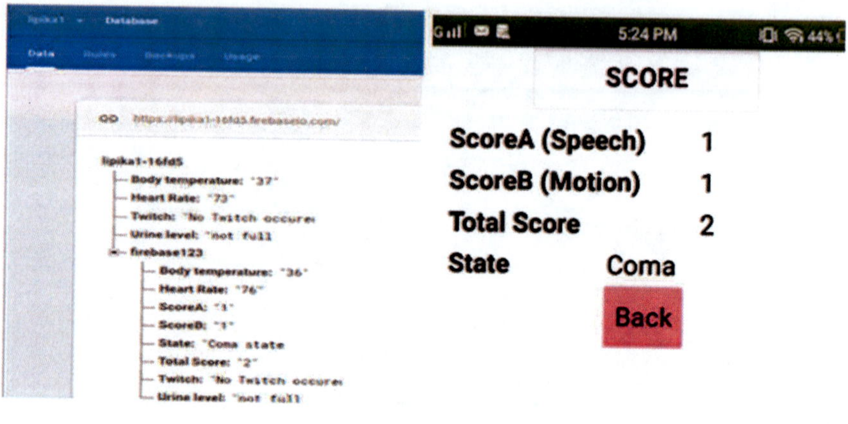

(a) (b)

Fig. 7 Real-time uploading of diagnostic test scores on **a** firebase, **b** App

5 Conclusion and Future Work

In this paper, it has been highlighted that digitalizing the whole process of monitoring and diagnosing the state of coma patient, makes the system simple and error-free. Moreover, patients will be provided with the appropriate treatment timely.

IoT helps in remote monitoring of the comatose patient through the developed mobile application and the real-time database. Also, the digital GCS test helps in diagnosing the state of the coma patient effectively. The proposed paper aims to eliminate the burden of continuous supervision and alerts the doctor or paramedical staff when attention is needed. This system will help assist the doctor in gauging the patient's condition.

The system can incorporate response of eye movement as an additional parameter in GCS test as the response of the eye to the external stimuli together with motor and verbal response will help in determining the state of coma more accurately. The system needs to be medically graded so that it can be used in the medical field for monitoring coma patients. Single radar-based sensor can be used to monitor various vital signs of the patient which will eliminate the requirement of multiple sensors. This will make the system compact and more efficient that can easily be integrated with the existing CMS of the hospital.

References

1. Lokesh B et al (2020) IoT based coma patient monitoring system using wearable sensors. Int J Adv Sci Technol 29(10s):4401–4409
2. Sankalp M (2011) Coma patient monitoring system using image processing. In: AIP conference proceedings
3. Islam M et al (2020) Development of smart healthcare monitoring system in IoT environment. SN Comput Sci 1:185
4. Krishnan D et al (2018) An IoT based patient health monitoring system. In: International conference on advances in computing and communication engineering (ICACCE), IEEE, 01-07
5. Kaur H et al (2016) Assessing nurses knowledge of glasgow coma scale in emergency and outpatient department. Nurs Res Pract Hindawi 2016:5
6. Lokesh B et al (2020) IoT based coma patient monitoring system using wearable sensors. Int J Adv Sci Technol SERSC 29(10s):4401–4409
7. Subha R et al (2020) Coma patient health monitoring system using IOT. In: International conference on advanced computing and communication systems (ICACCS). IEEE, 6, 1454–1457
8. Wan J et al (2018) Wearable IoT enabled real-time health monitoring system. J Wireless Com Netw. Springer, 298
9. Tamilselvi V et al (2020) IoT based health monitoring system. In: International conference on advanced computing and communication systems (ICACCS). IEEE, 6, 386–389
10. Trivedi S, Cheeran A (2017) Android based health parameter monitoring. In: International conference on intelligent computing and control systems (ICICCS). IEEE, 1145–1149
11. Kumar S et al (2017) Smart health monitoring system of patient through IoT. In: International conference on I-SMAC (IoT in social, mobile, analytics and cloud) (I-SMAC). IEEE, 551–556
12. Gregoski M et al (2012) Development and validation of a smartphone heart rate acquisition application for health promotion and wellness telehealth applications. Int J Telemed Appl. Hindawi,1–7
13. Oresko J et al (2010) A wearable smartphone-based platform for real-time cardiovascular disease detection via electrocardiogram processing. IEEE Trans Inf Technol Biomed 14:734–740
14. Acharya A, Patil S (2020) IoT based health care monitoring kit. In: International conference on computing methodologies and communication (ICCMC). IEEE, 4, 363–368
15. Banerjee S, Roy S (2016) Design of a photoplethysmography-based pulse rate detector. Int J Rec Trends Eng Res 2:302–306
16. Teasdale G et al (2014) The glasgow coma scale at 40 years: standing the test of time. Lancet Neurol 13(8):844–854
17. Alhassan A et al (2019) Knowledge of the glasgow coma scale among nurses in a tertiary hospital in Ghana. Nutr Res Pract 2019:7
18. Bender A et al (2015) Persistent vegetative state and minimally conscious state. DeutschesÄrzteblatt Int 2015(112):235–242

Image Forgery Detection by CNN and Pretrained VGG16 Model

Pranjal Raaj Gupta, Disha Sharma, and Nidhi Goel

Abstract Image forgery has been existing in various fields in some manner. Its severity can range from innocuous social media editing to forged medical records or legal documents. Especially at a time when most organizations have taken a virtual form, it is necessary to verify the quality and validity of records, bills, etc. In this research, various deep learning approaches have been used, such as CNN and pretrained CNN models. These are used to identify image distortions and to compare and interpret the outcomes. In this paper, CNN models and pretrained CNN models are designed to assess their efficacy in the recognition and detection of image datasets. The algorithms are applied to the dataset to calculate the outcome of the model. In proposed work, the obtained accuracy of CNN model and pretrained CNN model is 74.95% and 91.62%, respectively. These CNN models have been designed to observe the better performance of deep learning neural networks over the pre-existing active and passive approaches.

Keywords Digital forgery · CNN model · Pretrained CNN · Model analysis

1 Introduction

With the advancements in the use of digital images for multiple purposes, it has become a mandatory part of regular online communications among entities. Also, the ease of access to smart devices, spread of digital images has grown vastly over recent years. These advancements have made certain aspects of interaction more comprehensible and easily accessible. In some cases, it has been exploited to create fake replicas. It is done with the help of easily accessible and low-cost applications for digital image forgery.

Additionally, the growth of popularity of social media platforms like Facebook, Instagram, and Twitter has increased the spread of edited images exponentially. Therefore, it has become crucial to create forgery detection techniques [1] to be

P. R. Gupta (✉) · D. Sharma · N. Goel
IGDTUW, New Delhi, Delhi, India

G. Gupta et al. (eds.), *Proceedings of Academia-Industry Consortium for Data Science*, Advances in Intelligent Systems and Computing 1411,
https://doi.org/10.1007/978-981-16-6887-6_13

Table 1 Comparison table for both models

Models	Training loss	Training accuracy	Validation loss	Validation accuracy
CNN	0.5329	0.7712	0.5407	0.7495
VGG16 CNN	0.2281	0.9120	0.1885	0.9162

able to avoid frauds and misinformation. Several researchers have studied the above-mentioned obstacle and came up with a systematic classification of various techniques for image forgery detection [2].

Currently, the active image detection technique is used to avoid the forgery fraud by many organizations by watermarking [3] their issued documents. But this has been proven inefficient multiple times as the growth of editing applications is ever evolving and they are easily accessible at no or a low-cost. The passive techniques [4] that exist are, however, related to the physical attributes of the sample image. Techniques based on camera, pixel, geometry, format, and physics. These techniques do not always guarantee clear outputs and can be slow and inefficient over larger and more practical datasets. Deep learning models have been proven more efficient, accurate and faster than the conventionally used techniques.

The proposed solution involves the use of deep learning models [5] for detecting forged digital images and avoiding the forgery of the images. This involves testing the datasets over the primarily designed CNN model and a pretrained CNN model (VGG16) [6]. Then, the results of both the models are analyzed and compared in Table 1. The working of each model throughout the dataset has been presented using graphs.

2 Literature Survey

The authentication of the forged image is detected by the desire for accuracy and legitimacy of the image. Image forgery identification is becoming a difficult challenge to solve due to relentless developments in digital image processing technology. As a result, several researchers have previously focused on this subject using different methods such as active and passive techniques [2]. Detecting the image forgery using blind methods [1] is one such technique. It gives the comparison of various types of techniques to detect image manipulation.

Active forgery detection [2] is the standard approach used to detect image manipulation. Dynamic methods validate the validity of the image by means of concealing schemes. Watermarks or digital signatures are the data that are concealed in the picture at the time of creation. A lot of research has been conducted in various branches of this field [3]. Hidden data or signatures shall be retrieved and checked with the stored data for the authentication process. This technique includes methods such as digital watermarking and digital signature methodology. Digital marking involves a precise digest being incorporated into the original image at the time of capture. Digital signature involves special properties being incorporated into the original image at the time of capture.

However, passive forgery identification [2, 4] criteria are exactly at odds with active methods. Passive techniques authenticate the modification of the image without any direct awareness of the original image or its characteristics. The authentication process shall be carried out based on the facts in the image itself, without the use of any supplementary information. This methodology involves pixel-based techniques, i.e., the recognition of forgery in the original image by the study of pixels constituting the image. Also, format-based techniques, i.e., the detection of forgery in the image based on changes in the image format. Lastly, camera-based techniques, i.e., the detection of irregularities in the image using objects introduced by the camera lens, the imaging sensor, etc.

Recently, some amount of research has also been done using the convolutional neural networks, which are deep learning models specialized in image classification [7] and forgery detection using large datasets. These models use gradient descent optimization algorithms, used as black-box optimizers [8]. Along with this, some other methods are based on machine learning, domain adaption and multitask learning [9]. Researches based on transfer learning using pretrained model of CNN such as VGG16 [6], Inception [10], ResNet [11], Xception [12], etc., have also surfaced in recent years.

3 Proposed Solution

The proposed solution is based on deep learning neural network. Pretrained convolutional neural network is tuned according to the parameters. After tuning, both the models are trained and tested over a dataset. The dataset used is divided into two categories. The training set which constitutes 75% of the dataset, and the rest is categorized as the validation set.

3.1 Convolutional Neural Networks

CNN's are nonlinearly interconnected neural networks focused on the structure of the human visual system. It includes 3 layers: a convolutional layer, a pooling layer, and fully connected layers [13]. The first two layers of are very necessary layers of the CNN model. The first layer is used for extraction of features by integrating the sample field with various filters. To avoid over fitting [14], pooling layer restricts the dimension of the output feature map of the first layer. Number of neurons, parameters, and interactions are smaller in CNN model for these two layers. It makes CNN more effective compared to back propagation system with equivalent layers. The formula for the output image map is depicted in Eq. (1) [15].

$$\text{convolution}(N, K)_{xy} = f\left(B + \sum_{p=1}^{l}\sum_{q=1}^{m}\sum_{r=1}^{n} K_{pqr}.N_{x+p-1,y+q-1,r}\right) \tag{1}$$

where, K is kernel, N is no. of filters, x and y are axis used for convolution, f is function, B is bias value, l, m and n are the constraints for convolution. The layers are insistent on unique kernels, K along with their bias value. It then runs by computing the output of the preceding layer with each of the kernel. From Eq. (1), it can be inferred that kernel size, filter count, stride and padding are important parameters in CNN.

Different features of input are extracted through conversion of sample data. The transformation results in a linear alteration for the input sample in accordance with spatial information present. A common outlook for down-sampling involves the use of pooling layer. It decrypts the sample (like 2 × 2) and aggregates it into a one unit. The used aggregation method is the integration of the highest value (max pooling). Max pooling splits the outcome from the convolutional layer into multiple compact grids. It then uses the highest value of each grid for creation of reduced sample matrix. The formula used for the max pooling operation can be written as Eq. (2) where, l and m are constraint, x and y are input/output neuron

$$y = \max_{(p,q=1)}^{(l,m)} x p, q \tag{2}$$

Every neuron in the convolutional layer is converted into one-dimensional data and incorporated into completely connected layer. This is achieved by using the rectified linear unit (ReLU). The ReLU layer performs nonlinear activation on every element. If a given neuron y, it is transformed into single neuron x with the help of Eq. (3).

$$x = \max(0, y) \tag{3}$$

Figure 1 shows the flow chart of proposed CNN model. Training phase and test phase of the proposed method has been shown in Fig. 1. CNN training operates in the same way as a normal neural network, using backpropagation [16] or gradient descent [8]. The convolution operations used, make calculations mathematically complicated. Initially, the input is preprocessed (which includes resampling the images and converting digital images into binary images). Then, the processed data is set to both the layers of a convolutional neural network (feature extraction layer and classification layer). In the testing phase, the input samples are again preprocessed and passed through feature extraction layers of the CNN model. The output of both phases is simultaneously fed to the predictive model, which is an analyzing model that uses mathematical operations to predict certain outcomes from given data. Then, the final layer does a pattern recognition training which provides a prediction depending on the input sample. The softmax function compresses output of every unit in range of 0–1, like sigmoid function [17]. In the last stage, the image is classified as forged or pristine.

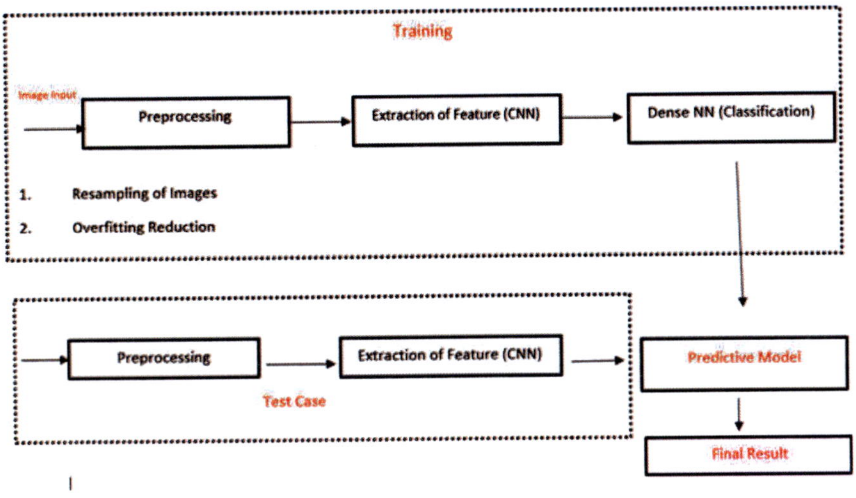

Fig. 1 Flow chart of the proposed CNN model

3.2 Pretrained Convolutional Neural Networks

Pretrained VGG16 [6] model is constructed from transfer learning [9] methodology. Transfer learning seeks to expand the insights from a single task (source task) to a separate but equivalent task (target task). It prevents having to reinvent the wheel and helps in creating ML frameworks in less time. There are three steps used here to implement transfer learning:

- Development of an open source pretrained model by a third party: A pretrained model has been built and trained by another entity to solve a problem that is similar to the current one. Usually, it uses a very large dataset, such as ImageNet or Wikipedia Corpus.
- Repurposing the model: The information gained is reused, which involves layers, attributes, weights, and prejudices. The pretrained models are stored and are conveniently available in several deep learning libraries, such as Tensorflow, Keras application, Pytorch hub, etc.
- Fine tuning for the problem: Fine tuning is done to achieve a higher level of accuracy and to be able to generate output in a correct format.

3.3 VGG16 Pretrained Model

The architecture of CNN VGG-very-deep-16 [15] is promising, so it is used as a pre-training model for the original image and forged image recognition. VGG16 [6] was first tested on the 2012 ImageNet (92.7% accuracy in a dataset of over 14 million images belonging to 1000 classes). It is a network composed of a convolutional,

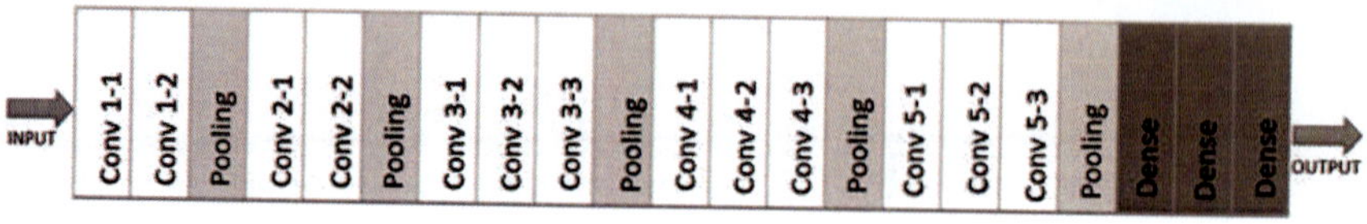

Fig. 2 Architecture of the VGG16 model

pool, and totally linked network (FC). The convolutional layer uses a 3D filter, and the pooling layer operates with factor 2. The architecture of VGG16 [6] is shown in Fig. 2.

It uses 1 × 1 convolutional filter that can be seen as a linear alteration of the input. The stride for this layer is set at 1 pixel. The padding of the convolution layer input is such that the spatial resolution is white folded afterward. The padding is fixed at 1 pixel. Spatial pooling [18] is done using 5 maximum pool layers, accompanied by several convolutional layers. Max pooling is implemented across a 2 × 2 window keeping the stride at 2 pixels. Lastly, three dense layers are added to network. The first two layers, each have 4096 channels and the third layer conducts a 1000-way ILSVRC classification. The softmax layer is added as the final layer to the model. All hidden layers are structured using the ReLU.

4 Results and Discussion

Dataset used: The dataset used for the testing and validation of the proposed model contains a total of 1950 images. These images are further classified into 1050 pristine images and 900 masked and forged images. The dataset resolution is changed to 224 × 224 pixels. The dataset is split into two, namely the training set and the validation set. The training set is constituted of 75% of the total dataset, while the remaining 25% is categorized as validation set.

4.1 Model Summary for CNN Model

The combination of convolution, activation, and max pooling layers has been designed to detect the forged images using the deep learning CNN architecture. In Fig. 3, layer-wise output dimensions and parameters for the input image sample are mentioned.

Layer (type)	Output Shape	Param #
conv2d_31 (Conv2D)	(None, 218, 222, 64)	1408
activation_49 (Activation)	(None, 218, 222, 64)	0
max_pooling2d_31 (MaxPooling	(None, 109, 111, 64)	0
conv2d_32 (Conv2D)	(None, 103, 109, 128)	172160
activation_50 (Activation)	(None, 103, 109, 128)	0
max_pooling2d_32 (MaxPooling	(None, 51, 54, 128)	0
conv2d_33 (Conv2D)	(None, 49, 52, 512)	590336
activation_51 (Activation)	(None, 49, 52, 512)	0
max_pooling2d_33 (MaxPooling	(None, 24, 26, 512)	0
dropout_10 (Dropout)	(None, 24, 26, 512)	0
flatten_10 (Flatten)	(None, 319488)	0
dense_19 (Dense)	(None, 64)	20447296
activation_52 (Activation)	(None, 64)	0
dense_20 (Dense)	(None, 1)	65
activation_53 (Activation)	(None, 1)	0

Total params: 21,211,265
Trainable params: 21,211,265
Non-trainable params: 0

Fig. 3 Model summary for CNN model

4.2 Accuracy and Loss of CNN Model

The dataset is divided into 20 epochs for the entire process. In the training phase, the input is initially fed to the model to make the model learn the parameters using various samples. The validation phase is required to test the learnings of the model. By analyzing the training and validation data, it can be determined if the model is under-fitted [14], over-fitted [14], or perfectly fitted [14]. The graphical outputs for the training, validation loss and accuracy are shown in Fig. 4. The training and validation data lie in comparative ranges. Therefore, model is efficient.

4.3 Confusion Matrices

Confusion matrices [19] are used to derive the output of the proposed models. These can be used to derive the output terms like precision [20] and recall [20] for the designed machine learning model. Precision and recall helps in determining the

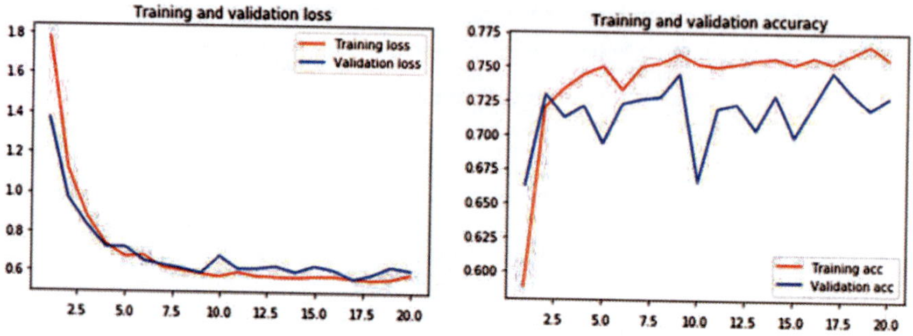

Fig. 4 Training and validation loss and accuracy for CNN

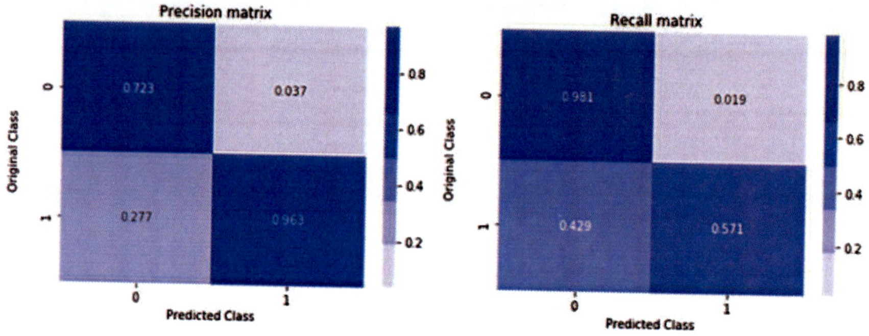

Fig. 5 Confusion matrices for CNN model

accuracy and deviation of the model over the entire datasets. Figure 5 depicts the confusion matrices for the proposed CNN model.

4.4 Pretrained CNN Model Summary

The pre-designed architecture for the VGG16 [6] pretrained CNN model including 16 layered combinations of convolution, pooling, and dense layers is depicted in Fig. 6.

More convolution, activation, and pooling layers are added according to the specific requirements to fine tune the models and to produce greater accuracy. The retouching layers added to VGG 16 [6] model are depicted in Fig. 7.

Layer (type)	Output Shape	Param #
input_1 (InputLayer)	(None, 224, 224, 3)	0
block1_conv1 (Conv2D)	(None, 224, 224, 64)	1792
block1_conv2 (Conv2D)	(None, 224, 224, 64)	36928
block1_pool (MaxPooling2D)	(None, 112, 112, 64)	0
block2_conv1 (Conv2D)	(None, 112, 112, 128)	73856
block2_conv2 (Conv2D)	(None, 112, 112, 128)	147584
block2_pool (MaxPooling2D)	(None, 56, 56, 128)	0
block3_conv1 (Conv2D)	(None, 56, 56, 256)	295168
block3_conv2 (Conv2D)	(None, 56, 56, 256)	590080
block3_conv3 (Conv2D)	(None, 56, 56, 256)	590080
block3_pool (MaxPooling2D)	(None, 28, 28, 256)	0
block4_conv1 (Conv2D)	(None, 28, 28, 512)	1180160
block4_conv2 (Conv2D)	(None, 28, 28, 512)	2359808
block4_conv3 (Conv2D)	(None, 28, 28, 512)	2359808
block4_pool (MaxPooling2D)	(None, 14, 14, 512)	0
block5_conv1 (Conv2D)	(None, 14, 14, 512)	2359808
block5_conv2 (Conv2D)	(None, 14, 14, 512)	2359808
block5_conv3 (Conv2D)	(None, 14, 14, 512)	2359808
block5_pool (MaxPooling2D)	(None, 7, 7, 512)	0

Total params: 14,714,688
Trainable params: 7,079,424
Non-trainable params: 7,635,264

Fig. 6 VGG16 layer architecture

4.5 *Accuracy and Loss of VGG16 Model*

In Fig. 8, it is seen that the validation accuracy and validation loss is following the training phase output data. Hence, it can be inferred that the proposed model is suitable. For this analysis, the dataset is divided into 30 epochs.

```
Layer (type)                 Output Shape              Param #
=================================================================
vgg16 (Model)                (None, 7, 7, 512)         14714688
_________________________________________________________________
conv2d_1 (Conv2D)            (None, 7, 7, 512)         262656
_________________________________________________________________
activation_1 (Activation)    (None, 7, 7, 512)         0
_________________________________________________________________
max_pooling2d_1 (MaxPooling2 (None, 3, 3, 512)         0
_________________________________________________________________
dropout_1 (Dropout)          (None, 3, 3, 512)         0
_________________________________________________________________
flatten_1 (Flatten)          (None, 4608)              0
_________________________________________________________________
dense_1 (Dense)              (None, 1024)              4719616
_________________________________________________________________
dropout_2 (Dropout)          (None, 1024)              0
_________________________________________________________________
dense_2 (Dense)              (None, 1)                 1025
=================================================================
Total params: 19,697,985
Trainable params: 12,062,721
Non-trainable params: 7,635,264
```

Fig. 7 Fine-tuned model after adding more layers to VGG16 architecture

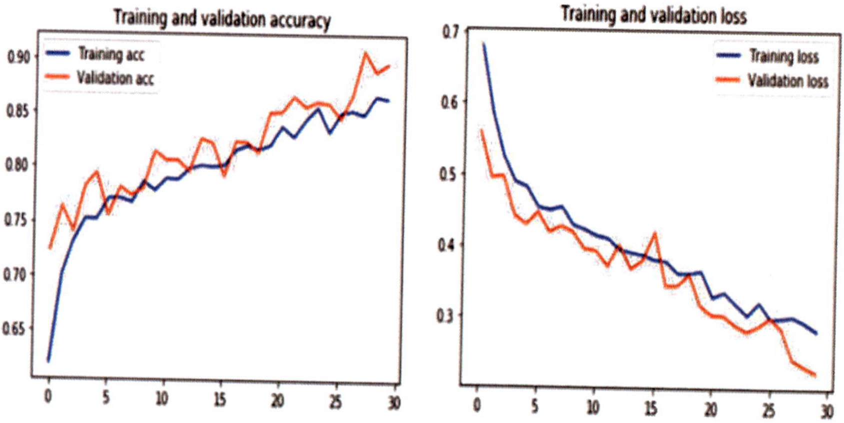

Fig. 8 Training and validation accuracy and loss for VGG16

4.6 *Confusion Matrices of VGG16 Model*

Figure 9 displays the confusion matrices for the refined VGG16 [6] model used in proposed work. As discussed in Sect. 4.3, the precision and recall for the proposed pretrained model can be derived using the resultant confusion matrices.

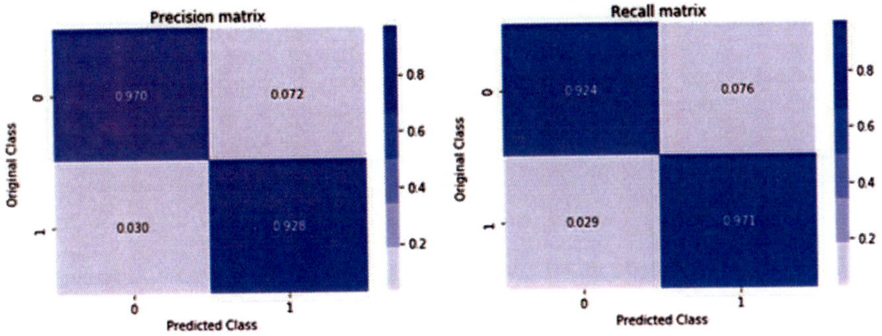

Fig. 9 Confusion matrices for VGG16

5 Conclusion

In this research, a CNN model and a fine-tuned VGG16 [6] model is tested and validated over the same dataset of digital images. These models have been designed to observe the better performance of deep learning neural networks over the pre-existing active and passive approaches.

As inferred from Table 1, it can be seen that CNN produces a validation accuracy of 74.95%, while the fine-tuned VGG16 produces a validation accuracy of 91.62%. It can be concluded that in most cases aforementioned neural networks yield higher accuracy results than the conventional forgery detection methods. In most cases, active and passive approaches are observed to have an accuracy of below 80% when tested over the similar datasets [21]. CNN models and its pretrained version also do not require human supervision and thus can be automated. Furthermore, CNN models are having better forgery localization abilities and can detect digital image forgeries with lesser computational complexity and high robustness [21].

In future research, other image classification models can be analyzed and compared with CNN models to derive better observational outputs. These models may include long short-term memory (LSTM) [22] and generative adversarial network (GAN) [23]. This might help in achieving more detailed comparative results to determine the better suitable model for image forgery detection.

References

1. Gill NK et al (2017) A review paper on digital image forgery detection techniques. In: 8th international conference on computing, communication and networking technologies (ICCCNT), 1–7
2. Kumar et al (2018) A systematic study of image forgery detection. J Comput Theor Nanosci
3. Potdar VM et al (2005) A survey of digital image watermarking techniques. In: 3rd IEEE international conference on industrial informatics, 709–716
4. Panda S, Mishra M (2005) Passive techniques of digital image forgery detection: developments and challenge. Springer

5. Goodfellow I et al (2016) Deep learning book review health inform resources. The MIT Press
6. Simonyan K, Zisserman A (2015) Very deep convolutional networks for large-scale image recognition. In: 3rd international conference on learning representations (ICLR)
7. Singhal S, Ranga V (2021) Passive authentication image forgery detection using multilayer CNN. Springer
8. Ruder S (2017) An overview of gradient descent optimization algorithms
9. Pan SJ, Yang Q (2010) A survey on transfer learning. IEEE Trans Knowl Data Eng 1345–1359
10. Szegedy C et al (2016) Rethinking the inception architecture for computer vision. In: IEEE conference on computer vision and pattern recognition (CVPR)
11. He K et al (2016) Deep residual learning for image recognition. In: IEEE conference on computer vision and pattern recognition (CVPR)
12. Fran C et al (2017) Deep learning with depth wise separable convolutions. In: IEEE conference on computer vision and pattern recognition (CVPR)
13. Albawi S et al (2017) Understanding of a convolutional neural network. In: International conference on engineering and technology (ICET), 1–6
14. Jabbar H, Khan RZ (2015) Methods to avoid over-fitting and under-fitting in supervised machine learning (comparative study). Comput Sci Commun Instrum Dev, 163–172
15. Sudiatmika IBK et al (2019) Image forgery detection using error level analysis and deep learning, 653–659
16. Leung H, Haykin S (1991) The complex backpropagation algorithm. IEEE Trans Sig Process 2101–2104
17. Ito Y (1991) Representation of functions by superpositions of a step or sigmoid function and their applications to neural network theory. Neural Netw: 385–394
18. Wang Z, Shang X (206) Spatial pooling strategies for perceptual image quality assessment. In: International conference on image processing, 2945–2948
19. Visa et al (2011) Confusion matrix-based feature selection. MAICS 710, 120–127
20. Davis J, Goadrich M (2006) The relationship between precision-recall and ROC curves. In: 23rd international conference on machine learning, 233–240
21. Basavarajappa et al (2016) Digital image forgery detection techniques, ACCENTS Trans Inf Secur
22. Bappy JH et al (2019) Hybrid LSTM and encoder-decoder architecture for detection of image forgeries. IEEE Trans Image Process: 3286–3300
23. Abdalla et al (2019) Copy-move forgery detection and localization using a generative adversarial network and convolutional neural-network, 286

A Study on Brain Tumor Segmentation in Noisy Magnetic Resonance Images

Shiv Naresh Shivhare and Nitin Kumar

Abstract Performance of brain tumor segmentation from Magnetic Resonance (MR) images has been significantly improved in the last decade by various advanced machine learning techniques. However, it remains challenging when MR images suffer from bias field and various types of noise, while acquisition process. Noise typically gets introduced in MR images because of uneven measurements and selection of several statistical factors of radio frequency pulse sequences such as T1 and T2 relaxation times, repetition time, time to echo, etc. In this work, we present a study of brain tumor segmentation on noisy MR images by adding Gaussian noise on different brain MR modalities. Random Forest (RF) and Support Vector Machine (SVM) classifiers are utilized to segment various labels of brain tumor, *viz.*, complete brain tumor region, brain tumor core region, and enhancing tumor region. Classifiers are trained after applying appropriate noise removal filtering algorithms such as Median Filter and Wiener Filter, followed by extracting relevant statistical and texture image features. The proposed work is evaluated on the real glioma MR images of benchmark BRATS dataset. Experimental results demonstrate favorable segmentation performance, i.e., 95% for segmenting complete brain tumor region in terms of Dice Similarity Coefficient (DSC).

Keywords Brain tumor segmentation · Magnetic resonance imaging · Noise removal

1 Introduction

Recently, brain tumor image detection and segmentation have attracted significant attention since Brain Tumor Segmentation (BRATS) challenge started providing the benchmark dataset [1]. Brain tumor segmentation is carried out by processing MR

S. N. Shivhare (✉) · N. Kumar
Department of Computer Science and Engineering, National Institute of Technology Uttarakhand, Srinagar Garhwal, India
e-mail: nitin@nituk.ac.in

S. N. Shivhare
School of Computer Science, University of Petroleum and Energy Studies, Dehradun, India

G. Gupta et al. (eds.), *Proceedings of Academia-Industry Consortium for Data Science*, Advances in Intelligent Systems and Computing 1411,
https://doi.org/10.1007/978-981-16-6887-6_14

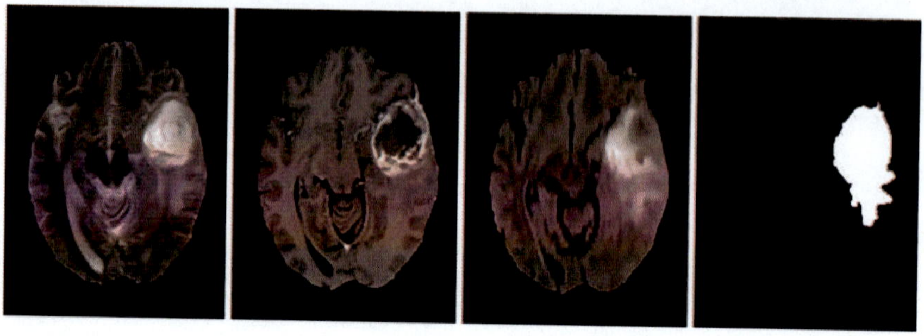

Fig. 1 Axial view of MRI modalities of a subject from BRATS dataset [9]. From left to right: T2-weighted MRI, T1C MRI, FLAIR MRI, and the corresponding ground truth as complete tumor region

images of different modalities such as T1-weighted, contrast enhanced T1-weighted (T1C), T2-weighted, and Fluid Attenuated Inversion Recovery (FLAIR). Brain tumor image analysis and tumor segmentation is challenging because of mainly two reasons, (i) irregular anatomical structure of brain abnormality, (ii) presence of noisy signals over MR images. In the past decade, several effective methods have been proposed to recognize and detect the brain tumor structure [2–8]. However, handling with noise on real MR images is still a major issue for brain image analysis and tumor segmentation.

Since 2012, BRATS community has devoted significant efforts through conducting annual challenge and providing the benchmark brain tumor dataset. The advantages of BRATS dataset are that it is preprocessed, and skulls are also removed to maintain subjects anonymity [1]. Each subject's brain is scanned and captured in the form of four MR image modalities, i.e., T1-weighted, T2-weighted, T1C, and FLAIR as shown in Fig. 1. MRI in the form of its various modalities expresses extensive global and local level information of brain's healthy and neoplastic tissues. Clinical experts or radiologists usually analyse these MR images by manual observations to diagnose brain tumor disease and plan its treatment. However, automation in this regard is needed to meet the following requirements: (i) accurate and efficient tumor diagnosis, (ii) availability of growing number of skilled radiologists, and (iii) prevention of human error.

Noise in MR images is the unwanted signal information which originates from several modes including electronic interference due to human body movement, and certain acquisition errors, i.e., T1 and T2 relaxation times, repetition time, time to echo, etc. [10]. MRI images acquired through any old standard device generally are prone to Rician distribution noise, salt and pepper noise and Gaussian noise. The major challenge with noisy MR images is that noise may results in false extraction of intensity and texture-based features that lead to incorrect and poor segmentation results. This motivates us to work towards developing an automatic, highly accurate, and computationally fast brain tumor segmentation method which is more robust, while handling with noisy MR images. In order to perform experiments on noisy images, Gaussian white noise is added on each grayscale image (slice) of MRI modalities. A noise reduction algorithm is targeted to remove noise intensity, while preserving its resolution and maintaining significant visual cues. Additive Gaussian noise is removed by two significant and state-of-the-art algorithms before extracting features, i.e., median filter and Wiener filter.

State-of-the-art methods of brain tumor detection and segmentation can be divided into generative methods and discriminative methods. Generative class of tumor segmentation methods depends on the comparative analysis of tumorous brain with the anatomy of normal or healthy brain tumor [2, 3]. Contrarily, discriminative class of segmentation methods rely less on prior anatomical appearance of brain. Discriminative class of methods mainly depend on various feature-based methods such as Artificial Neural Networks (ANN), Active Contour Model (ACM) [11], classification and clustering-based, methods [4–8] etc. In a generative method, Prastawa et al. [2] utilized T2-weighted MR images to detect complete tumor region by outlier detection approach and image registration based on healthy brain atlas. Further, edema is separated from complete tumor region using statistical clustering followed by applying a deformable model. Moreover, Menze et al. [3] proposed an automated segmentation method based on generative approach. Tumor region is detected by channel-specific segmentation of different MR modalities using an explicit model of image formation. Based on the discriminative class of methods, Sachdeva et al. [11] extracted several intensity and texture features and segmented six different types of brain tumor using active contour model. Bauer et al. [4] trained Support Vector Machine (SVM) classifier with multi-spectral intensities and texture features to segment brain images into active tumor, necrosis, edema as tumor regions and white matter, gray matter and Cerebrospinal Fluid (CSF) as brain's healthy tissues. Finally, Conditional Random Field (CRF) is exploited for regularization. Furthermore, Zhao et al. [5] proposed supervised classification-based tumor detection method by extracting significant 2-dimensional and 3-dimensional gradient and context-sensitive features. Pei et al. [6] combined texture and another relevant cell density features corresponding to tumor region to train the Random Forest (RF) classifier. Further the obtained results are also fused with an existing tumor segmentation method to ensure better tumor segmentation results.

Convolutional Neural Networks (CNN) have been very popular and widely used in the field of medical image analysis and computer vision due to their high performing characteristics. However, it requires large amount of time and computational resources to train such large and complex networks. Segmentation accuracy is one of the major requirement, and therefore, many variants of CNN have been proposed in the literature for highly accurate brain tumor segmentation. Kamnitsas et al. [7] presented 11-layer deep, 3-dimensional CNN to detect brain tumor tissues and segment its sub-compartments. CRF is also utilized as post-processing step to minimize false positive cases. In a similar research, Havaei et al. [8] also introduced another deep CNN architecture using two-phase training procedure that deals with the class-imbalance problem of BRATS dataset. Two CNN networks are used in cascaded manner to take advantages of concatenated feature maps. This method works faster than several existing methods.

Beevi et al. [12] proposed a method to segment noisy MR images using Fuzzy C-means (FCM) clustering. Images are de-noised by exploiting Principal Component Analysis (PCA) based local pixel grouping method. The method demonstrates the segmentation of normal brain MR images. Another noisy image segmentation is proposed by Guo et al. [13] that uses an improved version of FCM in association with guided filtering algorithm. Guided filtering performs image smoothing and enhancement by preserving edge information. Method is applied on the healthy and noisy synthetic MR images to perform segmentation.

This work presents a study on brain tumor segmentation considering noisy MRI images. The major contribution of this work is as follows:

1. Additive White Gaussian Noise is applied on different modalities of brain MR images to observe the behavior of state-of-the-art learning algorithms, i.e., RF and SVM on noisy images.
2. Median filter and Wiener filter algorithms are exploited to denoise MR images before extracting significant statistical, texture, and Histogram of Oriented Gradients (HOG) features to predict the required brain tumor structures, i.e., complete tumor, tumor core, and enhancing tumor regions.

The remaining paper is structured as follows: Sect. 2 elucidates the step-wise procedure of the proposed segmentation method. Section 3 demonstrates the experimental details and obtained results. At last, we conclude by presenting the prominent outcomes of this work with future scope in Sect. 4.

2 Proposed Method

Here, a fully automated method is proposed to segment brain tumor regions from noisy MR images. The flowchart of the proposed segmentation method is shown in Fig. 2 which demonstrates the step-wise procedure. The overall method works in three modules. In the first module T2, T1C, and FLAIR MR images are taken as input, pre-processed, and denoised. Features extraction and their dimensionality reduction is carried out in second module. Finally, in the third module, the extracted features are trained via RF and SVM classifiers to segment brain tumor regions. The detailed explanation is given as follows:

2.1 Preprocessing

Each 2-dimensional slice of MR modalities is pre-processed to extract Region of Interest (ROI) by discarding unwanted boundary pixels which do not represent any significant brain tissue information. Additive White Gaussian Noise (AWGN) is applied on each MRI slice to consider the natural, random, and normally distributed noise with zero mean and variance as 0.01. The added distributed noise is usually

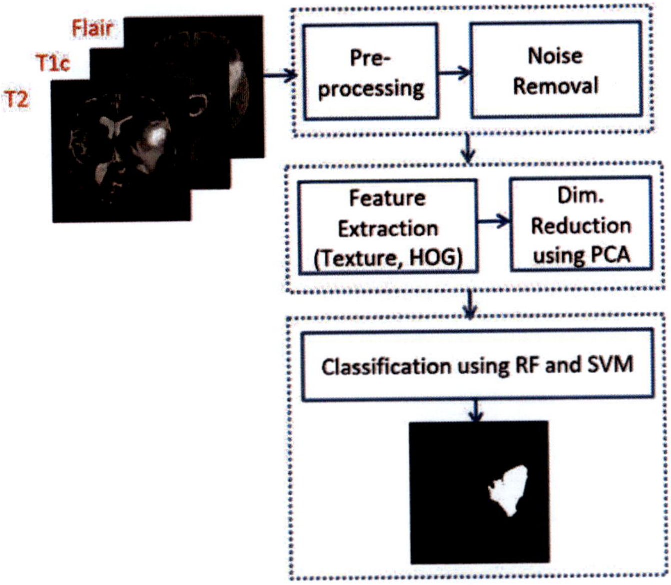

Fig. 2 Flowchart of the proposed segmentation method

removed by reducing noise intensity and preserving its resolution. Further, all the images are denoised by using median filter and Wiener filter algorithms. Being a non-linear filtering algorithm, median filter gives favorable denoising performance on AWGN [14]. The median value from the moving window of size 5×5 is considered as the denoised value of the corresponding central pixel. Wiener filter is another state-of-the-art denoising algorithm that denoises the signals based on the statistical computation by minimizing the mean squared error between the estimated random process and the desired process [10]. The ROI is detected by identifying the high energy regions using corner and edge information [15]. ROI is extracted by automated cropping the central patch of size 60×60 from each 2-dimensional MR slice to focus mainly on tumor tissues. Selecting such central patch also handles the class imbalance problem of the dataset in which the portion of healthy brain tissues are much greater than that of tumor tissues.

2.2 *Feature Extraction*

In this step, pixel-wise features from the identified ROI are extracted. First-order statistical, second-order texture features, and HOG features are extracted to capture the patterns of tumorous brain tissues and establish the relationship between healthy and abnormal brain tissues. First-order statistics features, i.e., mean, variance, skewness, and kurtosis have taken into consideration and computed for each input ROI.

First-order features are generally based on the distribution of pixel intensities and the histogram of the image. Let N_g be the number of grayscale levels, $N(I)$ represents the number of pixels having value I, and M be the total number of pixels in that image region. The probability of pixel occurrence can be computed as $P(I) = \frac{N(I)}{M}$. Features that demonstrate the characteristics of image region's histogram can be computed as follows:

$$\text{mean} = \sum_{I=0}^{N_g-1} I^1 P(I) \tag{1}$$

$$\mu_r = \sum_{I=0}^{N_g-1} (I - \text{mean})^r P(I) \tag{2}$$

where $r = 2, 3, 4$ representing the measures of central moments, i.e., variance, skewness, and kurtosis, respectively.

Second-order texture features describe the relationship between any two pixels of an image region based on their relative distance (d) and orientation ϕ. Texture features are calculated by analyzing the Gray Level Co-occurrence Matrix (GLCM) [16]. The relationship between the two pixels at a distance, $d = 1, 2$ is computed in four different orientations, i.e., 0°, 45°, 90° and 135°. Contrast, homogeneity, correlation, Angular Second Moment (ASM), and entropy features are extracted. Contrast feature can be defined as the deviation in the local cues such as grayscale values. Homogeneity is also termed as Inverse Difference Moment (IDM) and value of homogeneity feature maximizes when the value of the contrast feature minimizes. Correlation is the texture feature that describes how much linearly dependent the grayscale values of the image region are. ASM describes the smoothness cue of the image region. It measures the homogeneity of the image region with respect to the local grayscale cues as smaller for a homogeneous image region. Entropy measures the degree of randomness. Lower the smoothness of image region, higher the randomness and so higher the entropy. Let $P(I_1, I_2)$ be the matrix describing the occurrences of pair-wise grayscale values and positioned at a specific distance d and orientation ϕ in the original image region. Features are defined [16] as follows:

$$\text{Contrast} = \sum_{I_1, I_2}^{N_g-1} |I_1 - I_2|^2 \log P(I_1, I_2) \tag{3}$$

$$\text{Homogeneity} = \sum_{I_1, I_2}^{N_g-1} P(I_1, I_2) \frac{1}{1 + |I_1 - I_2|^2} \tag{4}$$

$$\text{Correlation} = \sum_{I_1, I_2}^{N_g-1} P(I_1, I_2) \frac{(I_1 - \mu_1)(I_2 - \mu_2)}{\sigma_1 \sigma_2} \tag{5}$$

$$\text{Angular Second Moment} = \sum_{I_1, I_2}^{N_g - 1} P(I_1, I_2)^2 \quad (6)$$

$$\text{Entropy} = -\sum_{I_1, I_2}^{N_g - 1} P(I_1, I_2)\ \log\ P(I_1, I_2) \quad (7)$$

where μ_1, μ_2, σ_1, and σ_2 represent means and standard deviations corresponding to I_1 and I_2, respectively.

HOG [17] features are also extracted for each ROI in addition to the aforementioned features. Gradient in horizontal direction fires on the vertical lines, whereas gradient in vertical direction fires on horizontal lines whenever there are abrupt changes in pixel grayscale values. Combining gradients of both the directions gives the clue about the appearance of tumorous brain tissues in MR images. Features are extracted per pixel basis for the central pixel of a 9×9 sized moving image window. Initially oriented gradient is computed by exploiting the 1-dimensional Sobel filter mask, i.e., $[-1\ \ 0\ \ 1]$ in horizontal direction and $[-1\ \ 0\ \ 1]^T$ in vertical direction. Let G_{mag} and G_{dir} be the magnitude and orientation of the image window gradient which are computed as follows: $G_{\text{mag}} = \sqrt{{G_h}^2 + {G_v}^2}$ and $G_{\text{dir}} = \tan^{-1}\left(\frac{G_v}{G_h}\right)$. Where G_h and G_v represent the gradients in horizontal and vertical directions, respectively. A histogram with 9 bins is calculated from the obtained gradient of the image window. One advantage of calculating histogram over small 9×9 moving image window is that it is less sensitive to any kind of noise. Histogram of gradient is calculated as a vector of size 9 where each cell represents angles from 0° to 160° with a difference of 20° (Table 1).

Table 1 Overview of features extracted from each 9×9 moving image window

Feature class	Feature type	No. of features
First order statistical	Mean	4
	Variance	
	Skewness	
	Kurtosis	
Second order texture	Contrast	40 (For each combination of $d = 1, 2$ and $\phi = 0°, 45°, 90°$ and $135°$)
	Homogeneity	
	Correlation	
	Angular second moment	
	Entropy	
Histogram of oriented gradients (HOG)	Histogram with 9 bins	9
	Total features	53

Thus, total 53 features (4 first-order statistical features + 40 second-order texture features + 9 HOG features) are extracted for each ROI of 2-dimensional MRI modality. Finally, features of the aforementioned three MRI modalities are combined which makes total 159 features.

2.3 *Dimensionality Reduction Using PCA*

High dimensionality of feature set affects the training speed of machine learning algorithms. However, less number of significant features also degrades the performance of classifiers. PCA is a widely used and efficient statistical method to reduce the dimensionality of a feature set [18]. PCA reduces the dimensions of an input vector without affecting its recognition ability by computing the eigenvectors of its covariance matrix in the form of principal components which are linearly uncorrelated. Each principal component (from first to last) is having higher variance than its successor component, whereas each component is orthogonal to its predecessor component. The original feature matrix can be represented in terms of *score* matrix in the principal component space. *Score* matrix is the representation of the original feature matrix corresponding to the computed principal components. In our experiments, we select first 16 out of total 159 components to train the classifier as they possess more than 95% of variance.

2.4 *Tumor Classification Using RF and SVM*

Breiman [19] introduced RF classifier as an ensemble of decision tree predictors in order to solve the known problems of decision trees such as slow learning, generalization accuracy and overfitting behavior. Each decision tree of RF depends on the values of a randomly chosen vector which is sampled independently of the same distribution for all the decision trees of RF. The classification results of all decision tree with randomly selected input are then combined to get the final classification using majority vote basis. Cases of training set are sampled randomly with replacement strategy in order for growing decision trees. Let M be the number of inputs, choose m such that $M >> m$. For each node, randomly select m inputs from the available M number of inputs. Out of the selected values of m, the optimal value is chosen for node splitting. Thus, training speed is achieved by random selection of inputs and greater classification is ensured by ensemble of numerous tree classifiers.

RF also improves the generalization performance by measuring the strength of individual trees and also by measuring correlation between individual random trees. Let ψ_k be the identically distributed random vector for k^{th} decision tree of RF. As per the characteristics, ψ_k should not depend on other previous random vectors such as $\psi_1, \psi_2, ..., \psi_{k-1}$. However, they share the same distribution property. The margin function for combining the decision tree classifiers, i.e., $\{t_1(x), t_2(x), ..., t_k(x)\}$ where the training set is taken randomly from the distribution of the random vector Y, X, is defined as follows [19]:

$$mf(X, Y) = \bar{x_k} I(t_k(X) = Y) - \max_{j \neq Y}(\bar{x_k} I(t_k(X) = j)) \tag{8}$$

where $I(\cdot)$ represents the indicator function and margin function computes that upto the extent such that the average prediction for the correct class exceeds the average predictions of any other class. The error of generalization can be computed in terms of margin function as follows:

$$PE^* = P_{X,Y}(mf(X, Y) < 0) \tag{9}$$

where $P_{X,Y}$ represents the probability over X, Y space. When the number of decision tree classifiers increases, for all possible sequences ψ, ..., the generalization error converges to as follows:

$$P_{X,Y}(P_\psi(t(X, \psi) = Y) - \max_{j \neq Y} P_\psi(t(X, \psi) = j) < 0) \tag{10}$$

Above Eq. 10 describes the robustness of RF toward the overfitting problem of decision tree classifiers. However, it also limits the generalization error as follows [19]:

$$PE^* \leq \bar{\rho} \frac{1 - s^2}{s^2} \tag{11}$$

where $\bar{\rho}$ and s represent average correlation between classifiers and strength of ensemble, respectively, which control the generalization error of RF. In our experiments, 50 tree estimators are empirically selected to train RF on the extracted and selected features.

SVM [20] is a supervised machine learning algorithm which originally developed to solve binary classification problem with the help of a hyperplane. Therefore, the datapoints, through which the hyperplane is drawn, are known as *Support Vectors*. The equation of hyperplane can be defined as follows:

$$W^T X + b = 0 \tag{12}$$

where W^T, X, and b represent the transpose of weight vector, input vector, and bias term, respectively. From the Eq. 12, support vectors can be identified as $W^T X + b = 1$ for one class and $W^T X + b = -1$ for another class. Thus, the problem of finding the optimal hyperplane can be transformed as problem of maximizing the margin or minimizing the norm of W. It can be stated as follows:

$$\begin{aligned} &\min_{W,b} \|W\| \\ &\text{subject to} \quad y_i(W^T x_i + b) \geq 1, \qquad \text{for} \quad i = 1, 2, ..., k \end{aligned}$$

where y_i represents the respective class label. Various kernel functions such as linear, radial basis function, sigmoid, and polynomial are defined to identify and optimize the margins of a non-linear problem. In our experiments, linear kernel is exploited for 5-class classification problem.

3 Experimental Results

Experimental study based on the proposed segmentation method have been carried out over real glioma MR images taken from publicly available benchmark dataset[1] (BRATS-2013). All the experiments are conducted over HP laptop machine with 64-bit Windows 10, Intel i-5 (2.40 GHz) and 8GB of RAM. BRATS-2013 dataset [9] containing real images of 20 high-grade and 10 low-grade glioma subjects is used to conduct experiments. Each subject's data is available in 3-dimensional form that contains 176 skull-stripped slices of size 216 × 160. Out of total 176 slices, several do not contain any useful tumor information. Therefore, in preprocessing, such slices are removed. Each subject's data is available in four sets of relevant MR modalities, i.e., T1-weighted, contrast T1-weighted , T2-weighted, and FLAIR MRI. All MRI modalities of 20 high-grade Glioma subjects are utilized in this work, except to T1-weighted MR modality.

Class-wise indexing of the dataset is as follows: healthy tissue—0, necrosis—1, edema—2, non-enhancing solid core—3, enhancing tumor—4. However, objective of this work is to predict and segment three brain tumor regions, i.e., complete tumor region, tumor core, and enhancing tumor. Complete tumor region is the combination of all tumor sub-compartments. Tumor core is the region that combines the most affected brain tissues, i.e., necrosis, enhancing tumor, and non-enhancing solid core. However, edema is not included in tumor core region. Finally, enhancing tumor region only contains tumor tissues which highlight after applying contrast-enhancing element, i.e., Gad. chemical agent.

3.1 Performance Evaluation

Dice Similarity Coefficient (DSC), Sensitivity, and Positive Predictive Value (PPV) have been considered as the prominent performance measures to validate the efficacy of image segmentation methods [1]. PPV measures the count of true positive predictions over all the predictions. Sensitivity represents true positive rate of predictions that means both predicted and actual results are positive (i.e., tumor class). DSC is the score which is computed as the harmonic mean of PPV and Sensitivity. DSC measures the overlapping region of predicted and actual segmentation results. The mathematical formulations are given as below:

---

[1] https://www.smir.ch/BRATS/Start2013.

$$\text{DSC} = \frac{2 * \text{TP}}{(\text{TP} + \text{FP}) + (\text{TP} + \text{FN})} \tag{13}$$

$$\text{Sensitivity} = \frac{\text{TP}}{\text{TP} + \text{FN}} \tag{14}$$

$$\text{PPV} = \frac{\text{TP}}{\text{TP} + \text{FP}} \tag{15}$$

where TP, TN, FP, and FN represents true positive, true negative, false positive, and false negative, respectively.

The performance of RF and SVM to segment brain tumor regions is presented in Tables 2 and 3 respectively. Table 2 shows segmentation results with RF classifier in terms of all the aforementioned brain tumor regions for noisy data, denoised data with median filter, and denoised data with Wiener filter. Similarly, Table 3 demonstrates the segmentation results with SVM multi-class classifier for noisy data, denoised data with median filter, and denoised data with Wiener filter. It can be observed that in association with both the classifiers, Wiener filter does better denoising as we get better segmentation results. It is also observed that except for few of the cases, RF classifier produces better segmentation results than that of SVM classifier. Overall, both RF and SVM classifiers produces high DSC scores of 92, 94, and 95% to segment complete tumor region over the noisy data, denoised data with median filter, and denoised data with Wiener filter, respectively.

The experimental results of the proposed method are compared with the state-of-the-art methods such as Demirhan et al. [21] as given in Tables 2 and 3 for the subjects of high-grade glioma. The results of the compared method are reproduced on the same settings. Segmentation results of Demirhan et al. [21] are obtained by training the suggested model on BRATS-2013 training dataset. As BRATS dataset provides us the real glioma images which are skull stripped and registered with T1C MRI, preprocessing steps were skipped, while implementing the model.

3.2 *Discussion*

A study on noisy brain MR images is presented by conducting experiments on BRATS dataset. Three MR modalities are chosen for this work, i.e., contrast-enhanced and T1-weighted, T2-weighted, and FLAIR because of their effects on certain signal types and visual characteristics over specific types of brain tissues. Contrast-enhanced T1-weighted MRI in this way, highlights the enhancing tumor tissues. T2-weighted MR modality has such visual characteristic due to which it is used to segment edema by manual practitioners. Similarly, FLAIR MR modality images are used to cross check the extension of edema class and to discriminate it against brain ventricle and other fluid filled structures. For all the MR modalities, central patches of size 60×60 from 2-dimensional slices are extracted and the further processing such

Table 2 Experimental results of using random forest (RF) as classifier on the subjects of high-grade (HG) glioma of BRATS-2013 dataset

Results on	DSC			Sensitivity			PPV		
	Comp.	Core	Enh.	Comp.	Core	Enh.	Comp.	Core	Enh.
Noisy data	0.92	0.88	0.69	0.92	0.87	0.74	0.93	0.85	0.63
Noise removed data using median filter	0.94	0.87	0.68	0.93	0.89	0.75	0.94	0.85	0.66
Noise removed data using Wiener filter	0.95	0.88	0.69	0.94	0.90	0.75	0.94	0.87	0.67
Demirhan et al. [21]	0.61	0.46	0.42	0.59	0.57	0.42	0.66	0.57	0.35

Table 3 Experimental results of using support vector machine (SVM) as classifier on the subjects of high-grade (HG) glioma of BRATS-2013 dataset

Results on	DSC			Sensitivity			PPV		
	Comp.	Core	Enh.	Comp.	Core	Enh.	Comp.	Core	Enh.
Noisy data	0.92	0.84	0.64	0.89	0.82	0.68	0.92	0.83	0.58
Noise removed data using median filter	0.94	0.85	0.65	0.91	0.84	0.71	0.94	0.85	0.61
Noise removed data using Wiener filter	0.95	0.86	0.67	0.94	0.87	0.72	0.95	0.87	0.62
Demirhan et al. [21]	0.61	0.46	0.42	0.59	0.57	0.42	0.66	0.57	0.35

as features extraction, and classifier training are carried out based on these central patches only. One advantage of extracting features from these ROI is that they do not suffer with class imbalance problem as majority of the healthy brain tissues are removed in the process of ROI selection. AWGN is applied with zero mean and 0.01 variance to ensure the natural and random distribution of noise. State-of-the-art filtering algorithms to deal with Gaussian noise, i.e., median filter and Wiener filter are applied on noisy images. Median value is selected from the 5×5 moving window. Size of the neighborhood window is selected empirically by conducting experiments.

Total 50 decision tree estimators have been chosen by empirical study and used to classify with RF. All other experimental details related to features extraction are given in Sect. 2.2.

4 Conclusion and Future Work

Noise in brain MR images is considered as a major challenge affecting segmentation performance. We analyse the behaviour and effect of natural and randomly distributed noise by adding it manually on MR images. RF and SVM classifiers are trained on both noisy as well as denoised MR images which were denoised by median filter and Wiener filter. Denoising with Wiener filter along with RF classifier produces favorable results as compared to other combinations of filter and classifier. Due to the random selection of inputs of decision trees and its explorative characteristics, RF gives a robust segmentation method in association with the other modules of the proposed method. In future, we will apply different kinds of noise on brain MR images and study the behavior after applying other noise removal techniques.

References

1. Menze BH, Jakab A, Bauer S, Kalpathy-Cramer J, Farahani K, Kirby J, Burren Y, Porz N, Slotboom J, Wiest R et al (2015) The multimodal brain tumor image segmentation benchmark (brats). IEEE Trans Med Imaging 34(10):1993
2. Prastawa M, Bullitt E, Ho S, Gerig G (2004) A brain tumor segmentation framework based on outlier detection. Med Image Anal 8(3):275–283
3. Menze BH, Leemput KV, Lashkari D, Weber M-A, Ayache N, Golland P (2010) A generative model for brain tumor segmentation in multi-modal images. In: International conference on medical image computing and computer-assisted intervention. Springer, pages 151–159
4. Bauer S, Nolte LP, Reyes M (2011) Fully automatic segmentation of brain tumor images using support vector machine classification in combination with hierarchical conditional random field regularization. In: International conference on medical image computing and computer-assisted intervention. Springer, pages 354–361
5. Zhao J, Meng Z, Wei L, Sun C, Zou Q, Su R (2019) Supervised brain tumor segmentation based on gradient and context-sensitive features. Front Neurosci 13
6. Pei L, Bakas S, Vossough A, Reza SMS, Davatzikos C, Iftekharuddin KM (2020) Longitudinal brain tumor segmentation prediction in mri using feature and label fusion. Biomed Signal Process Control 55:101648
7. Kamnitsas K, Ledig C, Newcombe VFJ, Simpson JP, Kane AD, Menon DK, Rueckert D, Glocker B (2017) Efficient multi-scale 3d cnn with fully connected crf for accurate brain lesion segmentation. Med Image Anal 36:61–78
8. Havaei M, Davy A, Warde-Farley D, Biard A, Courville A, Bengio Y, Pal C, Jodoin P-M, Larochelle H (2017) Brain tumor segmentation with deep neural networks. Med Image Anal 35:18–31
9. Kistler M, Bonaretti S, Pfahrer M, Niklaus R, Büchler P (2013) The virtual skeleton database: an open access repository for biomedical research and collaboration. J Med Internet Res 15(11):e245

10. Fan L, Zhang F, Fan H, Zhang C (2019) Brief review of image denoising techniques. Vis Comput Indus Biomed Art 2(1):7
11. Jainy S, Vinod K, Indra G, Niranjan K, Chirag KA (2012) A novel content-based active contour model for brain tumor segmentation. Magn Reson imaging 30(5):694–715
12. Beevi Z, Sathik M (2009) A robust segmentation approach for noisy medical images using fuzzy clustering with spatial probability. Methods 29(37):38
13. Guo L, Chen L, Philip Chen CL, Zhou J (2018) Integrating guided filter into fuzzy clustering for noisy image segmentation. Digital Signal Process 83:235–248
14. Mohd Sagheer SV, George SN (2020) A review on medical image denoising algorithms. Biomed Signal Process Control 61:102036
15. Harris C, Stephens M (1988) A combined corner and edge detector. In: Alvey vision conference, volume 15, pages 147–151. Citeseer
16. Haralick RM, Shanmugam K, Hak Dinstein I (1973) Textural features for image classification. IEEE Trans Syst Man Cybernet 6:610–621
17. Dalal N, Triggs B (205) Histograms of oriented gradients for human detection. In: 2005 IEEE computer society conference on computer vision and pattern recognition (CVPR'05), volume 1. IEEE, pages 886–893
18. Jolliffe IT (2002) Choosing a subset of principal components or variables. Principal component analysis, pages 111–149
19. Breiman L (2001) Random forests. Machine learning 45(1):5–32
20. Burges Christopher JC (1998) A tutorial on support vector machines for pattern recognition. Data Min Knowl Disc 2(2):121–167
21. Demirhan A, Törü M, Güler I (2014) Segmentation of tumor and edema along with healthy tissues of brain using wavelets and neural networks. IEEE J Biomed Health Inform 19(4):1451–1458

Designing New Tools for Curing Cardiovascular Diseases to Avoid Angioplasty, Coronary Bypass and Open Heart Surgery: Simulation and Modeling

Hozzatul Islam Sayed

Abstract In only 2016, across the globe 17.9 million people died from cardiovascular diseases (CVDs) where 75% of CVDs deaths turn up from less and average income country like Bangladesh. In real-world perspective, cardiovascular surgery (CVS) is very expensive. Billions of dollar expense behind the CVS globally each year. According to statistics, three out of four people are affected in CVDs in Bangladesh, and one die in every 38 s in USA. We have developed a new idea for curing the CVDs via simulation. We have also developed algorithms, designed cost-effective new drugs and stimulus focusing the blocked vessels through angiogram and collected ECG data for observing blood flow, mass and heat transfer via simulation of heart and vessels. We used bio-fluid mechanics property (Navier–Stokes equations of Newtonian flow), BVP model and Hodgkin–Huxley (1952) mathematical model for heart simulation and transport phenomena of blood. We developed two new algorithms to design a primer individually. We have shown the genetic engineering techniques for pushing the condition on primer. Traditional CVS are time-consuming, risky and entirely not effective or accurate especially over seventy years old peoples. We wish to turn the method of traditional CVS into a simulation-based algorithm which is a time-saving, cost-effective and more accurate technique for curing CVDs.

Keywords Cardiovascular diseases (CVDs) · Cardiovascular surgery (CVS) · Stimulus · Action potential · Blood flow velocity · BVP model · Laminar flow

1 Introduction

Approximately 422.7 million people (95% incertitude average: 415.53–427.87 million subjects) and 17.92 million people (95% incertitude average: 17.59–18.28 million CVDs) died from the several cardiovascular diseases (CVDs) in 2016 alone. The data shows between 1990 and 2015 Downfall in the period measured CVDs

---

H. I. Sayed (✉)
Biomedical Engineering Department, Jashore University of Science and Technology, Jashore–7408, Khulna, Bangladesh

G. Gupta et al. (eds.), *Proceedings of Academia-Industry Consortium for Data Science*, Advances in Intelligent Systems and Computing 1411,
https://doi.org/10.1007/978-981-16-6887-6_15

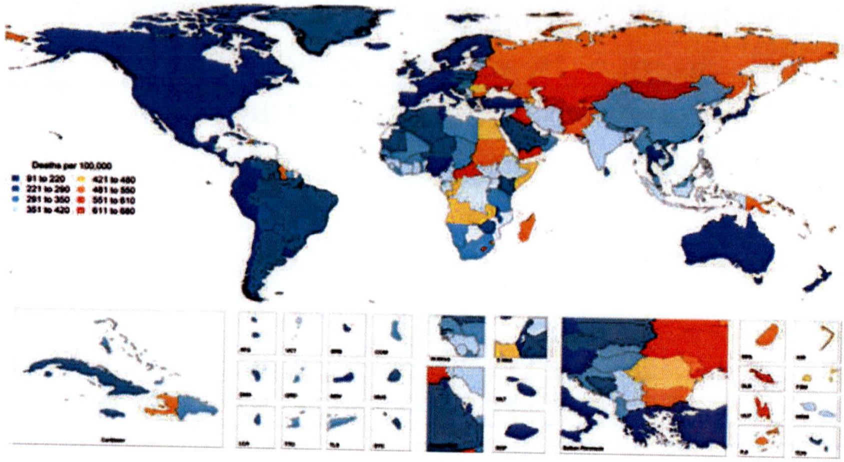

Fig. 1 Period standardized death rate of CVDs in 2016 global map. This map showed the how many peoples carry or develop cardiovascular diseases in per hundred thousand peoples. In the case of Bangladesh, in per 100,000 people is 351–420 peoples

death rate take place in the strong revenue and average revenue countries [1–3], The aim of this work is to build up a unique tool to cure major cardiovascular diseases (only for angina, arteriosclerosis, coronary atheroma as well as myocardial infraction, bradycardia and heart failure) to avoid angioplasty, coronary bypass and open heart surgery. Designing two new stimulus (hormone) using genetic engineering techniques (Based on developing algorithms for primer) (Fig. 1).

2 Background Study

For showing the how blood circulates within a stenotic artery used by Coupled fluid–network communication model, which is a sinusoidal function guided the carving of plaque:

$$d_a = \frac{Y - B}{2}.\cos x\left(1 + \frac{1}{\cos x}\right) \quad \text{and} \quad d_b = \frac{Y - B}{2}.\cos x\left(1 + \frac{1}{\cos x}\right) - y \quad (1)$$

They assumed the bloodstream to be laminar, Newtonian, viscous and incompressible. Here, incompressible Navier–Stokes equations in free Lagrangian–Eulerian system are used as the guiding equations:

$$\rho\frac{\partial v}{\partial t}\nabla\left[(-q)l + \eta\left(\nabla v + (\nabla v)^R\right)\right] + \rho(v - \psi).\nabla)v = F \quad \text{and} - \nabla.v = 0 \quad (2)$$

In this case, BVP model is used in stimulus simulation through ring (blood vessels)

$$\dot{z} = b(e + z - z^3/3 + x) \tag{3}$$

$$d = -\frac{(z - b + vx)}{c} \tag{4}$$

So,

$$1 - 2v/3 < b < 1, \quad 0 < v < 1, \quad v < c^2$$

Simulation of cardiac cell using Hodgkin_huxeley_1952 mathematical model and applying Basic Hemodynamics property [4–9], This method are applying for only computer based simulation by using Sim-Vascular and physio-Designer simulation software [10, 11], as well as other computational work.

3 Experimental Protocol

3.1 Required Tools

This study is based on fully computational method. No hardware is required for this study. Some high configuration software are used for simulation such as Physio-Designer-physiological system Simulation, Sim-Vascular—vascular Simulation, bioRender- biological template design and MATLAB for designing algorithm.

3.2 Prototype Construction and Theory

This simulation collected from Shadden lab (university of California—Berkeley) [12]. This study shows flow simulation, particle dynamics and 3D CAD model of the arch of aorta. In that case, fibrous thickness covers 50% area of actual cardiovascular artery area. Blood flow becomes turbulent because of the 50% fibrous thickness, and also, homeostasis of the blood flow changes [13, 14], (Fig. 2).

4 Analysis

4.1 Computational Methods

We developed two new algorithms, one is increased blood flow velocity from normal states, and other is decaying cholesterol from blocked vessels.

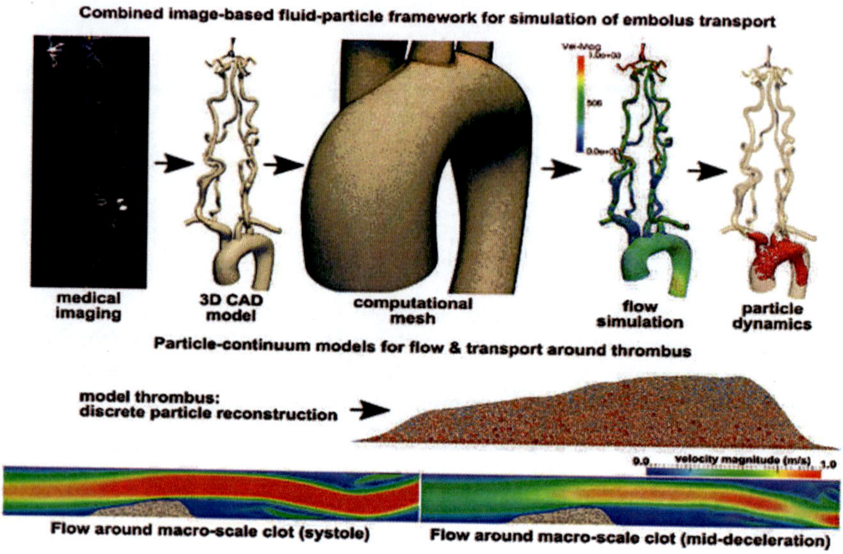

Fig. 2 Simulation of blocked cardiovascular artery where figure shows medical imaging image, a 3D designed model, flow simulation and particle dynamic simulation and also shows the fibrous cap thickness and laminar flow simulation where due to cap thickness the flow is not still laminar

4.2 For Increasing the Velocity of Blood Stream (Primary Stimulus)

Order> 1: Prescribe the aorta, capillaries, time, rate of the heart, etc.
Order> 2: Set up the stimulus over the heart for 0.7 s per heart cycle.
Order> 3: Set up the heart beat 85 times/min and for 123,429 times pumping within 86,400 s or 24 h.
Order> 4: Set up the velocity of blood flow through aorta ‘Aorta = 33 cm/s’ and through capillary ‘capillaries = 0.027 cm/s’.
Order> 5: Set up the 1.1 s/heart cycle heart rate for the 50% plaque.
Order> 6: Set up the 55 times/min heart beat and 78,545 pumping within 86,400 s or 24 h for 50% plaque.
Order> 7: Set up the velocity of blood flow through aorta ‘Aorta = 20 cm/s’ and through capillary ‘capillaries = 0.015 cm/s’ for 50% plaque.
Order> 8: Putting time = 1.1 s–0.7 s = 0.4 s, aorta = 33–20 = 13 cm/s, capillaries = 0.027–0.015 = 0.012 cm/s, then run the loop for raising the flow velocity.
Order> 9: ‘0.4 s = 13 && 0.012, ++’ (aorta && capillaries) cm/s, this loop returns after 0.4 s interval for 216,000 times and repeats in 86,400 s or 24 h.
Order> 10: ‘heart beat = 55 + 30, heart beat 30++ ’ (for every 0.4 s), ‘216,000 times running loops in 86,400 s’.
Order> 11: After 24 h or 86,400 s, the loops were stopped by itself.

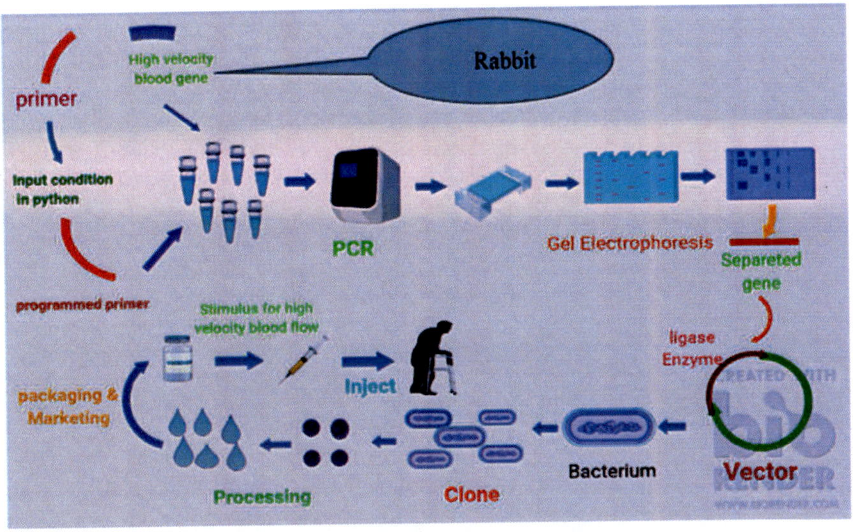

Fig. 3 Producing methods of new stimulus for high velocity blood flow (Algorithm I). The figure shows the whole processing methods of the stimulus for rising blood flow velocity

This algorithm is implemented on the Bio-Nano chips for designing primer. This chip is attached with primer and conducts the stimulus in blood. Study shows that rabbit contains high velocity blood gene. Then, the rabbit gene was hybridized with programmed primer. In this way, stimulus is made by using traditional genetic engineering techniques (Fig. 3).

4.3 *Algorithms for Decaying Cholesterol*

Order> 1: Prescribes the cartdw = intima & choldec molec = chemical substances.
Order> 2: Set up the choldec_molec_paas-diam = = <2.05 mm–<3.75 mm.
Order> 3: Set up the general carotid lumen diameter = 6.15 mm.
Order> 4: Set up the stimulus over via heart 0.7 s/heart cycle.
Order> 5: Set up the heart beat 85 times/min and 123,429 times pumping within 86,400 s or 24 h.
Order> 6: Set up the velocity of blood flow through aorta 'Aorta = 33 cm/s' and through capillary 'capillaries = 0.027 cm/s'.
Order> 7: Set up the cartd_lumen_diameter value for 50% plaques.
Order> 8: Set up the fibrous cape thickness = 2 mm for 50% plaques.
Order> 9: If carotid lumen diameter at 50% plaque <= half of the general carotid diameter (3.75 mm) or fibrous cap thickness <= one third (1/3) of the general carotid lumen diameter (2.05 mm) == choldec molec_paas_diam then choldec_molec attach to the cartdw.

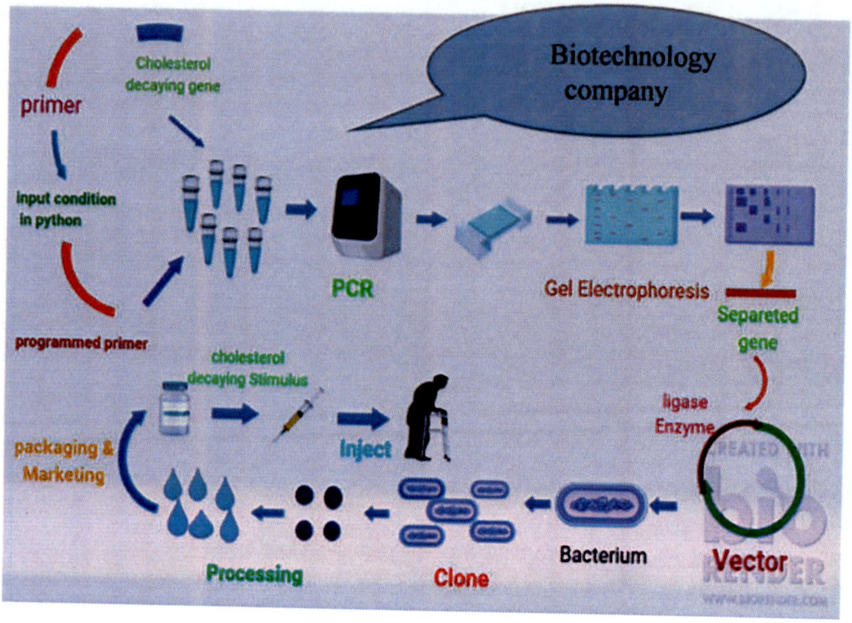

Fig. 4 Producing methods of cholesterol decaying stimulus which is for second algorithm. This stimulus decays the cholesterol from the most inner layer of the carotid intima. This figure shows how we generate and process this stimulus

Order> 10: Attached to cartdw 'choldec_molec = 172,800 s or 48 h'.
Order> 11: Detached after 172,800 + 1 = choldec_molec.

Similarly, this algorithm is implemented on a Bio-Nano chip as described earlier. The process of making stimulus is same as the previous method. But cholesterol decaying gene is not identified yet (Fig. 4).

This molecule operates when carotid lumen diameter is less than 3.75 mm. This molecule binds in most inner layer of the blood vessel (intima). After binding this molecule, decaying of the cholesterol from intima is started of the blood vessels [15], and N.B: condition, statement and input–output value will be changed for secondary stimulus and different types of stenosis (say 70%, 80%, etc.). Also, lumen diameter can be changed. In this case, it is developed for 50% stenosis. Appropriate command and condition applied by the professional programmer will not have an indent, either.

5 Results

We simulate the mass and heat flow for both blocked and healthy arteries. Note that the value of z for blocked artery is 0.5. As we see in Fig. 5 in blocked artery, the mass and heat flow does not entirely spread at particular time.

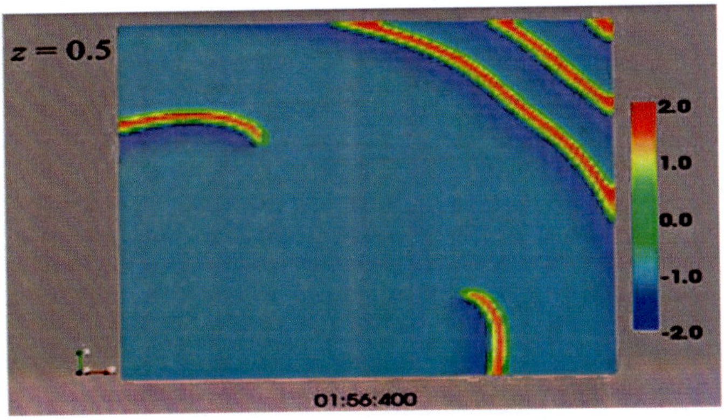

Fig. 5 Mass and heat transfer simulation (flow) in blocked cardiovascular artery. This is the blood flow simulation figure for blocked artery patient. Where the mass and heat does not fully flow through the lumen

In contrast, in case of healthy artery, the phenomenon is totally reverse. Mass and heat flow is entirely spread in artery at particular time. Note that, in this case, the value of z is 0.9. We can see it from Fig. 6. Then, we simulate the heart for blocked artery. We see the heat and mass flow does not manage full area of the heart chamber that means blood flow does not cover full area of the heart chambers. It can be shown in Fig. 7.

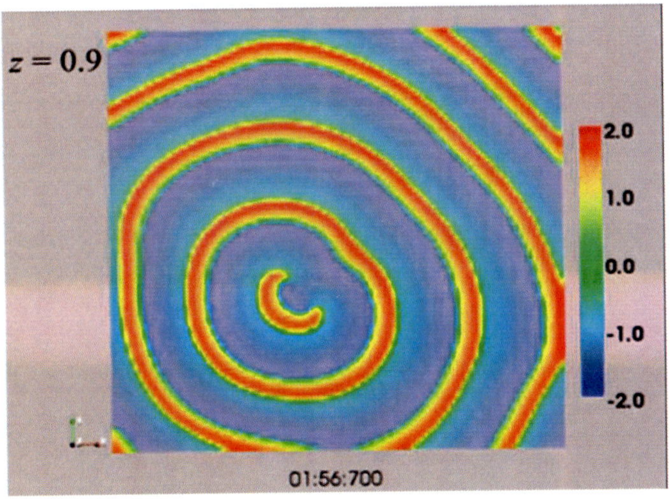

Fig. 6 Mass and heat transfer simulation (flow) in healthy cardiovascular artery patient. For healthy patient case, simulation showed the mass and heat transfer for blood flow with full area through artery lumen

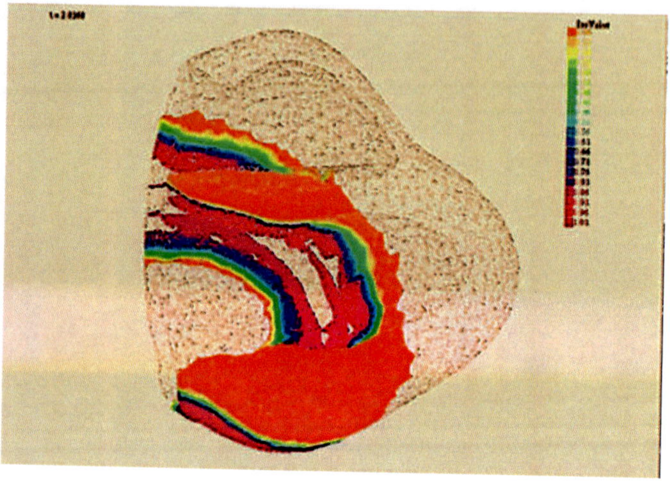

Fig. 7 This figure shows the blood flow simulation in blocked artery patient heart. Here, both mass and heat transfer does not fully cover area when blood flows through heart. The lacings of blood of the heart develops the myocardial infarction in blocked cardiovascular artery patient

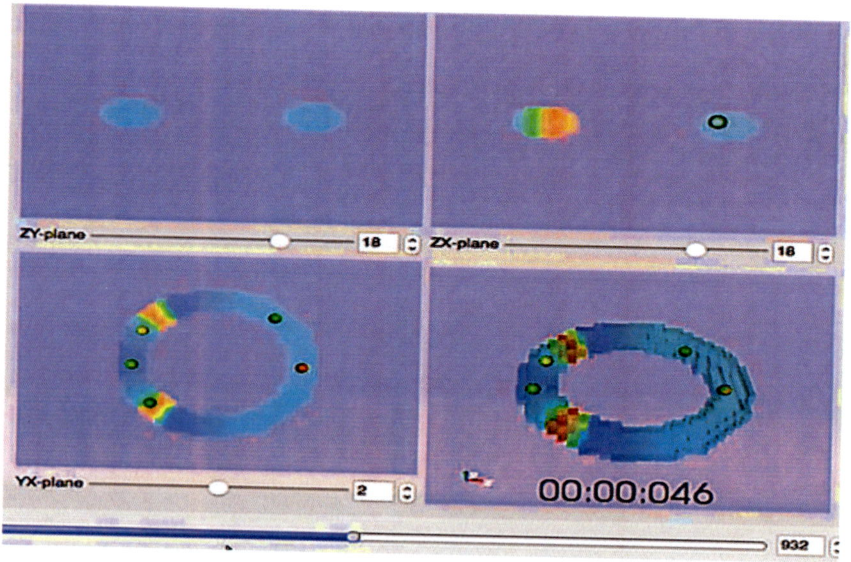

Fig. 8 This figure completely showed the simulation of primary stimulus. When primary stimulus pushes into blood, then this stimulus increases the blood velocity, and it is two times greater than normal blood flow velocity

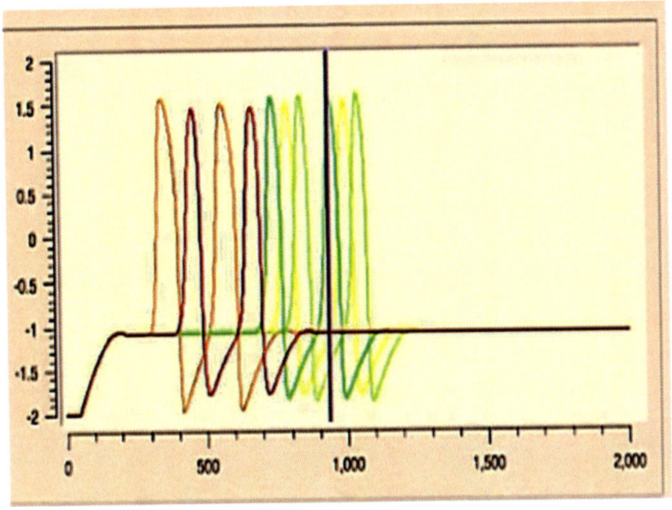

Fig. 9 Generated action potential for primary stimulus. These action potentials are two times higher than normal action potentials. That means the stimulus works and rises blood flow velocity

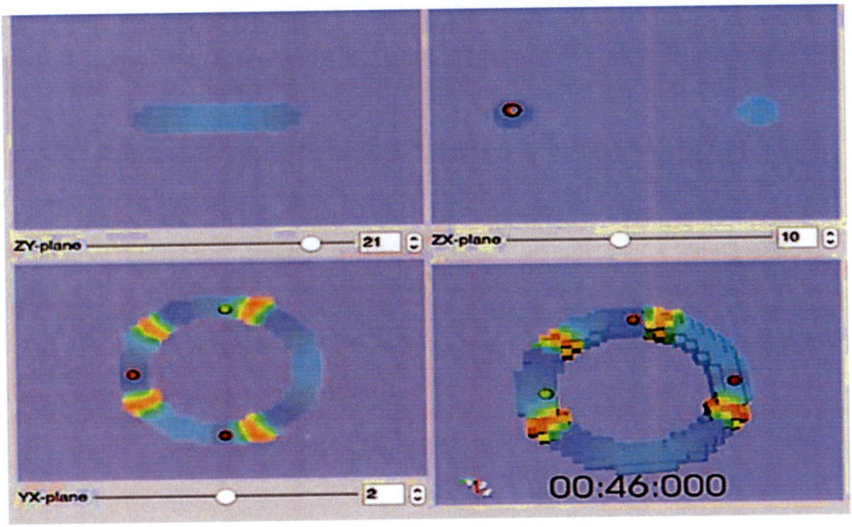

Fig. 10 Similarly, this simulation is for secondary stimulus. At the same time, flow speed of secondary stimulus is two times higher than primary stimulus and four times higher than normal blood flow velocity

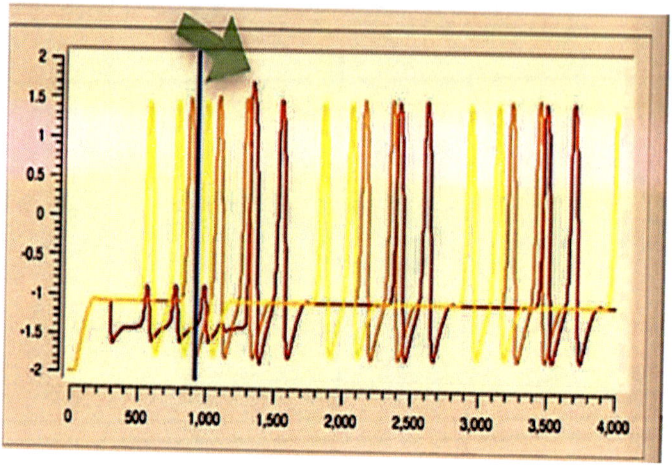

Fig. 11 Generated action potential for secondary stimulus. So similarly, in the case of secondary stimulus, the action potentials are higher than primary stimulus. That means it works and increases the blood velocity two times higher than the previous stimulus

Then, we simulate the heart for blocked artery, and we see the heat and mass flow does not manage full area of the heart chamber that means blood flow does not cover full area of the heart chambers as we see in this figure.

If we push the stimulus in patient, then the resultant simulation is shown in Fig. 8. The velocity simulation showed that how action potentials are rapidly increased at particular time as we defined in the algorithm. That means this stimulus increases the blood flow velocity (Figs. 9 and 10).

Large amplitude and frequency of action potential mean the stimulus is responsible for increasing the blood velocity. This phenomena occur when fibrous thickness is greater such 80–85%. This stimulus is known as secondary stimulus. This stimulus pushes when fibrous thickness is greater. From Fig. 11, we can see the generation of greater amplitude and large number of action potentials. That means increased the greater blood velocity than primary stimulus. It works within 24 hours.

6 Future Scopes and Challenges

1. Using machine learning and deep learning algorithms for a better understanding of the machine, we could build up this algorithm. This algorithm can be developed by using machine learning and deep learning algorithms for a better understanding of the machine. Implementing this algorithm in machine can be set or designing value and regulate the blood flow velocity and cholesterol level.

2. Complete understanding of Bio-Nano chips and the algorithms are how to implements. In addition, also try to identify the aspects, binding kinetics and Study the binding process of the materials and what types of gene express.
3. Study the gene expression aspects and materials attributes.
4. By using different machine (supervised/unsupervised) and deep learning algorithm, we could build up 'blood flow velocity and cholesterol level control' software.
5. Also, we can develop drug rather than discovering drug between associated ligand and protein by occurring better molecular docking, molecular dynamic simulation and using various types of advances in pharmaceutical technology within consuming few times.

7 Discussion

Cardiovascular diseases (CVDs) alone kill 2.56 lakh people in Bangladesh (WHO 2018). Angioplasty, coronary artery bypass (CAB) surgery and open heart surgery are very expensive especially for middle- or low-income country like Bangladesh. These techniques are riskier, time-consuming and entirely not effective or accurate especially over seventy years old people [16, 17], But only in Bangladesh, normal CAB surgery costs 0.3 million Bangladeshi taka (BDT), and angioplasty (coronary and carotid jointly) costs 0.2 million BDT. Worldwide billions of dollar expense behind this surgery every year [18–20]. Our developed approaches are cost effective than traditional CVS, and this ensures no injury in the body. This technique paves the way as the most effective for low- or middle-income countries as well.

8 Conclusion

Major cardiovascular diseases occur in Bangladeshi people due to high blood pressure. So, this can be a new and effective idea for controlling the blood pressure by using stimulus. This algorithm can cure mentioned CVDs without taking any risks to the CVS patients all over the world. It is cost effective and time saving. The risk of this techniques is very low. Also, those approaches are not harmful for elderly peoples.

References

1. WHO Homepage, https://www.who.int/health-topics/cardiovascular-disease/
2. WHO Homepage, https://www.who.int/en/news-room/fact-sheets/detail/cardiovascular-disease-(cvds)

3. Roth GA, Johnson C, Abajobir A, Abd-Allah F, Abera SF (2017) Global, regional, and national burden of cardiovascular diseases for 10 causes, 1990 to 2015. J Am Coll Cardiol 70(1):1–25
4. Wu S-N (2004) Simulations of the cardiac action potential based on the Hodgkin_huxley kinetics. Chin J Physiol 47(1):15–22
5. Fitzhugh R (1961) From the National Institutes of Health, Bethesda: Impulses and physiological states in theoretical models of nerve membrane. Biophys J 1(6):455–466
6. Hodgkin AL, Huxley AF (1952) A quantitative description of membrane cur-rent and its application to conduction and excitation in nerve. J Physiol 117(4):500–544
7. McDONALD DA (1952) The velocity of blood flow in the rabbit aorta studied with high-speed cinematography. J Physiol 118(3):328–339
8. Zafar H, Sharif F, Lehy MJ (2014) Measurement of the blood flow rate and velocity in coronary artery stenosis using intracoronary frequency domain optical coherence tomography: Validation against fractional flow reserve. IJC Heart Vasculature 5:68–71
9. Krejza J, Arkuszewski M, Kasner SE, Weigele J, Usty-mowicz A, Hurst RW, Cucchiara BL, Messe SR (2006) Carotid artery diameter in men and women and the relation to body and neck size. AHA J (Stroke) 37(4):1103–1105
10. Physio-designer Homepage, https://www.physiodesigner.org
11. Sim-Vascular Homepage, http://simvascular.github.io/
12. Shadden Lab Homepage, https://shaddenlab.berkeley.edu/research.html
13. Li Z-Y, Howarth SPS, Tang T, Gillard JH (2006) How critical is fibrous cap thickness to carotid plaque stability? A flow–plaque interaction model. AHA Strokes J 37(5):1195–1199
14. Razavi SE, Zanbouri R (2011) Simulation of blood flow coronary artery with consecutive stenosis & coronary-coronary bypass. Bioimpacts 1(2):99–104
15. Khanam F, Hossain MB, Mistry SK, Afsana K, Rahman M (2019) Prevalence and risk factors of cardiovascular diseases among Bangladeshi adults: findings from a cross-sectional study. J Epidemiol Global Health 9(3):176–184
16. Ibrahim Cardiac Hospital Homepage, https://www.ibrahimcardiac.org.bd/index.php?option=com_content&view=article&id=138&m_i=73
17. Chowdury MZI, Haque M, Farhana Z, Mustufa Anik A, Hoque Chowdury A, Haque SM, Wal Marjana L-L, Bristi PD, Al Mamun A (2018) Prevalence of cardiovascular disease among Bangladeshi adult population: a systematic review & meta analysis of the studies. J Vasc Health Risk Manage 14:165–181
18. Square hospital Homepage, https://www.squarehospital.com/post/71/Cardiologypackage
19. New Age Bangladesh Homepage, http://www.newagebd.net/article/51904/cardiac-patients-rising-in-bangladesh-25-lakh-die-annually
20. Cost helper Homepage, https://health.costhelper.com/heart-surgery.html

A Deep Learning Approach for the Obstacle Problem

Majid Darehmiraki

Abstract Due to the increasing use of deep learning in science and engineering, as well as the high ability of this method to solve computational problems, in this article, we use it for the obstacle problem (OP). Since many physical problems are modelled as variational inequality and are of obstacle types, therefore, we present an efficient solution to the obstacle problem. Here, two neural network models were employed for the obstacle problem. In the first method, we first discrete the OP using finite element method to turn it into a quadratic programming problem (QPP). Then, we extract the optimality conditions for the obtained QPP. Finally, a neural network model is employed to solve the QPP, which equilibrium point is the same as the solution to the optimization problem. We show the suggested neural network has global convergence. The second method includes a new approach for the OP using the dynamical functional particle method and the deep learning method.

Keywords Obstacle · Neural network · Deep learning · Dynamical functional particle method · Operational matrix

1 Introduction

There are many OPs in physics and engineering. For example, these types of problems have wide applications in constrained heating, fluid filtration in porous media, elasto-plasticity, control theory, or economic mathematics and so on. There are also many problems such as the filtration dam problem [19], the Stefan problem [24], the subsonic flow problem [24], American options pricing model [9] that can be modelled by the OP. Also, it should be noted where the OP is member of the elliptic variational inequalities class [11]. The classical OP illustrated by the graph of

M. Darehmiraki (✉)
Department of Mathematics, Behbahan Khatam Alanbia University of Technology, Khouzestan, Iran
e-mail: darehmiraki@bkatu.ac.ir

G. Gupta et al. (eds.), *Proceedings of Academia-Industry Consortium for Data Science*, Advances in Intelligent Systems and Computing 1411,
https://doi.org/10.1007/978-981-16-6887-6_16

$w : \Omega \subset \mathcal{R}^n \longrightarrow \mathcal{R}$ under an outer force $f \in L^2(\Omega)$ on the frontier $\partial\Omega$ and constrained to lie over an obstacle function, $\theta \in H^1(\Omega) \cap \bar{C} : \Omega \longrightarrow \mathcal{R}$ is investigated here:

$$\min \frac{1}{2}\int_{\Omega} |\nabla w| dx - \int_{\Omega} f w \mathrm{d}x,$$
$$S.t. \quad w \in \mathcal{K}, \tag{1}$$

where $\mathcal{K} = \{w \in H_0^1(\Omega) | \theta \leq w \in \Omega\}$ is a convex and closed admissible set. Another representation of the OP is to find $w \in \mathcal{K}$ subject to:

$$a(w, v - w) \geq (f, v - w)_{\Omega} \quad \forall v \in \mathcal{K}, \tag{2}$$

where $a(w, v) = \int_{\Omega} \nabla w . \nabla v dx$, and $(f, v)_{\Omega} = \int_{\Omega} f v \mathrm{d}x$. It is also modelled as follows

Finding the displacement w of a film constrained to stay above an obstacle given by φ (with $\varphi \leq 0$ at $\partial\Omega$)

$$-\Delta w - f \geq 0, \;\; w \geq \varphi, \;\; (w - \varphi)(f + \Delta w) = 0 \, in \, \Omega, \quad w = 0 \, in \, \partial\Omega. \tag{3}$$

where Ω is a convex polygon.

Various methods for different types of OPs have been suggested so far, some of which are mentioned here. Wang et al. employed discontinuous Galerkin methods to solve elliptic variational inequality problems [22]. Also, Lewi et al. [12] used symmetric dual-wind discontinuous Galerkin methods for the second-order elliptic OP. Authors of [4] using the Lagrange FEM provide a fully estimable approach for parabolic variational inequality problem. A hybrid method was proposed to solve the OP in [23]. The level set method and some of its modified versions have been used for the OP [8, 13]. A primal-dual hybrid gradient was used for solving reformulation of the obstacle problem [25]. The weak Galerkin finite element method was investigated to solve the second kind of the elliptic variational inequality and the obstacle problem in [16]. Authors of [18] proposed a robust algorithm based on the dynamical functional particle method for the unilateral obstacle problem. The Galerkin least square finite element method based on an augmented Lagrangian formulation was employed for the obstacle problem in [2]. Calder and Yezzi [3] hybrided methods momentum, Nesterov's accelerated gradient and Polyak's heavy ball for the OP. Recently, a first-order reformulation of the OP was solved using a least-squares FEM in [6].

Tank and Hopfield [20] first solved optimization problems by neural networks for the first time in two years. Later, researchers used many of these tools to solve various optimization problems. In the following, we mention some of them. Qin et al. [17] proposed a recurrent one-layer NN for the pseudoconvex optimization problem. Also, NNs were used to solve a especial class of pseudoconvex optimization problems in [1]. Nonlinear programming was solved by projection neural networks in [21].

To our knowledge, no results have been obtained about NN models for obstacle problems. In this work, we employ neural network estimates for the unilateral obstacle problem. The finite element is used for the discretization. FEM can be used to write the obstacle problem in the form of the quadratic programming problem [5] as follows:

$$\begin{gathered} \min \ \frac{1}{2} w^T K w - w^T F, \\ \text{S.t.} \\ T w \geq \bar{\theta}, \end{gathered} \tag{4}$$

where K is the symmetric, positive definite stiffness matrix, F is load vector, $\bar{\theta}$ is the vector with information about the given obstacle surface, and T represents a transformation matrix. The primary purpose of this paper is to first extract the optimality of the model (4) and then provide an impressive neural network model for solving it. In here, a NN model for solving resulted QP (4) based on optimality conditions is provided and the existence of the solution of it is also investigated.

1.1 Outline

The paper is structured as follows. In Sect. 2 a NN model is presented for the resulted QP problems. In Sect. 3 we analyse the convergence and stability of the proposed network and numerical studies are presented in Sect. 4.

2 Neural Network Model

Before introducing a NN model for the problem (4) , we need to remember the following three definitions.

Definition 1 If $h(t, u^*) = 0$ for all $t \geq 0$ then u^* is called an equilibrium point of $\dot{u}(t) = h(t, u)$.

Definition 2 Assume $D \subset \mathcal{R}^n$ be an open neighbourhood of u^*. A function $P : \mathcal{R}^n \longrightarrow \mathcal{R}$, that is continuously differentiable, is called a Lyapunov function at the state u^* for a system $\dot{u}(t) = h(t, u)$ if:

$$\begin{aligned} &P(u^*) = 0, \quad P(u) > 0, \ \forall u \in D \backslash u^*, \\ &\frac{\mathrm{d}P(u(t))}{\mathrm{d}t} = [\nabla P(u(t))]^T h(t, u) \leq 0, \ \forall u \in D. \end{aligned}$$

Lemma 1

1. *If there exists a Lyapunov function over some neighbourhood $\hat{\Omega}$ of u^* then the point u^* of a system $u' = \mathcal{H}(x)$, that is isolated equilibrium, is called Lyapunov stable .*
2. *If there is a Lyapunov function over some neighbourhood $\hat{\Omega}$ of u^* such that $\frac{dP(u(t))}{dt} < 0$, $\forall u \in \hat{\Omega} \setminus \{u^*\}$ then the point u^* of a system $u' = \mathcal{H}(u)$, that is isolated equilibrium, is called asymptotically stable .*

In order to provide a NN model for the OP we have to use the NCP-functions. So let's define these functions in following. The nonlinear complementarity problem (NCP) is corresponding to finding a vector $u \in \mathcal{R}^n$ so that

$$u \geq 0,\; H(u) \geq 0,\; (u, H(u)) = 0,$$

where $H : \mathcal{R}^n \longrightarrow \mathcal{R}^n$ must be a continuously differentiable function [14].

Definition 3 A function $\Psi : \mathcal{R}^2 \longrightarrow \mathcal{R}$, that has following conditions, is an NCP-function,

$$\Psi(s, z) = 0 \Longleftrightarrow s \geq 0,\; z \geq 0,\; sz = 0. \tag{5}$$

The FB function is one of the famous NCP-functions that stated as

$$\Psi(s, z) = (s + z) - \sqrt{s^2 + z^2}, \tag{6}$$

for more explanation see [15] and references there in.

Now, in order to present a NN model for the problem (4), we first need to derive the optimality conditions for it. The KKT optimality condition for Problem (4) can be stated as follows

$$Kw - F + Ty = 0 \tag{7}$$

$$(\bar{\theta} - Tw) \geq 0,\; y \geq 0,\; y^T(\bar{\varphi} - Tw) = 0, \tag{8}$$

where the Lagrangian multiplier vector y^T is corresponding to w. Since K is a positive definite matrix, so the best solution of the problem (4) is the corresponding to the solution of Eqs. (7) and (8). The Eq. (8) using FB function can be rewritten as following:

$$\Psi(\bar{\theta} - Tw, y) = 0. \tag{9}$$

For continence, consider the function ξ as follows

$$\xi(z) = \begin{cases} Kw - F + Ty \\ \Psi(\bar{\theta} - Tw, y) \end{cases}, \tag{10}$$

where $z = (w, y)^T$. Now, we propose the following NN for the problem (4)

$$\frac{dz(t)}{dt} = -\gamma\xi(z), \qquad z(t_0) = z_0, \tag{11}$$

with the convergence rate $\gamma \geq 0$.

3 Stability and Convergence Analysis

Here, we intend to examine the convergence and the stability of the NN proposed in (11). To do this, let's first state a several lemmas that are needed.

Lemma 2 *Let $s : \mathcal{R}_+^n \longrightarrow \mathcal{R}_+^n$, and is defined by $s(x) = \sqrt{x}$, $\forall x \in \mathcal{R}_+^n$. The matrix Jacobian for $s(x)$ at $x = (x_1, x_2)^T \in \mathcal{R}_+^n$, $x \neq 0$ is computed as follows*

$$\nabla s(x) = \frac{1}{2}L_w^{-1} = \begin{cases} \dfrac{1}{2\sqrt{x_1}I}, & x_2 = 0, \\ \begin{pmatrix} \dfrac{1}{2\sqrt{x_2}I} & cx_2^T/\|x_2\| \\ cx_2/\|x_2\| & aI + (b-a)x_2x_2^T/\|x_2\| \end{pmatrix}, & x_2 \neq 0, \end{cases}$$

where $\quad w = \sqrt{x},\ a = \dfrac{\sqrt{\sigma_2} - \sqrt{\sigma_1}}{\sigma_2 - \sigma_1},\ b = \dfrac{1}{4}(\dfrac{1}{\sqrt{\sigma_2}} + \dfrac{1}{\sigma_1}),\ c = \dfrac{1}{4}(\dfrac{1}{\sqrt{\sigma_2}} - \dfrac{1}{\sigma_1})$
with $\sigma_i = x_1 + (-1)^i\|x_2\|$, $i = 1, 2$, and $\nabla s(x)$ is positive definite for all $x \in \mathcal{R}_+^n$.

Lemma 3 *For all $(x, y)^T \in \mathcal{R}_+^n \times \mathcal{R}_+^n$, $x^2 + y^2 \in \mathcal{R}_+^n$, then $\Psi_{FB}(x, y)$ is continuously differentiable at $(x, y)^T$ and $V_i \in \nabla\Psi_{FB}(x, y)$, $i = 1, 2, \ldots, n$ satisfies in the following condition*

$$V_i = [I - P_W^{-1}L_x,\ I - P_W^{-1}P_y],$$

where $w = \sqrt{z}$, $z = x^2 + y^2$. I is an identity matrix and $L_s = \begin{pmatrix} s_1 & s_2^T \\ s_2 & s_1 I_{n-1} \end{pmatrix}$ for $s = (s_1, s_2)^T \in \mathcal{R} \times \mathcal{R}^{n-1}$ where $s = x$ or y.

Lemma 4 *Let $x \in \mathcal{R}_+^n$, $y \in \mathcal{R}_+^n$, $w = \sqrt{x^2 + y^2}$, then we have*

$$(P_w - P_x)(P_w - P_y) > 0,\ (P_w - P_x) > 0,\ (P_w - P_y) > 0.$$

Theorem 1 *If the matrix K be positive semidefinite matrix, then the Jacobian matrix $\nabla\xi$ of mapping ξ defined in (10) is nonsingular and vice versa .*

Proof Due to the Lemma 2, the following conditions apply to the Jacobian matrix of $\xi(z)$

$$\nabla\xi(z) = \begin{pmatrix} K & T^T \\ (-\text{diag}\{\nabla_{(\bar{\varphi}-Tw)_k}\Psi_{FB}((\bar{\varphi}-Tw)_k, y_k)\}_{k=1}^n) & (-\text{diag}\{\nabla_{y_k}\Psi_{FB}((\bar{\varphi}-Tw)_k, y_k)\}_{k=1}^n) \end{pmatrix}$$
$$= \begin{pmatrix} K & T^T \\ (-\text{diag}\{I + P_{v_{mi}}^{-1}P_{(\bar{\varphi}-Tu)_{mi}}\}_{i=1}^n)T & -\text{diag}\{I - P_{v_{mi}}^{-1}P_{y_{mi}}\}_{i=1}^n) \end{pmatrix}$$
$$= \begin{pmatrix} K & T^T \\ -[I + P_v^{-1}L_{(\bar{\varphi}-Tw)}]T & I - P_v^{-1}P_y \end{pmatrix} \quad (12)$$

where $v = \sqrt{(\bar{\theta} - Tw)^2 + y_i^2}$, $i = 1, 2, \ldots, n$, $P_v = diag\{P_{v_{m1}}, P_{v_{m2}}, \ldots, P_{v_{mn}}\}$, $P_{(\bar{\theta}-Tw)} = diag\{P_{(\bar{\theta}-Tw)_{m1}}\ P_{(\bar{\theta}-Tw)_{m2}}, \ldots, P_{(\bar{\theta}-Tw)_{mn}}\}$, $P_y = \text{diag}\{P_{y_{m1}}, P_{y_{m2}}, \ldots, P_{y_{mn}}\}$. Therefore, $\nabla\xi(z)$ is nonsingular if and only if the following matrix is nonsingular.

$$\zeta(z) = \begin{pmatrix} K & T^T \\ -[P_v + P_{(\bar{\varphi}-Tw)}]T & P_v - P_y \end{pmatrix}^T$$

Now, we use inducing contradiction to prove it. Suppose $\eta = (\eta_1, \eta_2)^T \in \mathcal{R}^n\mathcal{R}^n$, and $\eta \neq 0$. We have $\zeta(z)\eta = 0$ or

$$K\eta_1 - T^T - [P_v + P_{(\bar{\theta}-Tw)}]\eta_2 = 0, \quad (13)$$
$$T\eta_1 + (P_v - P_y)\eta_2 = 0. \quad (14)$$

From (14), we obtain

$$\eta_2^T [P_w + P_{(\bar{\theta}-Tw)}]T\eta_1 + \eta_2^T [P_v + P_{(\bar{\theta}-Tw)}]T(P_v - P_y)\eta_2 = 0 \quad (15)$$

From (15) and Lemma 4, we have

$$\eta_1^T T^T [P_v + P_{(\bar{\theta}-Tw)}]\eta_2 < 0.$$

Pre-multiplied (13) by η_1^T, it yields

$$\eta_1^T K\eta_1 - \eta_1^T T^T [P_v + P_{(\bar{\theta}-Tw)}]\eta_2 = 0.$$

So, we have

$$\eta_1^T K\eta_1 < 0.$$

This contradicts the assumption that the matrix K was semi-positive definite so we have $\eta_1 = 0$. With respect to (13), we get $(P_v - P_y)\eta_2 = 0$, i.e.

$$\eta_2^T(P_v - P_y)\eta_2 = 0 \quad (16)$$

According Lemma 4 and (16), we get $\eta_2 = 0$. This proves the nonsingularity of the matrix ξ.

In the following theorem, the relationship between the equilibrium point of model (11) and the solution to the problem (4) is investigated.

Theorem 2 *Let* $(w^*, y^*)^T \in \mathcal{R}^n \times \mathcal{R}^n$ *satisfies the KKT equations (7), (8). Then the NN (11) has the equilibrium point in* $z \in \mathcal{R}^n \times \mathcal{R}^n$*. Vice versa, if NN (11) has the equilibrium point in* $z^* = (w^*, y^*)$ *and the nonsingularity of the Jacobian matrix of* $\nabla\xi(z^*)$ *in (12) is proven, then* (w^*, y^*) *is the KKT point of the problem (4).*

Proof Let (w^*, y^*) satisfy KKT conditions (7)–(8). Consider the Lyapunov function as follows

$$E(z) = \frac{1}{2}\|\xi(z)\|^2. \tag{17}$$

From (11), we have $\xi(z^*) = 0$. In addition

$$\nabla E(z) = \nabla\xi(z)^T \xi(z), \tag{18}$$

where $\nabla\xi(z)$ is the Jacobian matrix of $\xi(z)$ and its nonsingularity was showed in Theorem 1. Using (18), we conclude that $\nabla E(z^*) = 0$. This means that the equilibrium point of the NN model (11) is z^*. On the contrary, with respect to Eqs. (10) and (18) the proof is clear.

Due to Theorem 2 and also uniqueness of the solution to the problem (4), therefore, the proposed NN (11) has an unique equilibrium point.

Theorem 3 *The isolated equilibrium point of the dynamical model (11),* (w^*, y^*)*, is asymptotically stable.*

Proof We know that $E(z) > 0$ and $E(z^*) = 0$. Furthermore, the existence of a neighborhood $\hat{\Omega} \subset \mathcal{R}^n \times \mathcal{R}^n$ of z^* such that

$$\nabla E(z^*) = 0, \; and \, \nabla E(z) \neq 0, \quad \forall z \in \hat{\Omega} \setminus \{z^*\},$$

is obvious because z^* is an isolated equilibrium point of (11).

Now we have for any $z \in \hat{\Omega} \setminus \{z^*\}$, $E(z) > 0$. Because otherwise, according to Equations (17) and (18), the assumption that z is an isolated equilibrium point is contradicted. Moreover, due to (17) we have

$$\frac{\mathrm{d}E(z(t))}{\mathrm{d}t} = [\nabla E(z(t))]^T \frac{\mathrm{d}z(t)}{\mathrm{d}t} = -k\|\nabla E(z(t))\|^2 \leq 0, \tag{19}$$

and

$$\frac{\mathrm{d}E(z(t))}{\mathrm{d}t} < 0, \; \forall z(t) \in \hat{\Omega} \; and \; z(t) \neq z^*.$$

According to Lemma 1, the theorem is proved.

Theorem 4 *If $z = z(t, z_0)$ be a trajectory of (11) with the initial point $z_0 = z(0, z_0)$ and the bounded level set $L(z_0) = \{z \in \mathcal{R}^n \times \mathcal{R}^n : E(z) < E(z_0)\}$. So*

1. *$\Upsilon^+(z_0) = \{z(t, z_0)|t \geq 0\}$ is bounded.*
2. *There exist $\mathfrak{z}$ so that $\lim_{t\to\infty} z(t, z_0) = \mathfrak{z}$.*

Proof

1. Suppose z^* be a point so that $\nabla E(z^*) = 0$, in other word z^* be an equilibrium point of the network in (11). Therefore the derivative of $E(z)$ along the trajectory $z(t, z_0)$, $(t \geq 0)$ is as follows

$$\frac{\mathrm{d}E(z(t))}{\mathrm{d}t} = [\nabla E(z(t))]^T \frac{\mathrm{d}z}{\mathrm{d}t} = -k\|\nabla E(z(t))\|^2 \leq 0. \tag{20}$$

Therefore, $E(z)$ in the mentioned trajectory is monotone nonincreasing. Now since $\Upsilon^+(z_0) \subset L(y_0)$ then $\Upsilon^+(z_0)$ is bounded.

2. The set $\Upsilon^+(z_0) = \{z(t, z_0)|t \geq 0\}$ is bounded with respect to (1). Since the sequence $\{\bar{t}_n\}$, $0 \leq \bar{t}_1 \leq \cdots \leq \bar{t}_n \leq \cdots$, $\bar{t}_n \to +\infty$ is strictly monotone increasing so the sequence $z(\bar{t}_n, z_0)$ is bounded. So there exist a subsequence $\{t_n\} \subset \{\bar{t}_n\}$, $t_n \to +\infty$ and a limiting point $\mathfrak{z}$ such that

$$\lim_{n\to+\infty} z(t_n, z_0) = \mathfrak{z}.$$

Also $\mathfrak{z}$ satisfies

$$\frac{\mathrm{d}E(z(t))}{\mathrm{d}t} = 0.$$

and this means that $\mathfrak{z}$ is ω-limit point of $\Upsilon^+(z_0)$. We have $z(t, z_0) \longrightarrow \bar{z} \in \mathcal{M}$ as $t \to \infty$, where $\mathcal{M}$ is the largest invariant set in $\mathcal{K} = \{z(t, z_0)| \frac{\mathrm{d}E(z(t, z_0))}{\mathrm{d}t} = 0\}$ thanks to the LaSalle Invariant set Theorem [7]. With respect to (11) and (20), it obtained that $\frac{\mathrm{d}u}{\mathrm{d}t} = 0$ and $\frac{\mathrm{d}y}{\mathrm{d}t} = 0 \Leftrightarrow \frac{\mathrm{d}E(z(t))}{\mathrm{d}t} = 0$; thus $\mathfrak{z} \in D^*$ by $\mathcal{M} \subset \mathcal{K} \subset D^*$. Therefore, by selecting any desired initial point z_0, we see that the trajectory $z(t, z_0)$ of (11) tends to $\mathfrak{z}$.

Finally we get the below result [10] thanks to Theorems 3 and 4.

Corollary 1 *The globally asymptotically stability of the NN (11) with the unique equilibrium point $z^* = (w^{*T}, y^{*T})$ is easily resulted.*

4 Deep Learning Approach

Ran et al. [18] showed that relations stated in the Eq. (3) are equivalent to

$$-\Delta u - f - \frac{1}{\epsilon}[\varphi - u - \epsilon(\Delta u + f)]_+ = 0, \tag{21}$$

where $\epsilon > 0$ is any constant. This convert relations (3) to a secondary problem becomes that it is easier to solve. The dynamical functional particle method is a useful idea to solve the original equation $F(u(x)) = 0$ where F is a functional, $u : X \to \mathcal{R}^n$. In this method, instead of solving the functional equation, the following dynamic system is solved

$$\ddot{u} + \tau\dot{u} = F(u), \tag{22}$$

where $\dot{u},\ \ddot{u} \to 0$ when t increases, in this way $u(x, t)$ approaches to $u(x)$, the solution of $F(u(x)) = 0$. Now if we put

$$F(u) = -\Delta u - f - \frac{1}{\epsilon}[\varphi - u - \epsilon(\Delta u + f)]_+$$

then in order to get the solution of the OP it is sufficient that solve the dynamical system (22). Of course, instead of the Laplace operator, we have to use the stiffness matrix, where is positive definite and symmetric. Now if we approximate the function u with the help of Bernstein polynomials (or Wavelets) and use the first and second derivative operators (B_1 and B_2 respectively) of these functions, then we will have

$$B_2P + \tau B_1P = F(P). \tag{23}$$

The algorithm approximates $P(x)$ by a neural network $\hat{P}(x, \theta)$, where θ is the stacked vector of the NN parameters. We propose a loss function of the form

$$L_C(\theta) = \frac{1}{N}\sum_{i=1}^{N}(B_2P(x_i) + \tau B_1P = F(P(x_i)))^2 \tag{24}$$

References

1. Bian W, Ma L, Qin S, Xue X (2018) Neural network for nonsmooth pseudoconvex optimization with general convex constraints. Neural Netw 101:1–14. https://doi.org/10.1016/j.neunet.2018.01.008
2. Burman E, Hansbo P, Larson MG, Stenberg R (2017) Galerkin least squares finite element method for the obstacle problem. Comput Methods Appl Mech Eng 313:362–374. https://doi.org/10.1016/j.cma.2016.09.025

3. Calder J, Yezzi A (2019) PDE acceleration: a convergence rate analysis and applications to obstacle problems. Res Math Sci 6(4):35. https://doi.org/10.1007/s40687-019-0197-x
4. Dabaghi J, Martin V, Vohralík M (2020) A posteriori estimates distinguishing the error components and adaptive stopping criteria for numerical approximations of parabolic variational inequalities. Comput Methods Appl Mech Eng 367:113105. https://doi.org/10.1016/j.cma.2020.113105
5. Doukhovni I, Givoli D (1996) Quadratic programming algorithms for obstacle problems. Commun Numer Methods Eng 12(4):249-256. doi.org/10.1002/(SICI)1099-0887(199604)12:4<249::AID-CNM976>3.0.CO;2-6
6. Fuhrer T (2020) First-order least-squares method for the obstacle problem. Numer Math 144(1):55–88. https://doi.org/10.1007/s00211-019-01084-0
7. Hale JK (1969) Ordinary differential equations. Wiley, New York
8. Hoppe RH (1987) Multigrid algorithms for variational inequalities. SIAM J Numer Anal 24(5):1046–1065. https://doi.org/10.1137/0724069
9. Howison S, Dewynne J (1995) The mathematics of financial derivatives. Cambridge University Press, Cambridge
10. Huang X, Lou X, Cui B (2016) A novel neural network for solving convex quadratic programming problems subject to equality and inequality constraints. Neurocomputing 214:23–31. https://doi.org/10.1016/j.neucom.2016.05.032
11. Glowinski R, Oden JT (1985) Numerical methods for nonlinear variational problems
12. Lewis T, Rapp A, Zhang Y (2020) Convergence analysis of symmetric dual-wind discontinuous Galerkin approximation methods for the obstacle problem. J Math Anal Appl 123840. https://doi.org/10.1016/j.jmaa.2020.123840
13. Majava K, Tai XC (2004) A level set method for solving free boundary problems associated with obstacles. Int J Numer Anal Model 1(2):157–171
14. Mansoori A, Effati S (2019) An efficient neurodynamic model to solve nonlinear programming problems with fuzzy parameters. Neurocomputing 334:125–133. https://doi.org/10.1016/j.neucom.2019.01.012
15. Nazemi A, Sabeghi A (2019) A novel gradient-based neural network for solving convex second-order cone constrained variational inequality problems. J Comput Appl Math 347:343–356. https://doi.org/10.1016/j.cam.2018.08.030
16. Peng H, Wang X, Zhai Q, Zhang R (2019) A weak Galerkin finite element method for the elliptic variational inequality. Numer Math Theory Methods Appl 12(3):923–941. https://doi.org/10.4208/nmtma.OA-2018-0124
17. Qin S, Yang X, Xue X, Song J (2016) A one-layer recurrent neural network for pseudoconvex optimization problems with equality and inequality constraints. IEEE Trans Cybern 47(10):3063–3074. https://doi.org/10.1109/TCYB.2016.2567449
18. Ran Q, Cheng X, Abide S (2020) A dynamical method for solving the obstacle problem. Numer Math Theory Methods Appl 13(3):1–19. https://doi.org/10.4208/nmtma.OA-2019-0109
19. Rodrigues JF (1987) Obstacle problems in mathematical physics. Elsevier
20. Tank D, Hopfield JJ (1986) Simple'neural'optimization networks: an A\D converter, signal decision circuit, and a linear programming circuit. IEEE Trans Circ Syst 33(5):533–541. https://doi.org/10.1109/TCS.1986.1085953
21. Xia Y, Wang J, Guo W (2019) Two projection neural networks with reduced model complexity for nonlinear programming. IEEE Trans Neural Netw Learn Syst 31(6):2020–2029. https://doi.org/10.1109/TNNLS.2019.2927639
22. Wang F, Han W, Cheng XL (2010) Discontinuous Galerkin methods for solving elliptic variational inequalities. SIAM J Numer Anal 48(2):708–733. https://doi.org/10.1137/09075891X
23. Yuan D, Li X, Lei C (2012) A fast numerical method for solving a regularized problem associated with obstacle problems. J Korean Math Soc 49(5):893–905. https://doi.org/10.4134/JKMS.2012.49.5.893
24. Zhang Y (2001) Multilevel projection algorithm for solving obstacle problems. Comput Math Appl 41(12):1505–1513. https://doi.org/10.1016/S0898-1221(01)00115-8
25. Zosso D, Osting B, Xia MM, Osher SJ (2017) An efficient primal-dual method for the obstacle problem. J Sci Comput 73(1):416–437. https://doi.org/10.1007/s10915-017-0420-0

Multi-attribute Group Decision Making Through the Development of Dombi Bonferroni Mean Operators Using Dual Hesitant Pythagorean Fuzzy Data

Nayana Deb, Arun Sarkar, and Animesh Biswas

Abstract Dual hesitant Pythagorean fuzzy set has been emerged as a successful fuzzy variant to deal with uncertain and vague information. Bonferroni mean possesses the capacity to deal with interrelated input arguments in multi-attribute group decision-making problems. Moreover, the process of aggregation becomes more supple using a dynamic Dombi parameter. Considering these two aspects, in this paper Bonferroni mean is merged with Dombi operations for dual hesitant Pythagorean fuzzy information processing. At first, Dombi operations on dual hesitant Pythagorean fuzzy sets are introduced. Then, utilizing Bonferroni mean operator with Dombi operations, some new aggregation operators for aggregating dual hesitant Pythagorean fuzzy information, viz., dual hesitant Pythagorean fuzzy Dombi Bonferroni mean, dual hesitant Pythagorean fuzzy Dombi geometric Bonferroni mean and their weighted operators, are proposed. Some special cases and important properties of the proposed operators have been discussed. Further, an approach for the application of the proposed operators in solving real-life group decision-making problem is presented and two illustrated numerical examples have been provided to show its validity.

Keywords Multi-attribute group decision making · Bonferroni mean · Dual hesitant Pythagorean fuzzy set · Dombi operations

1 Introduction

Fuzzy sets [1] are widely applied for processing vague data in different emerging areas, viz., pattern recognitions, expert systems, decision making, etc. Only membership degrees are considered in fuzzy sets. To include non-membership degrees simultaneously with membership degrees, the idea of intuitionistic fuzzy set (IFS) [2] came into account. The sum of membership degree and non-membership degree of it cannot

N. Deb · A. Biswas (✉)
Department of Mathematics, University of Kalyani, Kalyani 741235, India

A. Sarkar
Department of Mathematics, Heramba Chandra College, Kolkata 700029, India

G. Gupta et al. (eds.), *Proceedings of Academia-Industry Consortium for Data Science*, Advances in Intelligent Systems and Computing 1411,
https://doi.org/10.1007/978-981-16-6887-6_17

exceed 1. By extending the idea of IFS, Yager [3] introduced Pythagorean fuzzy (PF) set (PFS). It is characterized by both membership degree and non-membership degree, and the square sum of those degrees belongs to [0, 1]. After inception, PFSs are successfully applied to various practical multi-attribute decision-making (MADM) areas [4–6] to tackle vague data. Zhang and Xu [7] launched PF number (PFN) and proposed TOPSIS model to solve MADM problems. Peng and Yang [8] defined some basic algebraic operations of PFNs. Based on the operations on PFNs, different aggregation operators are developed. Using Einstein operations, Garg [9] introduced a series of aggregation operators to handle multicriteria decision-making (MCDM) problems. Wei [10] proposed some PF interaction aggregation operators. Utilizing Bonferroni mean (BM) [11], Zhang et al. [12] introduced generalized PF BM operator in multi-attribute group decision making (MAGDM). Combining Hamacher operations and geometric BM (GBM), Gao [13] generated PF prioritized aggregation operators for MCDM. Applying power aggregation operators and Maclaurin symmetric mean, Wei and Lu [14, 15] developed PF MADM methods. Liang et al. [16] introduced PF BM operator and weighted PF BM operator for MCDM under PF environments. Li et al. [17] proposed Hamy mean-based PF aggregation operators and utilized them in MAGDM relating to supplier selection. Some prioritized PF aggregation operators were developed by Khan et al. [18]. Further, based on Einstein operation, Rahman (2019) introduced a series of generalized induced PF aggregation operators. Rahman and Ali [19] developed PF Einstein hybrid geometric operator for solving PF decision-making problems. Recently, combining power aggregation operator and Schweizer and Sklar t-conorm and t-norm, Biswas and Deb [4] proposed several aggregation operators, viz., PF Schweizer and Sklar power average and weighted average along with their geometric operators.

Sometimes, it may seem difficult to estimate the membership value and non-membership values of a PFS by a single crisp value. To overcome such situations, Wei and Lu [20] developed dual hesitant PF (DHPF) sets (DHPFSs) by combining dual hesitant fuzzy sets [21] with PFS [3]. This helps the decision makers (DMs) in providing their evaluation values more flexibly choosing several possible crisp numbers in [0,1] in the form of membership degree and non-membership degree. DHPFSs extend the concept of various existing classical sets, viz., hesitant fuzzy sets, IFSs, PFSs, etc. Wei and Lu [20] also studied the basic operations and properties of DHPF elements (DHPFEs) and derived several DHPF aggregation operators. Tang and Wei [22] proposed MCDM method under DHPF environments. Recently, Tang et al. [23] utilized Heronian mean to develop new DHPF aggregation operators. Furthermore, Lu et al. [24] presented a DHPF bidirectional projection model for MADM contexts. To provide the DMs more scopes in expressing their evaluation information using DHPF data is considered and investigated in this paper.

Nowadays, interrelationship among aggregating parameters is usually needed to consider due to increase of complexity in MADM problems. BM introduced by Bonferroni [11] has the ability to tackle the interrelationship among attribute values. In hesitant fuzzy context, Zhu and Xu [25] investigated BM operator. Further, Zhu et al. [26] introduced GBM operator in hesitant fuzzy environment. Moreover, it is being studied by researchers on PFSs [12, 13, 16] to solve MADM problems. Keeping the benefit of the BM operator in mind, it is utilized in this paper, for aggregating the DHPF information.

In addition, Dombi operations [27] have now become an increasing demand in recent days due to its ability to make the decision process more flexible with the involvement of Dombi parameters. Dombi t-conorm and Dombi t-norm have the characteristic of Archimedean t-conorm and t-norm. Thus, Dombi operations have been successfully applied to many decision-making environments, viz., hesitant fuzzy [28], intuitionistic fuzzy [29], PF [30], etc.

From the best of authors' knowledge, the combination of BM operator with Dombi operations has not been introduced in DHPF environment. Employing the merits of both Dombi operations and BM, for developing the proposed methodology to solve MAGDM problems under DHPF environment, would be beneficial for the following reasons:

- Under DHPF environment, the DMs will have the wider range for expressing their evaluation values, and moreover, they can include their hesitancy along with their decision inputs.
- It would be very useful to consider BM as an aggregation operator for considering interrelationship among attributes.
- The presence of Dombi operations in the aggregation function would help decision-making process more flexible using parameters.

From this motivation, the main goal of this article is to generate several DHPF aggregation operators, viz., DHPF Dombi BM (DHPFDBM) operator, DHPF weighted Dombi BM (DHPFWDBM) operator, DHPF Dombi GBM (DHPFDGBM) operator, DHPF Dombi weighted GBM (DHPFDWGBM) operator for combining DHPF data. Further, a MAGDM approach based on these operators has been presented by aggregating DHPF information. Two illustrative examples are solved and compared with some existing approaches in order to validate the proposed method.

2 Preliminaries

In this section, some primary knowledge related to DHPFSs [20] is briefly discussed.

Definition 1 [3]. A PFS P on the fixed set X is defined as

$$\{\langle x, \mu_p(x), \nu_p(x)\rangle | x \in X\},$$

where $\mu_p : X \to [0, 1]$ and $\nu_p : X \to [0, 1]$ represent the membership grade and non-membership grade of x belongs to X in P satisfying $0 \le (\mu_p(x))^2 + (\nu_p(x))^2 \le 1$.

In addition, $\pi_p(x) = \sqrt{1 - (\mu_p(x))^2 - (\nu_p(x))^2}$ defines the degree of hesitation.

Definition 2 [20]. A DHPFS K on a fixed set X is described as

$$K = (\langle x, h_K(x), g_K(x)\rangle | x \in X),$$

in which $h_K(x)$ and $g_K(x)$ are two sets of real numbers in [0, 1], representing the possible membership and non-membership degrees, respectively, of element $x \in X$ in K satisfying: $0 \leq \gamma, \delta \leq 1$ and $0 \leq \left(\max_{\gamma \in h_K(x)}\{\gamma\}\right)^2 + \left(\max_{\delta \in g_K(x)}\{\delta\}\right)^2 \leq 1$, $\gamma \in h_K(x)$, $\delta \in g_K(x)$ for all $x \in X$.

In short, the pair $\kappa = (h_k(x), g_k(x))$ is called as DHPFE and denoted by $\kappa = (h, g)$.

To compare DHPFEs, score and accuracy functions are defined by Wei and Lu [20].

Definition 3 [20]. The score function $S(\kappa)$ of a DHPFE $\kappa = (h, g)$ is given by

$$S(\kappa) = \left((1/l_h)\sum\nolimits_{\gamma \in h}\gamma^2 - (1/l_g)\sum\nolimits_{\delta \in g}\delta^2\right) \tag{1}$$

The accuracy function of κ, indicated by $A(\kappa)$, is given by

$$A(\kappa) = \left((1/l_h)\sum\nolimits_{\gamma \in h}\gamma^2 + (1/l_g)\sum\nolimits_{\delta \in g}\delta^2\right), \tag{2}$$

where l_h and l_g are the number of elements in h and g, respectively.

Let $\kappa_i = (h_i, g_i)$, $(i = 1, 2)$ be any two DHPFEs, then the ordering of DHPFEs is done as follows:

1. If $S(\kappa_1) > S(\kappa_2)$, then κ_1 is superior to κ_2, denoted by $\kappa_1 \succ \kappa_2$.
2. If $S(\kappa_1) = S(\kappa_2)$, then
   - $A(\kappa_1) > A(\kappa_2)$, implies $\kappa_1 \succ \kappa_2$.
   - If $A(\kappa_1) = A(\kappa_2)$, implies κ_1 is equivalent to κ_2, denoted by $\kappa_1 \sim \kappa_2$.

The BM is defined by Bonferroni [11] as presented in the following:

Definition 4 [11]. Let a_i $(i = 1, 2, \ldots, n)$ be a set of nonnegative real numbers and $p, q \geq 0$. The BM of the collection $(a_1, a_2, \ldots., a_n)$ is defined by

$$BM^{p,q}(a_1, a_2, \ldots, a_n) = \left(\frac{1}{n(n-1)}\sum_{\substack{i,j=1 \\ i \neq j}}^{n} a_i^p a_j^q\right)^{\frac{1}{p+q}} \tag{3}$$

Definition 5 [31]. Let $a_i (i = 1, 2, \ldots, n)$ be a set of nonnegative real numbers with $p, q > 0$. The GBM of the collection $(a_1, a_2, \ldots., a_n)$ is given by

$$GBM^{p,q}(a_1, a_2, \ldots, a_n) = \frac{1}{p+q}\left(\prod_{\substack{i,j=1\\ i\neq j}}^{n}\left((pa_i) + \left(qa_j\right)\right)\right)^{\frac{1}{n(n-1)}} \tag{4}$$

Dombi [27] proposed Dombi t-conorms and t-norms operations, which are special types of Archimedean t-conorms and t-norms, respectively, having a variable parameter.

Definition 6 [27]. Assume that $\varphi > 0$ and $u, v \in [0, 1]$. Then, Dombi t-conorm $U_D(u, v)$ and t-norm $I_D(u, v)$ are presented as

$$U_D(u, v) = 1 - \frac{1}{1 + \left(\left(\frac{u}{1-u}\right)^{\varphi} + \left(\frac{v}{1-v}\right)^{\varphi}\right)^{1/\varphi}}$$

$$I_D(u,\ v) = \frac{1}{1 + \left(\left(\frac{1-u}{u}\right)^{\varphi} + \left(\frac{1-v}{v}\right)^{\varphi}\right)^{1/\varphi}}$$

.

3 Operations of DHPFEs Based on Dombi t-conorm and t-norm

Definition 7 Consider κ_1, κ_2 and κ any three DHPFEs. Then, the operational rules on DHPFEs using Dombi t-norm and t-conorm can be defined as: ($\varphi > 0$)

1. $\kappa_1 \oplus_D \kappa_2$

$$= \left(\bigcup_{\substack{\gamma_i \in h_i\\ i=1,2}}\left\{\sqrt{1 - \frac{1}{1 + \left(\left(\frac{(\gamma_1)^2}{1-(\gamma_1)^2}\right)^{\varphi} + \left(\frac{(\gamma_2)^2}{1-(\gamma_2)^2}\right)^{\varphi}\right)^{\frac{1}{\varphi}}}}\right\}, \bigcup_{\substack{\delta_i \in g_i\\ i=1,2}}\left\{\frac{1}{\sqrt{1 + \left(\left(\frac{1-(\delta_1)^2}{(\delta_1)^2}\right)^{\varphi} + \left(\frac{1-(\delta_2)^2}{(\delta_2)^2}\right)^{\varphi}\right)^{\frac{1}{\varphi}}}}\right\}\right)$$

2. $\kappa_1 \otimes_D \kappa_2$

$$= \left(\bigcup_{\substack{\gamma_i \in h_i\\ i=1,2}}\left\{\frac{1}{\sqrt{1 + \left(\left(\frac{1-(\gamma_1)^2}{(\gamma_1)^2}\right)^{\varphi} + \left(\frac{1-(\gamma_2)^2}{(\gamma_2)^2}\right)^{\varphi}\right)^{\frac{1}{\varphi}}}}\right\}, \bigcup_{\substack{\delta_i \in g_i\\ i=1,2}}\left\{\sqrt{1 - \frac{1}{1 + \left(\left(\frac{(\delta_1)^2}{1-(\delta_1)^2}\right)^{\varphi} + \left(\frac{(\delta_2)^2}{1-(\delta_2)^2}\right)^{\varphi}\right)^{\frac{1}{\varphi}}}}\right\}\right);$$

3. $$\lambda\kappa = \left(\bigcup_{\gamma \in h}\left\{\sqrt{1 - \frac{1}{1 + \left(\lambda\left(\frac{(\gamma)^2}{1-(\gamma)^2}\right)^{\varphi}\right)^{\frac{1}{\varphi}}}}\right\}, \bigcup_{\delta \in g}\left\{\frac{1}{\sqrt{1 + \left(\lambda\left(\frac{1-(\delta)^2}{(\delta)^2}\right)^{\varphi}\right)^{\frac{1}{\varphi}}}}\right\}\right), (\lambda > 0)$$

4. $\kappa^{\lambda} = \left(\bigcup_{\gamma \in h}\left\{\frac{1}{\sqrt{1+\left(\lambda\left(\frac{1-(\gamma)^2}{(\gamma)^2}\right)^{\varphi}\right)^{\frac{1}{\varphi}}}}\right\}, \bigcup_{\delta \in g}\left\{\sqrt{1-\frac{1}{1+\left(\lambda\left(\frac{(\delta)^2}{1-(\delta)^2}\right)^{\varphi}\right)^{\frac{1}{\varphi}}}}\right\}\right), (\lambda > 0).$

4 Development of DHPF Dombi BM Aggregation Operators

Dombi t-conorm and t-norm-based DHPF BM averaging and geometric aggregation operators are developed in this section.

Definition 8 Let $\kappa_i = (h_i, g_i)$ $(i = 1, 2, \ldots, n)$ be a collection of DHPFEs. Also, let $p, q \geq 0$ be any two numbers. Then, DHPFDBM operator is defined as

$$\mathrm{DHPFDBM}^{p,q}(\kappa_1, \kappa_2, \ldots, \kappa_n) = \left(\frac{1}{n(n-1)} \oplus_{D_{\substack{i,j=1\\ i\neq j}}}^{n} \left(\kappa_i^{p} \otimes_D \kappa_j^{q}\right)\right)^{\frac{1}{p+q}}$$

The $\mathrm{DHPFDBM}^{p,q}(\kappa_1, \kappa_2, \ldots, \kappa_n)$ is called DHPFDBM operator.

Theorem 1 Let $\kappa_i = (h_i, g_i)$, $(i = 1, 2, \ldots, n)$ be a set of DHPFEs with $p, q \geq 0$. Then, evaluation value based on DHPFDBM operator is still generated as a DHPFE and also

$$\mathrm{DHPFDBM}^{p,q}(\kappa_1, \kappa_2, \ldots, \kappa_n) = \left(\frac{1}{n(n-1)} \oplus_{D_{\substack{i,j=1\\ i\neq j}}}^{n} \left(\kappa_i^{p} \otimes_D \kappa_j^{q}\right)\right)^{\frac{1}{p+q}}$$

$$= \left(\bigcup_{\substack{\gamma_i \in h_i,\\ \gamma_j \in h_j}}\left\{\sqrt{1\Big/\left(1+\left(\frac{n(n-1)}{p+q}\left(1\Big/\left(\sum_{\substack{i,j=1\\ i\neq j}}^{n}\left(1\Big/\left(p\left(\frac{1-\gamma_i^2}{\gamma_i^2}\right)^{\varphi}+q\left(\frac{1-\gamma_j^2}{\gamma_j^2}\right)^{\varphi}\right)\right)\right)\right)\right)^{1/\varphi}\right)}\right\},\right.$$

$$\left.\bigcup_{\substack{\delta_i \in g_i,\\ \delta_j \in g_j}}\left\{\sqrt{1-\left(1\Big/\left(1+\left(\frac{n(n-1)}{p+q}\left(1\Big/\sum_{\substack{i,j=1\\ i\neq j}}^{n}\left(1\Big/\left(p\left(\frac{\delta_i^2}{1-\delta_i^2}\right)^{\varphi}+q\left(\frac{\delta_j^2}{1-\delta_j^2}\right)^{\varphi}\right)\right)\right)\right)^{\frac{1}{\varphi}}\right)\right)}\right\}\right) \tag{5}$$

.

Proof From Definition (7), it is found that

$$\kappa_i^{p} = \left(\bigcup_{\gamma_i \in h_i}\left\{1\Big/\sqrt{1+\left(p\left(\frac{1-\gamma_i^2}{\gamma_i^2}\right)^{\varphi}\right)^{\frac{1}{\varphi}}}\right\}, \bigcup_{\delta_i \in g_i}\left\{\sqrt{1-\left(1\Big/\left(1+\left(p\left(\frac{\delta_j^2}{1-\delta_j^2}\right)^{\varphi}\right)^{\frac{1}{\varphi}}\right)\right)}\right\}\right).$$

$$\kappa_j^q = \left(\underset{\gamma_j \in h_j}{\cup} \left\{ 1/\sqrt{1+\left(q\left(\frac{1-\gamma_j^2}{\gamma_j^2}\right)\right)^{\varphi}}^{\frac{1}{\varphi}} \right\}, \underset{\delta_j \in g_j}{\cup} \left\{ \sqrt{1-\left(1/\left(1+\left(q\left(\frac{\delta_j^2}{1-\delta_j^2}\right)\right)^{\varphi}\right)^{\frac{1}{\varphi}}\right)} \right\} \right).$$

And then, $\kappa_i^p \otimes_D \kappa_j^q = \left(\bigcup_{\substack{\gamma_i \in h_i, \\ \gamma_j \in h_j}} \left\{ 1/\sqrt{1+\left(p\left(\frac{1-\gamma_i^2}{\gamma_i^2}\right)^{\varphi} + q\left(\frac{1-\gamma_j^2}{\gamma_j^2}\right)^{\varphi}\right)^{1/\varphi}} \right\}, \right.$

$\left. \bigcup_{\substack{\delta_i \in g_i, \\ \delta_j \in g_j}} \left\{ \sqrt{1-1/\left(1+\left(p\left(\frac{\delta_i^2}{1-\delta_i^2}\right)^{\varphi} + q\left(\frac{\delta_j^2}{1-\delta_j^2}\right)^{\varphi}\right)^{1/\varphi}\right)} \right\} \right).$

In the following, using mathematical induction at first it is proved that

$$\oplus^n_{D\,\substack{i,j=1 \\ i \neq j}} \left(\kappa_i^p \otimes_D \kappa_j^q\right) = \left(\bigcup_{\substack{\gamma_i \in h_i, \\ \gamma_j \in h_j}} \left\{ \sqrt{1-\left(1/\left(1+\left(\sum_{\substack{i,j=1 \\ i \neq j}}^{n} \left(1/\left(p\left(\frac{1-\gamma_i^2}{\gamma_i^2}\right)^{\varphi} + q\left(\frac{1-\gamma_j^2}{\gamma_j^2}\right)^{\varphi}\right)\right)\right)^{1/\varphi}\right)\right)} \right\},_T \bigcup_{\substack{\delta_i \in g_i, \\ \delta_j \in g_j}} \left\{ \sqrt{1/\left(1+\left(\sum_{\substack{i,j=1 \\ i \neq j}}^{n} \left(1/\left(p\left(\frac{\delta_i^2}{1-\delta_i^2}\right)^{\varphi} + q\left(\frac{\delta_j^2}{1-\delta_j^2}\right)^{\varphi}\right)\right)\right)^{1/\varphi}\right)} \right\} \right) \tag{6}$$

It is obvious for $n = 2$.
Now, assume that (6) is true for $n = m$, i.e.,

$$\oplus^m_{D\,\substack{i,j=1 \\ i \neq j}} \left(\kappa_i^p \otimes_D \kappa_j^q\right) = \left(\bigcup_{\substack{\gamma_i \in h_i, \\ \gamma_j \in h_j}} \left\{ \sqrt{1-\left(1/\left(1+\left(\sum_{\substack{i,j=1 \\ i \neq j}}^{m} \left(1/\left(p\left(\frac{1-\gamma_i^2}{\gamma_i^2}\right)^{\varphi} + q\left(\frac{1-\gamma_j^2}{\gamma_j^2}\right)^{\varphi}\right)\right)\right)^{1/\varphi}\right)\right)} \right\}, \bigcup_{\substack{\delta_i \in g_i, \\ \delta_j \in g_j}} \left\{ \sqrt{1/\left(1+\left(\sum_{\substack{i,j=1 \\ i \neq j}}^{m} \left(1/\left(p\left(\frac{\delta_i^2}{1-\delta_i^2}\right)^{\varphi} + q\left(\frac{\delta_j^2}{1-\delta_j^2}\right)^{\varphi}\right)\right)\right)^{1/\varphi}\right)} \right\} \right) \tag{7}$$

Now when $n = m + 1$,

$$\oplus^{m+1}_{D\,\substack{i,j=1 \\ i \neq j}} \left(\kappa_i^p \otimes_D \kappa_j^q\right)$$

$$= \oplus_{D_{\substack{i,j=1\\i\neq j}}}^{m}\left(\kappa_i^{p}\otimes_D\kappa_j^{q}\right)\oplus_D\left(\oplus_{Di=1}^{m}\left(\kappa_i^{p}\otimes_D\kappa_{m+1}^{q}\right)\right)\oplus_D\left(\oplus_{Dj=1}^{m}\left(\kappa_{m+1}^{p}\otimes_D\kappa_j^{q}\right)\right) \tag{8}$$

Again, $\left(\oplus_{D_{i=1}^{m}}\left(\kappa_i^{p}\otimes_D\kappa_{m+1}^{q}\right)\right)$

$$= \left(\bigcup_{\gamma_i\in h_i}\left\{\sqrt{1-1/\left(1+\left(\sum_{i=1}^{m}\frac{1}{p\left(\frac{1-(\gamma_i)^2}{(\gamma_i)^2}\right)^{\varphi}+q\left(\frac{1-(\gamma_{m+1})^2}{(\gamma_{m+1})^2}\right)^{\varphi}}\right)^{1/\varphi}\right)}\right\},\right.$$

$$\left.\bigcup_{\delta_i\in g_i}\left\{\sqrt{1/\left(1+\left(\sum_{i=1}^{m}\frac{1}{p\left(\frac{(\delta_i)^2}{1-(\delta_i)^2}\right)^{\varphi}+q\left(\frac{(\delta_{m+1})^2}{1-(\delta_{m+1})^2}\right)^{\varphi}}\right)^{1/\varphi}\right)}\right\}\right) \tag{9}$$

Similarly,

$$\left(\oplus_{Dj=1}^{m}\left(\kappa_{m+1}^{p}\otimes_D\kappa_j^{q}\right)\right)$$

$$= \left(\bigcup_{\gamma_j\in h_j}\left\{\sqrt{1-\left(1/\left(1+\left(\sum_{j=1}^{m}\frac{1}{p\left(\frac{1-(\gamma_{m+1})^2}{(\gamma_{m+1})^2}\right)^{\varphi}+q\left(\frac{1-(\gamma_j)^2}{(\gamma_j)^2}\right)^{\varphi}}\right)^{1/\varphi}\right)\right)}\right\},\right. \tag{10}$$

$$\left.\bigcup_{\delta_j\in g_j}\left\{\sqrt{1/\left(1+\left(\sum_{j=1}^{m}\frac{1}{p\left(\frac{(\delta_{m+1})^2}{1-(\delta_{m+1})^2}\right)^{\varphi}+q\left(\frac{(\delta_j)^2}{1-(\delta_j)^2}\right)^{\varphi}}\right)^{1/\varphi}\right)}\right\}\right)$$

By (7), (9) and (10), calculate (8) as

$$\oplus_{D_{\substack{i,j=1\\i\neq j}}}^{m+1}\left(\kappa_i^{p}\otimes_D\kappa_j^{q}\right)$$

$$= \left(\bigcup_{\substack{\gamma_i\in h_i,\\ \gamma_j\in h_j}}\left\{\sqrt{1-\left(1/\left(1+\left(\sum_{\substack{i,j=1\\i\neq j}}^{m+1}\left(1/\left(p\left(\frac{1-\gamma_i^2}{\gamma_i^2}\right)^{\varphi}+q\left(\frac{1-\gamma_j^2}{\gamma_j^2}\right)^{\varphi}\right)\right)\right)^{1/\varphi}\right)\right)}\right\},\right.$$

$$\left.\bigcup_{\substack{\delta_i\in g_i,\\ \delta_j\in g_j}}\left\{\sqrt{1/\left(1+\left(\sum_{\substack{i,j=1\\i\neq j}}^{m+1}\left(1/\left(p\left(\frac{\delta_i^2}{1-\delta_i^2}\right)^{\varphi}+q\left(\frac{\delta_j^2}{1-\delta_j^2}\right)^{\varphi}\right)\right)\right)^{1/\varphi}\right)}\right\}\right) \tag{11}$$

i.e., (6) is true for all n.

Hence, $\mathrm{DHPFDBM}^{p,q}(\kappa_1,\ \kappa_2, \ldots, \kappa_n) = \left(\frac{1}{n(n-1)} \oplus^{n}_{D_{i,j=1} \atop i \neq j} \left(\kappa_i^p \otimes_D \kappa_j^q\right)\right)^{\frac{1}{p+q}}$

$$\left(\bigcup_{\gamma_i \in h_i, \atop \gamma_j \in h_j} \left\{\sqrt{1/\left(1+\left(\frac{n(n-1)}{p+q}\left(1/\left(\sum_{i,j=1 \atop i \neq j}^{n}\left(1/\left(p\left(\frac{1-\gamma_i^2}{\gamma_i^2}\right)^{\varphi}+q\left(\frac{1-\gamma_j^2}{\gamma_j^2}\right)^{\varphi}\right)\right)\right)\right)\right)^{\frac{1}{\varphi}}\right)}\right\}\right.$$

,

$$\left.\bigcup_{\delta_i \in g_i, \atop \delta_j \in g_j} \left\{\sqrt{1-\left(1/\left(1+\left(\frac{n(n-1)}{p+q}\left(1/\sum_{i,j=1 \atop i \neq j}^{n}\left(1/\left(p\left(\frac{\delta_i^2}{1-\delta_i^2}\right)^{\varphi}+q\left(\frac{\delta_j^2}{1-\delta_j^2}\right)^{\varphi}\right)\right)\right)\right)^{\frac{1}{\varphi}}\right)\right)}\right\}\right)$$

Thus, the theorem follows.

In regard to the parameters $p$and$q$, some special cases for $DHPFDBM^{p,q}$ operator are studied below.

(i) When $q \to 0$; $\varphi > 0$, then

$$DHPFDBM^{p,0}(\kappa_1, \kappa_2, \ldots, \kappa_n)$$

$$= \left(\bigcup_{\gamma_i \in h_i} \left\{\sqrt{1/\left(1+\left(\frac{n(n-1)}{p}\left(1/\left(\sum_{i=1}^{n}\left(1/\left(p\left(\frac{1-\gamma_i^2}{\gamma_i^2}\right)^{\varphi}\right)\right)\right)\right)\right)^{1/\varphi}\right)}\right\},\right.$$

$$\left.\bigcup_{\delta_i \in g_i} \left\{\sqrt{1-\left(1/\left(1+\left(\frac{n(n-1)}{p}\left(1/\sum_{i=1}^{n}\left(1/\left(p\left(\frac{\delta_i^2}{1-\delta_i^2}\right)^{\varphi}\right)\right)\right)\right)^{1/\varphi}\right)\right)}\right\}\right)$$

(ii) When $p \to 0$ and $\varphi > 0$, then

$\mathrm{DHPFDBM}^{0,q}(\kappa_1, \kappa_2, \ldots, \kappa_n)$

$$= \left(\bigcup_{\gamma_j \in h_j} \left\{\sqrt{1/\left(1+\left(\frac{n(n-1)}{q}\left(1/\left(\sum_{j=1}^{n}\left(1/\left(q\left(\frac{1-\gamma_j^2}{\gamma_j^2}\right)^{\varphi}\right)\right)\right)\right)\right)^{1/\varphi}\right)}\right\},\right.$$

$$\left.\bigcup_{\delta_j \in g_j} \left\{\sqrt{1-\left(1/\left(1+\left(\frac{n(n-1)}{q}\left(1/\sum_{j=1}^{n}\left(1/\left(q\left(\frac{\delta_j^2}{1-\delta_j^2}\right)^{\varphi}\right)\right)\right)\right)^{1/\varphi}\right)\right)}\right\}\right)$$

(iii) When $p = q = 1$ and $\varphi > 0$, then

$\mathrm{DHPFDBM}^{1,1}(\kappa_1, \kappa_2, \ldots, \kappa_n)$

$$=\left(\bigcup_{\substack{\gamma_i\in h_i,\\ \gamma_j\in h_j}}\left\{\sqrt{1/\left(1+\left(\frac{n(n-1)}{2}\left(1/\left(\sum_{\substack{i,j=1\\ i\neq j}}^{n}\left(1/\left(\left(\frac{1-\gamma_i^2}{\gamma_i^2}\right)^{\varphi}+\left(\frac{1-\gamma_j^2}{\gamma_j^2}\right)^{\varphi}\right)\right)\right)\right)\right)^{1/\varphi}\right)}\right\},\right.$$

$$\left.\bigcup_{\substack{\delta_i\in g_i,\\ \delta_j\in g_j}}\left\{\sqrt{1-\left(1/\left(1+\left(\frac{n(n-1)}{2}\left(1/\sum_{\substack{i,j=1\\ i\neq j}}^{n}\left(1/\left(\left(\frac{\delta_i^2}{1-\delta_i^2}\right)^{\varphi}+\left(\frac{\delta_j^2}{1-\delta_j^2}\right)^{\varphi}\right)\right)\right)\right)^{1/\varphi}\right)\right)}\right\}\right)$$

Theorem 2 (Idempotency)
Let $\kappa_i = (h_i, g_i)$ $(i = 1, 2, \ldots, n)$ be a set of DHPFEs. If $\kappa_i = \kappa = (h, g)$ for all $i = 1, 2, \ldots, n$, then it is found that

$$\text{DHPFDBM}^{p,q}(\kappa_1, \kappa_2, \ldots, \kappa_n) = \kappa.$$

Proof From the given Definition (8), $\text{DHPFDBM}^{p,q}(\kappa_1, \kappa_2, \ldots, \kappa_n) =$

$$\left(\bigcup_{\substack{\gamma_i\in h_i,\\ \gamma_j\in h_j}}\left\{\sqrt{1/\left(1+\left(\frac{n(n-1)}{p+q}\left(1/\left(\sum_{\substack{i,j=1\\ i\neq j}}^{n}\left(1/\left(p\left(\frac{1-\gamma_i^2}{\gamma_i^2}\right)^{\varphi}+q\left(\frac{1-\gamma_j^2}{\gamma_j^2}\right)^{\varphi}\right)\right)\right)\right)\right)^{\frac{1}{\varphi}}\right)}\right\},\right.$$

$$\left.\bigcup_{\substack{\delta_i\in g_i,\\ \delta_j\in g_j}}\left\{\sqrt{1-\left(1/\left(1+\left(\frac{n(n-1)}{p+q}\left(1/\sum_{\substack{i,j=1\\ i\neq j}}^{n}\left(1/\left(p\left(\frac{\delta_i^2}{1-\delta_i^2}\right)^{\varphi}+q\left(\frac{\delta_j^2}{1-\delta_j^2}\right)^{\varphi}\right)\right)\right)\right)^{\frac{1}{\varphi}}\right)\right)}\right\}\right).$$

Since $\kappa_i = \kappa$ $(i = 1, 2, \ldots, n)$, $\text{DHPFDBM}^{p,q}(\kappa_1, \kappa_2, \ldots, \kappa_n) =$

$$\left(\bigcup_{\gamma\in h}\left\{\sqrt{1/\left(1+\left(\frac{n(n-1)}{p+q}\left(1/\left(\frac{n(n-1)}{(p+q)\left(\frac{1-\gamma^2}{\gamma^2}\right)^{\varphi}}\right)\right)\right)^{1/\varphi}\right)}\right\},\right.$$

$$\left.\bigcup_{\delta\in g}\left\{\sqrt{1-\left(1/\left(1+\left(\frac{n(n-1)}{p+q}\left(1/\left(\frac{n(n-1)}{(p+q)\left(\frac{\delta^2}{1-\delta^2}\right)^{\varphi}}\right)\right)\right)^{1/\varphi}\right)\right)}\right\}\right)$$

$$=\left(\bigcup_{\gamma\in h}\{\gamma\}, \bigcup_{\delta\in g}\{\delta\}\right) = k.$$

Hence the proof.

Theorem 3 (Monotonicity)

Let $k_i = (h_i, g_i)(i = 1, 2, \ldots, n)$ and $k_i^{'} = \left(h_i^{'}, g_i^{'}\right)(i = 1, 2, \ldots, n)$ be two sets of DHPFEs; if $\gamma_i \le \gamma_i^{'}$ and $\delta_i^{'} \le \delta_i$ for all $(i = 1, 2, \ldots, n)$, then

$$\mathrm{DHPFDBM}^{p,q}(k_1, k_2, \ldots, k_n) \le \mathrm{DHPFDBM}^{p,q}\left(k_1^{'}, k_2^{'}, \ldots, k_n^{'}\right).$$

Proof Since $\gamma_i \le \gamma_i^{'}$ for all $i = 1, 2, \ldots, n$, $(\gamma_i)^2 \le \left(\gamma_i^{'}\right)^2$,

Then, $p\left(\frac{1-\gamma_i^2}{\gamma_i^2}\right)^k \ge p\left(\frac{1-\gamma_i^{'2}}{\gamma_i^{'2}}\right)^k$ and $q\left(\frac{1-\gamma_j^2}{\gamma_j^2}\right)^k \ge q\left(\frac{1-\gamma_j^{'2}}{\gamma_j^{'2}}\right)^k$.

So, $\left(p\left(\frac{1-\gamma_i^2}{\gamma_i^2}\right)^\varphi + q\left(\frac{1-\gamma_j^2}{\gamma_j^2}\right)^\varphi\right) \ge \left(p\left(\frac{1-\gamma_i^{'2}}{\gamma_i^{'2}}\right)^\varphi + q\left(\frac{1-\gamma_j^{'2}}{\gamma_j^{'2}}\right)^\varphi\right)$

$$\Rightarrow 1/\left(1+\left(\frac{n(n-1)}{p+q}\left(1/\left(\sum_{\substack{i,j=1\\ i\ne j}}^{n}\left(1/\left(p\left(\frac{1-\gamma_i^2}{\gamma_i^2}\right)^\varphi + q\left(\frac{1-\gamma_j^2}{\gamma_j^2}\right)^\varphi\right)\right)\right)\right)\right)^{\frac{1}{\varphi}}\right)$$

$$\le 1/\left(1+\left(\frac{n(n-1)}{p+q}\left(1/\left(\sum_{\substack{i,j=1\\ i\ne j}}^{n}\left(1/\left(p\left(\frac{1-\gamma_i^{'2}}{\gamma_i^{'2}}\right)^\varphi + q\left(\frac{1-\gamma_j^{'2}}{\gamma_j^{'2}}\right)^\varphi\right)\right)\right)\right)\right)^{\frac{1}{\varphi}}\right) \tag{12}$$

Again, $\delta_i^{'} \le \delta_i$ for all $i = 1, 2, \ldots, n$. So, $p\left(\frac{\delta_i^{'2}}{1-\delta_i^{'2}}\right)^\varphi \le p\left(\frac{\delta_i^2}{1-\delta_i^2}\right)^\varphi$ and $q\left(\frac{\delta_j^{'2}}{1-\delta_j^{'2}}\right)^\varphi \le q\left(\frac{\delta_j^2}{1-\delta_j^2}\right)^\varphi$ for all $j = 1, 2, \ldots, n$.

Thus, $p\left(\frac{\delta_i^{'2}}{1-\delta_i^{'2}}\right)^\varphi + q\left(\frac{\delta_j^{'2}}{1-\delta_j^{'2}}\right)^\varphi \le p\left(\frac{\delta_i^2}{1-\delta_i^2}\right)^\varphi + q\left(\frac{\delta_j^2}{1-\delta_j^2}\right)^\varphi$

$$\Rightarrow \sum_{\substack{i,j=1\\ i\ne j}}^{n}\left(1/\left(p\left(\frac{\delta_i^{'2}}{1-\delta_i^{'2}}\right)^\varphi + q\left(\frac{\delta_j^{'2}}{1-\delta_j^{'2}}\right)^\varphi\right)\right) \ge \sum_{\substack{i,j=1\\ i\ne j}}^{n}\left(1/\left(p\left(\frac{\delta_i^2}{1-\delta_i^2}\right)^\varphi + q\left(\frac{\delta_j^2}{1-\delta_j^2}\right)^\varphi\right)\right)$$

$$\Rightarrow 1-\left(1/\left(1+\left(\frac{n(n-1)}{p+q}\left(1/\sum_{\substack{i,j=1\\ i\ne j}}^{n}\left(1/\left(p\left(\frac{\delta_i^{'2}}{1-\delta_i^{'2}}\right)^\varphi + q\left(\frac{\delta_j^{'2}}{1-\delta_j^{'2}}\right)^\varphi\right)\right)\right)\right)^{\frac{1}{\varphi}}\right)\right)$$

$$\le 1-\left(1/\left(1+\left(\frac{n(n-1)}{p+q}\left(1/\sum_{\substack{i,j=1\\ i\ne j}}^{n}\left(1/\left(p\left(\frac{\delta_i^2}{1-\delta_i^2}\right)^\varphi + q\left(\frac{\delta_j^2}{1-\delta_j^2}\right)^\varphi\right)\right)\right)\right)^{\frac{1}{\varphi}}\right)\right) \tag{13}$$

By Definition (3) and Eqs. (12) and (13), theorem is proved.
Hence the theorem.

Theorem 4 (Boundary)
Let $k_i = (h_i, g_i)$, $(i = 1, 2, \ldots, n)$ be a collection of DHPFEs, and let $\gamma^* = \max\{\gamma_{i\,\max}\}$, where $\gamma_{i\,\max} = \max_{\gamma_i \in h_i}\{\gamma_i\}$ for all $i = 1, 2, \ldots, n$.
$\gamma_\# = \min\{\gamma_{i\,\min}\}$, where $\gamma_{i\,\min} = \min_{\gamma_i \in h_i}\{\gamma_i\}$ for all $i = 1, 2, \ldots, n$.

Again, let $\delta^* = \max\{\delta_{i\max}\}$, *where* $\delta_{i\max} = \max_{\delta_i \in g_i}\{\delta_i\}$ *for all* $i = 1, 2, \ldots, n$. $\delta_{\#} = \min\{\delta_{i\min}\}$, *where* $\delta_{i\min} = \min_{\delta_i \in g_i}\{\delta_i\}$ *for all* $i = 1, 2, \ldots, n$. *Now, if* $k_- = (\gamma_{\#}, \delta^*)$ *and* $k_+ = (\gamma^*, \delta_{\#})$, *then*

$$k_- \leq \text{DHPFDBM}^{p,q}(\kappa_1, \kappa_2, \ldots, \kappa_n) \leq k_+.$$

Proof Since $\gamma_{\#} \leq \gamma_i \leq \gamma^*$ and $\delta_{\#} \leq \delta_i \leq \delta^*$ for all $(i = 1, 2, \ldots, n)$, then $k_- \leq k_i$ for $i = 1, 2, \ldots, n$ and, therefore, from monotonicity,

$$\text{DHPFDBM}^{p,q}(k_-, k_-, \ldots, k_-) \leq \text{DHPFDBM}^{p,q}(k_1, k_2, \ldots, k_n).$$

and so by idempotency,

$$k_- \leq \text{DHPFDBM}^{p,q}(k_1, k_2, \ldots, k_n) \tag{14}$$

and

$$\text{DHPFDBM}^{p,q}(k_1, k_2, \ldots, k_n) \leq k_+. \tag{15}$$

Combining Eqs. (14) and (15), $k_- \leq \text{DHPFDBM}^{p,q}(k_1, k_2, \ldots, k_n) \leq k_+$.

Definition 9 Suppose k_i, $(i = 1, 2, \ldots, n)$ be a set of DHPFEs. Also, let $\omega = (\omega_1, \omega_2, \ldots, \omega_n)^T$ be the weight vector such that $\sum_{i=1}^{n} \omega_i = 1$, where $\omega_i \in [0, 1]$. Let $p, q \geq 0$ be any numbers. If

$$\text{DHPFDWBM}_{\omega}^{p,q}(k_1, k_2, \ldots, k_n) = \left(\frac{1}{n(n-1)} \oplus_{D_{i,j=1 \atop i \neq j}}^{n} \left((\omega_i k_i)^p \otimes_D (\omega_j k_j)^q\right)\right)^{\frac{1}{p+q}},$$

then $\text{DHPFDWBM}_{\omega}^{p,q}(k_1, k_2, \ldots, k_n)$ is called DHPFDWBM operator.

Theorem 5 Let $k_i = (h_i, g_i)(i = 1, 2, \ldots, n)$ be a set of DHPFEs, and corresponding weight vector is $\omega = (\omega_1, \omega_2, \ldots, \omega_n)^T$, which satisfies $\omega_i \in [0, 1]$ and $\sum_{i=1}^{n} \omega_i = 1$. Let $p, q \geq 0$ be any numbers. The evaluation value based on DHPFDWBM operator is also generated as a DHPFE and

$$\text{DHPFDWBM}_{\omega}^{p,q}(k_1, k_2, \ldots, k_n) = \left(\frac{1}{n(n-1)} \oplus_{D_{i,j=1 \atop i \neq j}}^{n} \left((\omega_i k_i)^p \otimes_D (\omega_j k_j)^q\right)\right)^{\frac{1}{p+q}}$$

$$= \left(\bigcup_{\substack{\gamma_i \in h_i, \\ \gamma_j \in h_j}} \left\{ \sqrt{1 \Big/ \left(1 + \left(\frac{n(n-1)}{p+q}\left(1 \Big/ \left(\sum_{\substack{i,j=1 \\ i \neq j}}^{n} \left(1 \Big/ \left(\frac{p}{\omega_i}\left(\frac{1-\gamma_i^2}{\gamma_i^2}\right)^{\varphi} + \frac{q}{\omega_j}\left(\frac{1-\gamma_j^2}{\gamma_j^2}\right)^{\varphi}\right)\right)\right)\right)\right)^{1/\varphi}\right)} \right\},\right.$$

$$\left.\bigcup_{\substack{\delta_i \in g_i, \\ \delta_j \in g_j}} \left\{ \sqrt{1 - \left(1 \Big/ \left(1 + \left(\frac{n(n-1)}{p+q}\left(1 \Big/ \sum_{\substack{i,j=1 \\ i \neq j}}^{n} \left(1 \Big/ \left(\frac{p}{\omega_i}\left(\frac{\delta_i^2}{1-\delta_i^2}\right)^{\varphi} + \frac{q}{\omega_j}\left(\frac{\delta_j^2}{1-\delta_j^2}\right)^{\varphi}\right)\right)\right)\right)^{\frac{1}{\varphi}}\right)\right)} \right\}\right) \tag{16}$$

Proof The proof of the theorem is as same as of Theorem 1.

DHPF Dombi GBM aggregation operator is proposed below.

Definition 10 Let $\kappa_i = (h_i, g_i)$ $(i = 1, 2, \ldots, n)$ be a set of DHPFEs. Also, let $p, q > 0$ be any two numbers. Then, DHPFDGBM operator is defined as $\text{DHPFDGBM}^{p,q}(k_1, k_2, \ldots, k_n) = \frac{1}{p+q}\left(\otimes^n_{D_{i,j=1},\, i\neq j}\left((pk_i) \oplus_D (qk_j)\right)\right)^{\frac{1}{n(n-1)}}$.

The $\text{DHPFDGBM}^{p,q}(\kappa_1, \kappa_2, \ldots, \kappa_n)$ is called DHPFDGBM operator.

Theorem 6 *Let $k_i = (h_i, g_i)(i = 1, 2, \ldots, n)$ be a collection of DHPFEs. Also, let $p, q > 0$ be any two numbers. Then, the evaluation value based on DHPFDGBM is also generated as a DHPFE and*

$$\text{DHPFDGBM}^{p,q}(k_1, k_2, \ldots, k_n) =$$

$$\left(\bigcup_{\substack{\gamma_i\in h_i,\\ \gamma_j\in h_j}}\left\{\sqrt{1-\left(1/\left(1+\left(\frac{n(n-1)}{p+q}\left(1/\sum_{\substack{i,j=1\\ i\neq j}}^{n}\left(1/\left(p\left(\frac{\gamma_i^2}{1-\gamma_i^2}\right)^{\varphi}+q\left(\frac{\gamma_j^2}{1-\gamma_j^2}\right)^{\varphi}\right)\right)\right)\right)^{1/\varphi}\right)\right)}\right\},\right.$$

$$\left.\bigcup_{\substack{\delta_i\in g_i,\\ \delta_j\in g_j}}\left\{\sqrt{1/\left(1+\left(\frac{n(n-1)}{p+q}\left(1/\left(\sum_{\substack{i,j=1\\ i\neq j}}^{n}\left(1/\left(p\left(\frac{1-\delta_i^2}{\delta_i^2}\right)^{\varphi}+q\left(\frac{1-\delta_j^2}{\delta_j^2}\right)^{\varphi}\right)\right)\right)\right)\right)^{1/\varphi}\right)}\right\}\right)$$

Proof Proof is similar.

Note: DHPFDGBM operator possesses the idempotency, boundary and monotonicity properties.

Definition 11 Suppose $k_i(i = 1, 2, \ldots, n)$ be a set of DHPFEs. Also, let $\omega = (\omega_1, \omega_2, \ldots, \omega_n)^T$ be corresponding weight vector with $\omega_i \in [0, 1]$, $\sum_{i=1}^n \omega_i = 1$. Let $p, q > 0$ be any two numbers. If

$$\text{DHPFDWGBM}_{\omega}^{p,q}(k_1, k_2, \ldots, k_n) = \frac{1}{p+q}\left(\otimes^n_{\substack{D_{i,j=1}\\ i\neq j}}\left((pk_i)^{\omega_i} \oplus_D (qk_j)^{\omega_j}\right)\right)^{\frac{1}{n(n-1)}},$$

then $\text{DHPFDWGBM}_{\omega}^{p,q}(k_1, k_2, \ldots, k_n)$ is said to be DHPFDWGBM operator.

Theorem 7 Let $k_i = (h_i, g_i)$ $(i = 1, 2, \ldots, n)$ be a collection of DHPFEs, whose weighted vector is $\omega = (\omega_1, \omega_2, \ldots, \omega_n)^T$, which satisfies $\omega_i \in [0, 1]$ and $\sum_{i=1}^n \omega_i = 1$. Consider $p, q > 0$ as any two numbers. Then, the evaluation value based on DHPFDWGBM operator is also generated as a DHPFE and

$$\text{DHPFDWGBM}_{\omega}^{p,q}(k_1, k_2, \ldots, k_n) =$$

$$\left(\bigcup_{\substack{\gamma_i \in h_i,\\ \gamma_j \in h_j}} \left\{ \sqrt{1 - \left(1 \Big/ \left(1 + \left(\frac{n(n-1)}{p+q}\left(1 \Big/ \sum_{\substack{i,j=1\\ i \neq j}}^{n} \left(1 \Big/ \left(\frac{p}{\omega_i}\left(\frac{\gamma_i^2}{1-\gamma_i^2}\right)^{\varphi} + \frac{q}{\omega_j}\left(\frac{\gamma_j^2}{1-\gamma_j^2}\right)^{\varphi}\right)\right)\right)\right)^{\frac{1}{\varphi}}\right)\right)} \right\},\right.$$

$$\left.\bigcup_{\substack{\delta_i \in g_i,\\ \delta_j \in g_j}} \left\{ \sqrt{1 \Big/ \left(1 + \left(\frac{n(n-1)}{p+q}\left(1 \Big/ \left(\sum_{\substack{i,j=1\\ i \neq j}}^{n} \left(1 \Big/ \left(\frac{p}{\omega_i}\left(\frac{1-\delta_i^2}{\delta_i^2}\right)^{\varphi} + \frac{q}{\omega_j}\left(\frac{1-\delta_j^2}{\delta_j^2}\right)^{\varphi}\right)\right)\right)\right)\right)^{\frac{1}{\varphi}}\right)} \right\}\right) \tag{17}$$

Proof Proof is similar.

5 A New MAGDM Method Under DHPF Context

This section presents the application of the newly defined DHPFDWBM and DHPFDWGBM operators to solve MAGDM problems.

Consider $A = \{A_i | i = 1, 2, \ldots, m\}$ as a collection of choices that is evaluated based on the attribute set $C = \{C_i | i = 1, 2, \ldots, n\}$ by the set of DMs, $E = \{e_1, e_2, \ldots, e_u\}$. Let $D^{(v)}_{m \times n} = \left[\tilde{d}^{(v)}_{ij}\right]_{m \times n}$ $(v = 1, 2, \ldots, u)$ be a DHPF decision matrix (DHPFDM), where $\tilde{d}^{(v)}_{ij} = \left(h_{ij}{}^{(v)}, g_{ij}{}^{(v)}\right)$ is a DHPFE. DM e_v designates his/her judgment value $\tilde{d}^{(v)}_{ij}$ for the alternative $A_i \in A$ on the attribute $C_j \in C$. Here, $h_{ij}{}^{(v)}$ and $g_{ij}{}^{(v)}$ denote the membership and non-membership degrees evaluating A_i on C_j by the DM e_v, $i = 1, 2, \ldots, m$; $j = 1, 2, \ldots, n$; $v = 1, 2, \ldots, u$.

Then, the newly defined DHPFDWBM (or DHPFDWGBM) operators are applied to yield the MAGDM approach using DHPF data. The successive steps of the proposed methodology are framed below.

Step 1. Basically, any attribute is either benefit or cost type. If the DHPFDM has cost type attributes, matrices $D^{(v)}_{m \times n} = \left[\tilde{d}^{(v)}_{ij}\right]_{m \times n}$ can be converted into the normalized DHPFDM, $R^{(v)}_{m \times n} = \left(\tilde{r}^{(v)}_{ij}\right)_{m \times n}$ in the following way,

$$\tilde{r}^{(v)}_{ij} = \begin{cases} \tilde{d}^{(v)}_{ij} & \text{for benefit attribute } C_j \\ \tilde{d}^{c(v)}_{ij} & \text{for cost attribute } C_j \end{cases} \tag{18}$$

$i = 1, 2, \ldots, m$ and $j = 1, 2, \ldots, n$, where $\tilde{d}^c_{ij}$ is the complement $\tilde{d}_{ij}$.

Step 2. For aggregating DHPFDMs $R^{(v)}_{m \times n}$ $(v = 1, 2, \ldots, u)$ into one collective DHPFDM $R_{m \times n} = \left(\tilde{r}_{ij}\right)_{m \times n}$, $i = 1, 2, \ldots, m$; $j = 1, 2, \ldots, n$, the following DHPFDWBM (and DHPFDWGBM) operators are used.

Thus using DHPFDWBM operator, $\tilde{r}_{ij} = \text{DHPFDWBM}\left(\tilde{r}^{(1)}_{ij}, \tilde{r}^{(2)}_{ij}, \ldots, \tilde{r}^{(u)}_{ij}\right)$ and DHPFDWGBM operator, $\tilde{r}_{ij} = \text{DHPFDWGBM}\left(\tilde{r}^{(1)}_{ij}, \tilde{r}^{(2)}_{ij}, \ldots, \tilde{r}^{(u)}_{ij}\right)$.

Step 3. Aggregate the DHPFEs $\tilde{r}_{ij}$ for each A_i applying DHPFDWBM (or DHPFDWGBM) operator.

$\tilde{r}_i = \text{DHPFDWBM}(\tilde{r}_{i1}, \tilde{r}_{i2}, \ldots\ldots, \tilde{r}_{in})$

or $\tilde{r}_i = \text{DHPFDWGBM}(\tilde{r}_{i1}, \tilde{r}_{i2}, \ldots, \tilde{r}_{in})$ for $i = 1, 2, \ldots, m$; $j = 1, 2, \ldots, n$.

Step 4. Evaluate the rank of each alternative utilizing the score function as presented in Definition 3.

6 Illustrative Examples

Two examples would be solved now, using the proposed methodology, to demonstrate the efficiency.

6.1 *Example 1*

For the purpose of showing application potentiality of the established method, a numerical example (adapted from Liu et al. [29]) about the investment to five possible companies based on, viz., A_1: car, A_2: food, A_3: computer, A_4: arms and A_5: TV has been quoted.

Three DMs, e_v, $(v = 1, 2, 3)$ maintaining the weight vector, $\omega = (0.35, 0.40, 0.25)$, put decision values by evaluating the five alternatives based on four attributes, viz., risk, growth, social–political impact and environmental impact, which are denoted by C_1,C_2, C_3, and C_4, respectively, with corresponding weight vector, $w = (0.2, 0.1, 0.3, 0.4)$.

The DMs, e_v $(v = 1, 2, 3)$, assess the five companies, A_i, based on the attributes C_j $(j = 1, 2, 3, 4)$, and express their decision values using DHPFEs. The DHPFDMs $D^v = \left[\tilde{d}^v_{ij}\right]_{5\times4}$ $(v = 1, 2, 3)$ are exhibited in Table 1.

Step 1. Normalize the decision matrices. All the attributes provided by the DMs being benefit type need not be normalized.

Step 2. Aggregate the values $\tilde{r}^{(v)}_{ij}$ $(v = 1, 2, 3)$ for each $i = 1, 2, 3, 4, 5$ and $j = 1, 2, 3, 4$, and construct the collective DHPF decision matrix, $R_{5\times4} = \left(\tilde{r}_{ij}\right)_{5\times4}$ using the DHPFDWBM operator (taking $p = q = 1$; $\varphi = 2$) in Table 2.

Step 3. Aggregate all DHPFEs $\tilde{r}_{ij}$ for each alternative A_i using the DHPFDWBM operator (taking $p = q = 1$; $\varphi = 2$) according to Eq. (16). And the derived results are

$$\tilde{r}_1 = (\{0.2664, 0.2623, 0.2555, 0.2515\}, \{0.6367, 0.6000, 0.6039, 0.5736, 0.6279\ 0.5925, 0.5964, 0.5670\})$$

$$\tilde{r}_2 = (\{0.3972, 0.3873, 0.3935, 0.3843, 0.3864, 0.3784, 0.3788, 0.3695, 0.3748,$$

Table 1 DHPFDM generated by DMs

DMs		C_1	C_2	C_3	C_4
e_1	A_1	$(\{0.6\}, \{0.5\})$	$(\{0.5\}, \{0.3\})$	$(\{0.2, 0.3\}, \{0.6\})$	$(\{0.5\}, \{0.6\})$
	A_2	$(\{0.9\}, \{0.2\})$	$(\{0.8\}, \{0.3\})$	$(\{0.7\}, \{0.3\})$	$\begin{pmatrix}\{0.6, 0.8\}, \\ \{0.2, 0.3\}\end{pmatrix}$
	A_3	$\begin{pmatrix}\{0.4, 0.5\}, \\ \{0.4, 0.6\}\end{pmatrix}$	$\begin{pmatrix}\{0.6, 0.7\}, \\ \{0.4, 0.5, 0.6\}\end{pmatrix}$	$(\{0.7\}, \{0.5\})$	$(\{0.6\}, \{0.5\})$
	A_4	$(\{0.9\}, \{0.3\})$	$(\{0.8\}, \{0.4\})$	$(\{0.4, 0.6\}, \{0.2\})$	$(\{0.7\}, \{0.4\})$
	A_5	$\begin{pmatrix}\{0.4, 0.5, 0.7\}, \\ \{0.3, 0.4\}\end{pmatrix}$	$(\{0.6\}, \{0.5\})$	$(\{0.5\}, \{0.6\})$	$\begin{pmatrix}\{0.4, 0.5\}, \\ \{0.5, 0.7\}\end{pmatrix}$
e_2	A_1	$\begin{pmatrix}\{0.4, 0.6\}, \\ \{0.5, 0.6\}\end{pmatrix}$	$(\{0.7\}, \{0.4\})$	$(\{0.5\}, \{0.6\})$	$(\{0.5\}, \{0.3, 0.5\})$
	A_2	$(\{0.8\}, \{0.3\})$	$(\{0.8\}, \{0.3\})$	$\begin{pmatrix}\{0.6, 0.7, 0.8\}, \\ \{0.3\}\end{pmatrix}$	$(\{0.9\}, \{0.2\})$
	A_3	$(\{0.5\}, \{0.6\})$	$(\{0.4\}, \{0.7\})$	$(\{0.5\}, \{0.6\})$	$(\{0.2, 0.3\}, \{0.6, 0.8\})$
	A_4	$(\{0.7\}, \{0.4\})$	$\begin{pmatrix}\{0.7, 0.8\}, \\ \{0.2, 0.3\}\end{pmatrix}$	$(\{0.4, 0.6\}, \{0.4\})$	$(\{0.8\}, \{0.3\})$
	A_5	$(\{0.7\}, \{0.5\})$	$(\{0.8\}, \{0.3\})$	$(\{0.6\}, \{0.5\})$	$(\{0.7\}, \{0.2, 0.3\})$
e_3	A_1	$(\{0.6\}, \{0.6\})$	$(\{0.7\}, \{0.4\})$	$(\{0.5\}, \{0.3, 0.6\})$	$(\{0.6\}, \{0.3\})$
	A_2	$(\{0.7\}, \{0.5\})$	$\begin{pmatrix}\{0.5, 0.6, 0.8\}, \\ \{0.2, 0.3\}\end{pmatrix}$	$(\{0.9\}, \{0.2\})$	$(\{0.8\}, \{0.3\})$
	A_3	$(\{0.4\}, \{0.7\})$	$(\{0.5\}, \{0.8\})$	$(\{0.6\}, \{0.5\})$	$\begin{pmatrix}\{0.3, 0.4\}, \\ \{0.3, 0.5, 0.7\}\end{pmatrix}$
	A_4	$\begin{pmatrix}\{0.5, 0.6, 0.7\}, \\ \{0.3, 0.4\}\end{pmatrix}$	$(\{0.6\}, \{0.4\})$	$(\{0.6\}, \{0.5\})$	$(\{0.7\}, \{0.2\})$
	A_5	$(\{0.6\}, \{0.3\})$	$(\{0.7\}, \{0.5\})$	$\begin{pmatrix}\{0.4, 0.5\}, \\ \{0.3, 0.4\}\end{pmatrix}$	$(\{0.7\}, \{0.4\})$

$0.3665, 0.3673, 0.3605, 0.3713, 0.3618, 0.3671, 0.3587, 0.3592, 0.3525\}, \{0.4918, 0.4764, 0.4717, 0.4576\})$

$\tilde{r}_3 = (\{0.2160, 0.2064, 0.2089, 0.1955, 0.2156, 0.2060, 0.2086, 0.1951, 0.2114, 0.2021, 0.2046, 0.1913, 0.2110, 0.2016, 0.2042, 0.1908\}, \{0.7770, 0.7401, 0.6720, 0.7494, 0.7217, 0.6668, 0.7551, 0.7211, 0.6591, 0.7298, 0.7042, 0.6544, 0.7475,$

Table 2 Collective DHPFDM

	C_1	C_2	C_3	C_4
A_1	({0.3335, 0.3853}, {0.4891, 0.5158})	({0.4199}, {0.3023})	({0.3153, 0.3330}, {0.5306, 0.6724})	({0.3656}, {0.3864, 0.4891})
A_2	({0.4350}, {0.4245})	({0.4833, 0.5187, 0.5904}, {0.3023, 0.3586})	({0.4925, 0.5155, 0.5273}, {0.3016})	({0.5565, 0.6389}, {0.3016, 0.3474})
A_3	({0.3128, 0.3335}, {0.5384, 0.6124})	({0.2559, 0.2584}, {0.5384, 0.6124})	({0.3434}, {0.4010})	({0.2150, 0.2659, 0.2559, 0.2951}, {0.4768, 0.6128, 0.6953, 0.4892, 0.6659, 0.7990})
A_4	({0.4405, 0.4870, 0.5461}, {0.4010, 0.4577})	({0.5278, 0.5591}, {0.5278, 0.5591})	({0.2804, 0.3059, 0.3115, 0.4136}, {0.4930})	({0.4199}, {0.2611})
A_5	({0.4229, 0.4570, 0.5187}, {0.3474, 0.3864})	({0.4419}, {0.3166})	({0.3128, 0.3382}, {0.4307, 0.5114})	({0.4419, 0.5155}, {0.4430, 0.4740, 0.4481, 0.4814})

$0.7153, 0.6530, 0.7238, 0.6983, 0.6483, 0.7282, 0.6983, 0.6411, 0.7062, 0.6825,$
$0.6369\})$

$$\tilde{r}_4 = (\{0.3518, 0.3288, 0.3278, 0.3235, 0.3450, 0.3214, 0.3203, 0.3159, 0.3414, 0.3179, 0.3168, 0.3122, 0.3353, 0.3116, 0.3105, 0.3057, 0.3315, 0.3076, 0.3065, 0.3015, 0.3258, 0.3020, 0.3008, 0.2957\}, \{0.5551, 0.5262, 0.5349, 0.5089\})$$

$$\tilde{r}_5 = (\{0.3208, 0.3085, 0.3154, 0.3028, 0.3046, 0.2971, 0.2987, 0.2911, 0.2956, 0.2898, 0.2894, 0.2836\}, \{0.5612, 0.5516, T0.5590, 0.5502, 0.5385, 0.5293, 0.5364, 0.5279, 0.5504, 0.5404, 0.5482, 0.5389, 0.5277, 0.5183, 0.5256, 0.5169\})$$

Step 4. Evaluate $S(\tilde{r}_i)$, the score for $\tilde{r}_i (i = 1, 2, 3, 4, 5)$ using Definition (3), and find the rankings depending upon the score values. The result is $S(\tilde{r}_1) = 0.2783$, $S(\tilde{r}_2) = 0.5151$, $S(\tilde{r}_3) = 0.1866$, $S(\tilde{r}_4) = 0.4081$, $S(\tilde{r}_5) = 0.3232$. Since $S(\tilde{r}_2) > S(\tilde{r}_4) > S(\tilde{r}_5) > S(\tilde{r}_1) > S(\tilde{r}_3)$, the ranking is $A_2 \succ A_4 \succ A_5 \succ A_1 \succ A_3$. And the best alternative that found is A_2.

Similarly, using the DHPFDWGBM operator the ranking $A_2 \succ A_4 \succ A_5 \succ A_1 \succ A_3$ is obtained. And thus for this case also A_2 is found as the most appropriate option.

Influence on the decision-making results for the presence *p* and *q*

Keeping the Dombi parameter $\varphi = 2$ fixed, different parameter values of p and q in (0, 10] are considered and the corresponding score values are depicted through Figs. 1, 2, 3, 4 and 5 (using DHPFDWBM operator) and Figs. 6, 7, 8, 9 and 10 (using DHPFDWGBM operator).

It is found that the ranking result remains consistent with the change of the parameters p and q i.e., varying the parameters the results are same. It is significant that for $p = q$ same score values are obtained for each of the options. For $p \neq q$, the comprehensive score values decrease using the DHPFDWBM operator, while p or q increases and using DHPFDWGBM operator the score values of the alternatives increase while p or q increases.

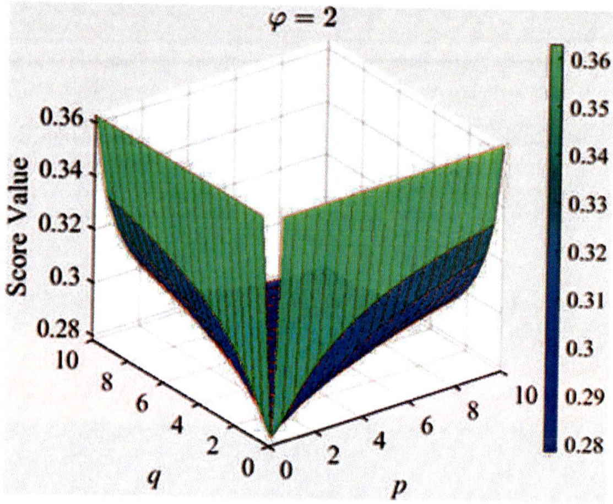

Fig. 1 Score values for A_1 using DHPFDWBM operator

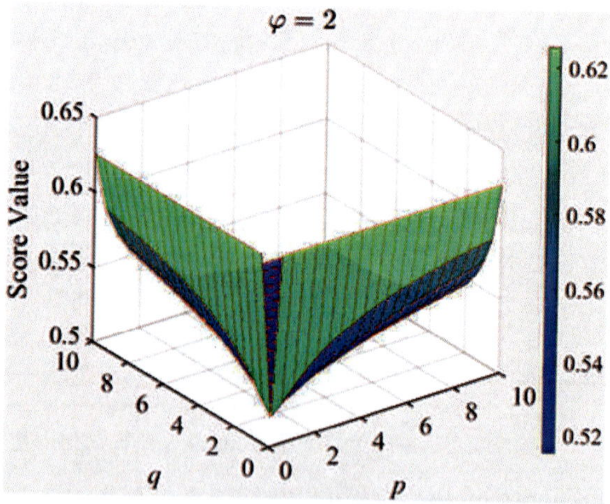

Fig. 2 Score values for A_2 using DHPFDWBM operator

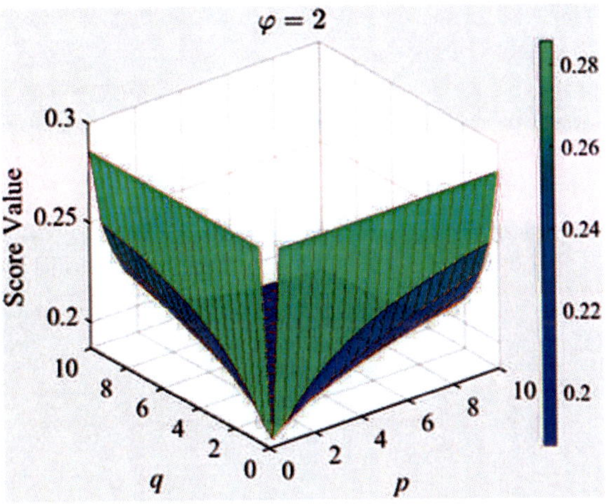

Fig. 3 Score values for A_3 using DHPFDWBM operator

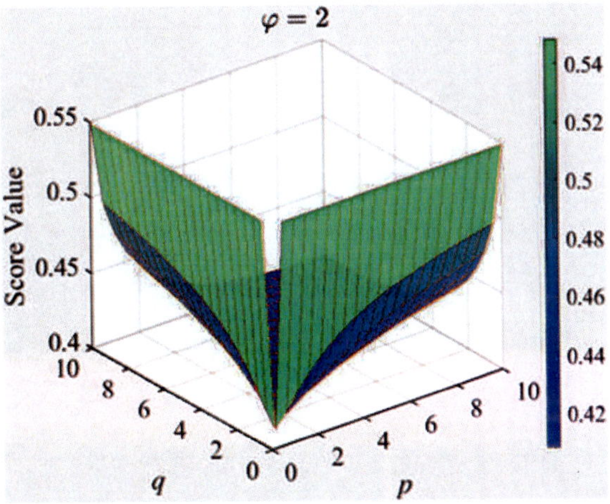

Fig. 4 Score values for A_4 using DHPFDWBM operator

Furthermore, considering BM parameter as $p = q = 1$, varying the Dombi parameter φ in $(0, 20]$, the corresponding score values related to respective alternatives are depicted through Figs. 11 and 12. It is clear that using DHPFDWBM operator the corresponding graphs representing the comprehensive score values are increasing monotonically.

Thus, the DMs may choose higher value of Dombi parameter for taking optimistic decision, and they may choose smaller value of Dombi parameter for making pessimistic decisions. And using DHPFDWGBM operator, the corresponding graphs representing the alternatives' comprehensive score values are exhibiting decreasing sense. Thus in this case also DMs can reflect their opinion through choosing the appropriate parameter value as per their requirement.

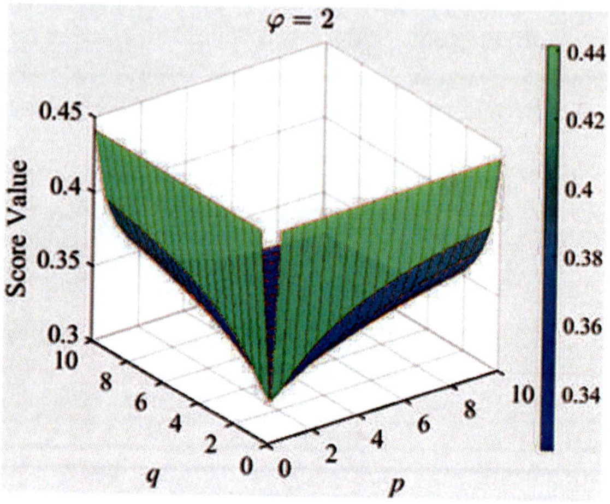

Fig. 5 Score values for A_5 using DHPFDWBM operator

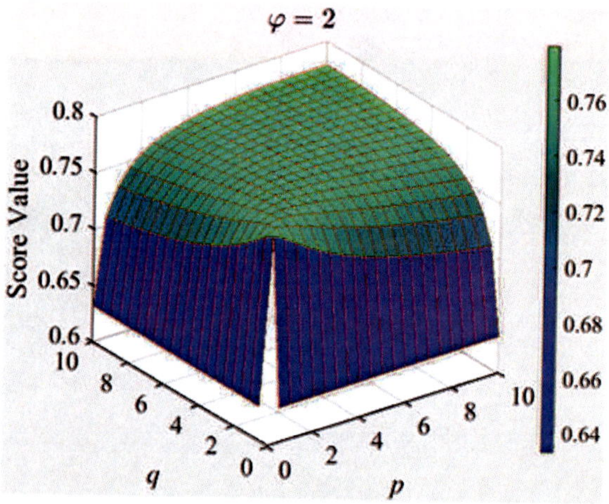

Fig. 6 Score values for A_1 using DHPFDWGBM operator

The above example was considered by Liu et al. [29] using IFSs. In this paper, the same example is revised and solved under DHPF environment and same ranking result is found, which displays feasibility and practicality of the developed method. But the proposed approach is more general in the sense that it can consider hesitancy of the DMs. Moreover, the range of evaluation value collected from DM is wider than by Liu et al.'s method [29]. Thus, scope of application during real-life decision making is enriched by the proposed method.

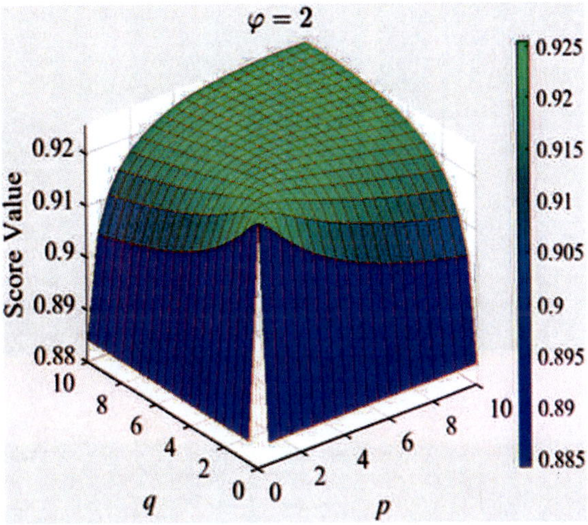

Fig. 7 Score values for A_2 using DHPFDWGBM operator

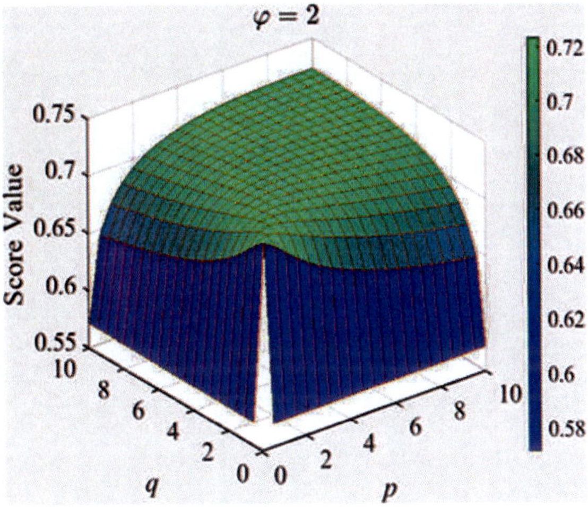

Fig. 8 Score values for A_3 using DHPFDWGBM operator

6.2 *Example 2*

The application potentiality of the newly developed method is further depicted by solving another example, previously considered by Garg [32]. The problem is stated below.

For the selection of the best market for investment from five available major markets, $A_i, (i = 1, 2, 3, 4, 5)$ are evaluated based on the four criteria $C_j, (j = 1, 2, 3, 4)$ with the weight vector $w = (0.20, 0.15, 0.35, 0.30)$. The evaluation value collected from DMs is considered as the same as presented in Table 1, Page No. 285 of the article published by Garg [32].

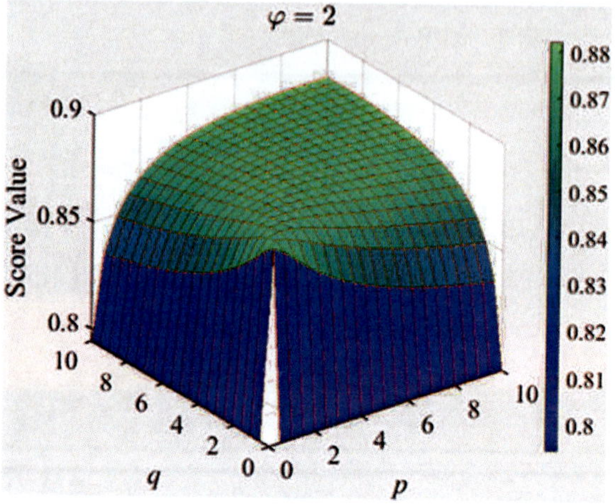

Fig. 9 Score values for 4 using DHPFDWGBM operator

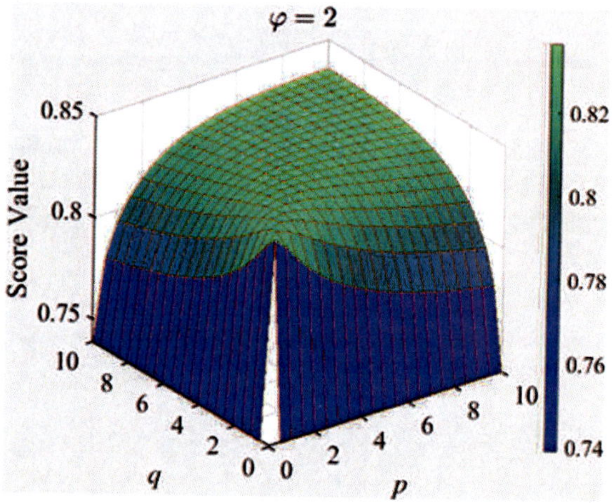

Fig. 10 Score values for A_5 using DHPFDWGBM operator

The aggregated result using the DHPFDWBM operator (considering $p = q = 1$; $\varphi = 2$) is found as: $S(\tilde{r}_1) = 0.3021$, $S(\tilde{r}_2) = 0.3603$, $S(\tilde{r}_3) = 0.4295$, $S(\tilde{r}_4) = 0.4755$, $S(\tilde{r}_5) = 0.4661$. Since $S(\tilde{r}_4) > S(\tilde{r}_5) > S(\tilde{r}_3) > S(\tilde{r}_2) > S(\tilde{r}_1)$, the ranking becomes $A_4 \succ A_5 \succ A_3 \succ A_2 \succ A_1$. So, the best alternative is identified as A_4.

Further, using DHPFDWGBM operator (considering $= q = 1$; $\varphi = 2$) the result is obtained as: $S(\tilde{r}_1) = 0.5505$, $S(\tilde{r}_2) = 0.6015$, $S(\tilde{r}_3) = 0.5526$, $S(\tilde{r}_4) = 0.5149$, $S(\tilde{r}_5) = 0.6113$. Thus, $A_5 \succ A_2 \succ A_3 \succ A_1 \succ A_4$. Hence, the most appropriate option is A_5.

Taking $p = q = 1$, and varying $\varphi \in (0, 20]$, corresponding score values related to the respective alternatives using DHPFDWBM and DHPFDWGBM operators are

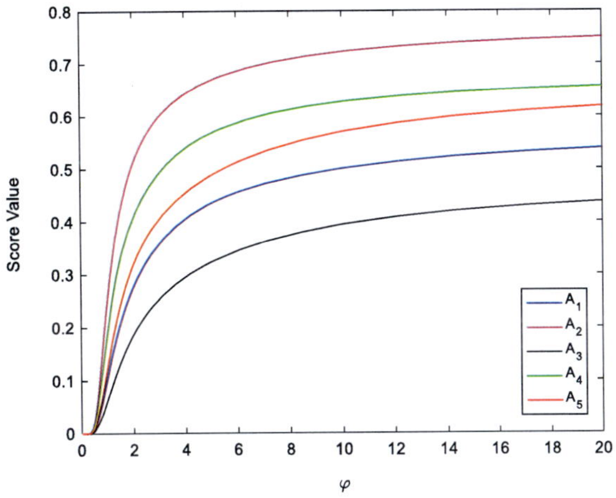

Fig. 11 Impact of using DHPFDWBM operator on the score values related to respective alternatives for $\varphi \in (0, 20]$

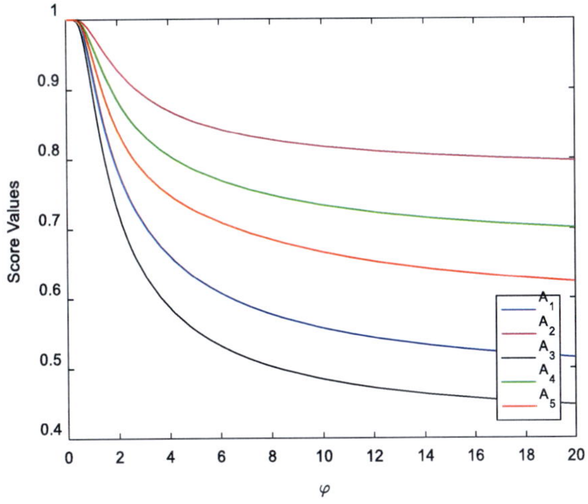

Fig. 12 Impact of using DHPFDWGBM operator on the score values related to respective alternatives for $\varphi \in (0, 20]$

represented in Figs. 13 and 14. For different values of the Dombi parameter, slight change in the ranking of the alternatives for both the operators is seen.

Using DHPFDWBM operator, for $\varphi \in (0, 1.642)$ the ranking is $A_5 \succ A_4 \succ A_3 \succ A_2 \succ A_1$; for $\varphi \in (1.642, 5.465)$, the ranking is $A_4 \succ A_5 \succ A_3 \succ A_2 \succ A_1$, and for $\varphi \in (5.465, 20]$, the ranking is $A_4 \succ A_3 \succ A_5 \succ A_2 \succ A_1$. Thus, the best alternative is either A_4 or A_5 using DHPFDWBM operator according to the choice of the Dombi parameter.

Using DHPFDWGBM operator, for $\varphi \in (0, 2.098)$ the ranking is $A_5 \succ A_2 \succ A_3 \succ A_1 \succ A_4$; for $\varphi \in (2.098, 11.56)$, the ranking is $A_2 \succ A_5 \succ A_4 \succ A_1 \succ A_3$, and for $\varphi \in (11.56, 20]$, the ranking is $A_2 \succ A_4 \succ A_5 \succ A_1 \succ A_3$. Thus, the best alternative is either A_2 or A_5 using DHPFDWGBM operator according to the choice of parameter.

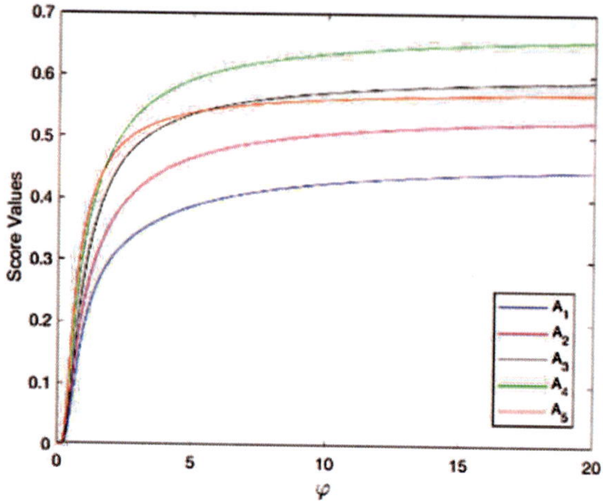

Fig. 13 Impact of using DHPFDWBM operator on the score values related to respective alternatives for $\varphi \in (0, 20]$

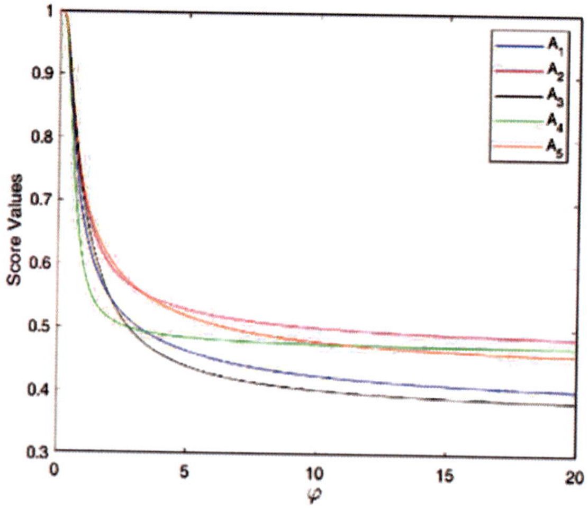

Fig. 14 Impact of using DHPFDWGBM operator on the score values related to respective alternatives for $\varphi \in (0, 20]$

It is to be mentioned here that using Garg's method [32] the ranking of the alternatives is $A_5 \succ A_4 \succ A_3 \succ A_2 \succ A_1$ in hesitant PF context. But, using the proposed methodology, changing the Dombi parameter according to their needs, the DMs can evaluate the alternatives more accurately and properly than the existing method [32] as discussed in the above. Thus, the proposed method has the ability to make the decision-making process more flexible using Dombi parameter φ.

Table 3 Comparison with existing operators

Methods	Consideration of interrelationship among inputs	Flexible using a variable parameter
From DHPFWA (Wei and Lu [20])	No	No
From DHPFWG (Wei and Lu [20])	No	No
From DHPFBM (Tang and Wei [33])	Yes	No
From DHPFGBM (Tang and Wei [33])	Yes	No
Using the proposed operators	Yes	Yes

7 Comparison with Existing Operators

The proposed DHPFDWBM and DHPFDWGBM operators have two important characteristics. Firstly, aggregation operators with BM can consider the interrelationship between input parameters; secondly, it can make the aggregation process more agile using Dombi operations. A comparative analysis of aggregation operators [20, 33] with the proposed operators on the basis of their characteristics, under DHPF MAGDM contexts, has been presented in Table 3. It is clear from Table 3 that DHPFWA [20] and DHPFWG [20] operators cannot consider interrelationship among input parameters. Although the DHPFBM [33] and DHPFGBM [33] operators have the ability to consider the interrelationship characteristics, none of them have a changeable parameter that can helps to generate a flexible aggregation process. Thus, the proposed operators are more powerful and advantageous for enabling these characteristics.

8 Conclusions

In this paper, two novel DHPF aggregation operators, viz., DHPFDBM and DHPFDGBM operators, and their weighted forms are developed. Some desirable characteristics and special cases related to those newly developed operators are also discussed. Then, the proposed DHPF operators are applied in developing a new methodology for MAGDM context. Two numerical examples with DHPF data are solved, to show suitability of the proposed operator in dealing real-life problems. Through the figures and the results, it has been visualized that the proposed operators are more superior than the existing ones. The novelty of the proposed methodology is ascertained by having the powerful BM within the aggregation function and utilizing the flexible Dombi operations in the process of aggregations. In the future research, the developed operators are to be extended and applied in information processing using q-rung orthopair fuzzy, dual hesitant q-rung orthopair fuzzy data, etc.

Acknowledgements The authors express their gratitude to the reviewers for their invaluable suggestions for improvement of the quality of the manuscript.

References

1. Zadeh LA (1965) Fuzzy sets. Inf Control 8:338–353
2. Atanassov KT (1986) Intuitionistic fuzzy sets. Fuzzy Sets Syst 20(1):87–96
3. Yager RR (2013) Pythagorean fuzzy subsets. In: Proceedings of joint IFSA world congress and NAFIPS annual meeting, Edmonton, Canada 57–61
4. Biswas A, Deb N (2020) Pythagorean fuzzy Schweizer and Sklar power aggregation operators for solving multi-attribute decision-making problems. Granul Comput. https://doi.org/10.1007/s41066-020-00243-1
5. Sarkar A, Biswas A (2019) Multicriteria decision-making using Archimedean aggregation operators in Pythagorean hesitant fuzzy environment. Int J Intell Syst 34(7):1361–1386
6. Sarkar B, Biswas A (2020) A unified method for Pythagorean fuzzy multicriteria group decision-making using entropy measure, linear programming and extended technique for ordering preference by similarity to ideal solution. Soft Comput 24:5333–5344
7. Zhang XL, Xu ZS (2014) Extension of TOPSIS to multiple criteria decision making with Pythagorean fuzzy sets. Int J Intell Syst 29:1061–1078
8. Peng X, Yang Y (2015) Some results for Pythagorean fuzzy sets. Int J Intell Syst 30(11):1133–1160
9. Garg H (2016) Generalized Pythagorean fuzzy geometric aggregation operators using einsteint-norm andt-conorm for multicriteria decision-making process. Int J Intell Syst 32(6):597–630
10. Wei GW (2017) Pythagorean fuzzy interaction aggregation operators and their application to multiple attribute decision making. J Intell Fuzzy Syst 33:2119–2132
11. Bonferroni C (1950) Sulle medie multiple di potenze. Bolletino Matematica Ital 5:267–270
12. Zhang R, Wang J, Zhu X, Xia M, Yu M (2017) Some generalized Pythagorean fuzzy Bonferroni mean aggregation operators with their application to multiattribute group decision-making. Complexity, Article ID 5937376
13. Gao H (2018) Pythagorean fuzzy Hamacher Prioritized aggregation operators in multiple attribute decision making. J Intell Fuzzy Syst 35:2229–2245
14. Wei G, Lu M (2018) Pythagorean Fuzzy Maclaurin symmetric mean operators in multiple attribute decision making. Int J Intell Syst 33(5):1043–1070
15. Wei G, Lu M (2018) Pythagorean fuzzy power aggregation operators in multiple attribute decision making. Int J Intell Syst 33:169–186
16. Liang D, Zhang Y, Xu Z, Darko AP (2018) Pythagorean fuzzy Bonferroni mean aggregation operator and its accelerative calculating algorithm with the multithreading. Int J Intell Syst 33(3):615–633
17. Li Z, Wei G, Lu M (2018) Pythagorean fuzzy hamy mean operators in multiple attribute group decision making and their application to supplier selection. Symmetry 10:505
18. Khan MSA, Abdullah S, Ali A, Amin F (2019) Pythagorean fuzzy prioritized aggregation operators and their application to multi-attribute group decision making. Granul Comput 4:249–263
19. Rahman K, Abdullah S, Hussain F (2020) Some generalised Einstein hybrid aggregation operators and their application to group decision-making using Pythagorean fuzzy numbers. Fuzzy Inf Eng 1–16

20. Wei G, Lu M (2017) Dual hesitant Pythagorean fuzzy Hamacher aggregation operators in multiple attribute decision making. Archiv Control Sci 27(3):365–395
21. Zhu B, Xu ZS, Xia MM (2012) Dual hesitant fuzzy sets. J Appl Math, Article ID 879629. https://doi.org/10.1155/2012/879629
22. Tang X, Wei G (2019) Dual hesitant Pythagorean fuzzy Bonferroni mean operators in multi-attribute decision making. Archiv Control Sci 29(2):339–386
23. Tang M, Wang J, Lu J, Wei G, Wei C, Wei Y (2019) Dual Hesitant Pythagorean fuzzy Heronian mean operators in multiple attribute decision making. Mathematics 7(4):344
24. Lu J, Tang X, Wei G, Wei C, Wei Y (2019) Bidirectional project method for dual hesitant Pythagorean fuzzy multiple attribute decision-making and their application to performance assessment of new rural construction. Int J Intell Syst 34:1920–1934
25. Zhu B, Xu ZS (2013) Hesitant fuzzy Bonferroni means for multi-criteria decision making. J Oper Res Soc 64(12):1831–1840
26. Zhu B, Xu Z, Xia M (2012) Hesitant fuzzy geometric Bonferroni means. Inf Sci 205:72–85
27. Dombi J (1982) A general class of fuzzy operators, the De-Morgan class of fuzzy operators and fuzziness induced by fuzzy operators. Fuzzy Sets Syst 8(2):149–163
28. He X (2017) Typhoon disaster assessment based on Dombi hesitant fuzzy information aggregation operators. Nat Hazards 90(3):1153–1175
29. Liu P, Liu J, Chen SM (2018) Some intuitionistic fuzzy Dombi Bonferroni mean operators and their application to multi-attribute group decision making. J Oper Res Soc 69(1):1–24
30. Jana C, Senapati T, Pal M (2019) Pythagorean fuzzy Dombi aggregation operators and its applications in multiple attribute decision-making. Int J Intell Syst 34(9):2019–2038
31. Xia M, Xu Z, Zhu B (2013) Geometric Bonferroni means with their application in multi-criteria decision making. Knowl-Based Syst 40:88–100
32. Garg H (2018) Hesitant Pythagorean fuzzy sets and their aggregation operators in multiple attribute decision-making. Int J Uncertain Quantif 8(3):267–289
33. Tang X, Wei G (2019) Dual hesitant pythagorean fuzzy Bonferroni mean operators in multi-attribute decision making. Arch Control Sci 29(2):339–386

Comparative Study of Pressurized Spherical Shell with Different Materials

Gaurav Verma, Pankaj Thakur, and Zoran Radaković

Abstract The paper deals with an analytic approach to a problem of transversely isotropic spherical shell under combined effect of interior and exterior pressure. The paper is based on spherical symmetry of shell. Radial and circumferential stresses are calculated by making different combinations of internal and external pressures with the help of generalized strain components. Stresses obtained for initial yielding as well as full plastic state are depicted graphically. Results are compared for transversely isotropic materials (beryl and magnesium) with isotropic material (brass). It has been observed that radial stresses as well as hoop stresses are maximum for isotropic material (brass) as compared to transversely isotropic materials (beryl and magnesium).

Keywords Elastic–plastic · Stresses · Spherical shell · Internal · External · Pressure

1 Introduction

Researchers have been shown a long-term interest in problems of spherical shell due to its simple geometry. Spherical shell problems are associated with stress boundary conditions to the interior and exterior part of it. There are many shell problems in domains of mechanics which attract scientific community for research activities. Blake problem is one of the problems related to spherical shell within an infinite medium in mechanics. Many authors Brandon et al. [1], Krawczuk et al. [2], Akhmedov et al. [3], Komijani et al. [4] and Mantari et al. [5] have worked on problem of spherical shells by using various techniques like benchmark solution, finite element method, shear deformation theory, etc. Thakur et al. [6, 7], Gupta et al.

G. Verma (✉)
Department of Mathematics, Hans Raj Mahila Maha Vidyalaya, Jalandhar, Punjab, India

P. Thakur
Department of Mathematics, ICFAI University, Baddi, Solan, Himachal Pradesh, India

Z. Radaković
Faculty of Mechanical Engineering, University of Belgrade, Belgrade, Serbia

G. Gupta et al. (eds.), *Proceedings of Academia-Industry Consortium for Data Science*, Advances in Intelligent Systems and Computing 1411,
https://doi.org/10.1007/978-981-16-6887-6_18

[8] and Sharma et al. [9–11] have analyzed the problems of rotating disc and cylinders with transversely isotropic nature. These authors utilized the idea of generalization of principal strain measures and transition theory to derive the results in discs and cylinders. These problems are based on existence of transition zone between the perfect elastic and prefect plastic state. By using same view of transition zone, we have derived results analytically in this paper for transversely isotropic spherical shell. The results obtained are shown graphically.

2 Spherical Shell Problem

A pressurized spherical shell with inner and outer radii a and b, respectively, is taken for study. The internal pressure p_1 and external pressure p_2 are simultaneously applicable on spherical body. The displacements in three directions are given in spherical coordinates (r, θ, ϕ) as

$$u = r(1 - g),\ v = 0,\ w = 0$$

where g is function of r.

The strain components of shell problem are written in generalized form as [12, 13]

$$e_{rr} = \tfrac{1}{n}\left[1 - \left(rg' + g\right)^n\right] = \tfrac{1}{n}\left[1 - g^n(1+Q)^n\right],$$

$$e_{\theta\theta} = e_{\phi\phi} = \frac{1}{n}\left[1 - g^n\right],$$

$$e_{r\theta} = e_{\theta\phi} = e_{\phi r} = 0, \quad (1)$$

where $rg' = gQ$.

The components of the stress for transversely isotropic shell are given as

$$\begin{aligned} T_{rr} &= c_{33}e_{rr} + 2c_{12}e_{\theta\theta} \\ T_{\theta\theta} &= T_{\phi\phi} = c_{13}e_{rr} + 2(c_{11} - c_{66})e_{\theta\theta} \\ T_{r\theta} &= T_{\theta\phi} = T_{r\phi} = 0 \end{aligned} \quad (2)$$

By using Eq. (1) in Eq. (2), the stresses are obtained as

$$\begin{aligned} T_{rr} &= \frac{c_{33}}{n}\left[1 - g^n(1+Q)^n\right] + \frac{2c_{13}}{n}\left[1 - g^n\right], \\ T_{\theta\theta} &= T_{\phi\phi} = \frac{c_{13}}{n}\left[1 - g^n(1+Q)^n\right] + \frac{2(c_{11} - c_{66})}{n}\left[1 - g^n\right], \\ T_{r\theta} &= T_{\theta\phi} = T_{r\phi} = 0 \end{aligned} \quad (3)$$

The problem of the spherical shell satisfies the following equilibrium condition

$$r\frac{dT_{rr}}{dr} - 2(T_{\theta\theta} - T_{rr}) = 0 \tag{4}$$

where T_{rr} and $T_{\theta\theta}$ are the radial and hoop stresses, respectively, and satisfy following conditions as

$$T_{rr} = -p_1 \quad \text{at} \quad r = a$$

$$T_{rr} = -p_2 \quad \text{at} \quad r = b \tag{5}$$

A nonlinear differential equation in g can be obtained by using Eq. (3) in (4) as

$$\begin{aligned} &Q(Q+1)^{n-1} g\frac{dQ}{dg} + Q(Q+1)^n + 2(1-\alpha_2)Q \\ &= \frac{2}{ng^n}\left[\alpha_2\left\{1 - g^n(Q+1)^n\right\} - \alpha_3(1-\alpha_2)\left(1-g^n\right)\right] = 0 \end{aligned} \tag{6}$$

where $\alpha_2 = (c_{33} - c_{13})/c_{33}$, $\alpha_3 = \frac{c_{11}+c_{12}-2c_{13}}{c_{33}(1-\alpha_2)}$.

3 Stress Analysis Using Principal Stress

In order to do stress analysis, we use the transition function R [14–22] at the critical point obtained from Eq. (6).

$$R = (3 - 2\alpha_2) - \frac{nT_{rr}}{c_{33}} = g^n\left[(Q+1)^n + 2(1-\alpha_2)\right] \tag{7}$$

Solving the logarithmic derivative of Eq. (7) with respect to r and taking the value of $\mathrm{d}Q/\mathrm{d}g$ from Eq. (6) at critical point $Q \to \pm\infty$, we get

$$\frac{\mathrm{d}\log R}{\mathrm{d}r} = \frac{-2(\alpha_2)}{r} \tag{8}$$

On solving ordinary differential Eq. (8), we get

$$R = d_1 r^{-2\alpha_2} \tag{9}$$

where d_1 is a constant.

Solving Eqs. (7) and (9), we get radial stress

$$T_{rr} = d_2 + \frac{d_3 r^{-m}}{n} \tag{10}$$

where $d_2 = \frac{c_{33}}{n}(3 - 2\alpha_2)$, $d_3 = d_1 c_{33}$, $m = 2\alpha_2$.

On applying stress conditions and Eq. (4), we get stresses as

$$\begin{aligned} T_{rr} &= -p_2 + \frac{p_1 - p_2}{(b/a)^m - 1}\left[1 - \left(\frac{b}{r}\right)^m\right], \\ T_{\theta\theta} - T_{rr} &= \frac{m}{2}\left(\frac{b}{r}\right)^m \left(\frac{p_1 - p_2}{(b/a)^m - 1}\right) \end{aligned} \tag{11}$$

Taking the value of $|T_{\theta\theta} - T_{rr}|$ at $r = a$.

$$|T_{\theta\theta} - T_{rr}|_{r=a} = \left|\frac{m}{2}\left(\frac{b}{a}\right)^m \left(\frac{p_1 - p_2}{(b/a)^m - 1}\right)\right| \equiv Y \tag{12}$$

where Y denotes yielding at $r = a$, and yielding in shell starts at the following pressure

$$P_i = \frac{p_1}{Y} - \frac{p}{Y} = \left\{\frac{2(b/a)^m - 1}{m(b/a)^m}\right\} \tag{13}$$

Stresses in terms of initial pressure can be expressed as

$$\begin{aligned} \frac{T_{rr}}{Y} &= -P_{i2} + \frac{P_i}{(b/a)^m - 1}\left[1 - \left(\frac{b}{r}\right)^m\right], \\ \frac{T_{\theta\theta}}{Y} &= -P_{i2} + \frac{P_i}{(b/a)^m - 1}\left[1 - \left(1 - \frac{m}{2}\right)\left(\frac{b}{r}\right)^m\right] \end{aligned} \tag{14}$$

Further, we take $\alpha_2 \to 0$ in Eq. (12) at $r = b$ which represents fully plastic state.

$$|T_{\theta\theta} - T_{rr}|_{r=b} = \left|\frac{m}{2}\left(\frac{p_1 - p_2}{(b/a)^m - 1}\right)\right| \equiv Y_1 \tag{15}$$

Therefore, fully plastic state in spherical shell starts at the following pressure

$$P_f = \frac{p_1}{Y_1} - \frac{p_2}{Y_1} = 2\log(b/a) \tag{16}$$

Stresses for fully plastic state are

$$\begin{aligned}\frac{T_{rr}}{Y_1} &= -P_{f2} - P_f\frac{\log(r/b)}{\log(a/b)},\\ \frac{T_{\theta\theta}}{Y_1} &= -P_{f2} - P_f\frac{(1+2\log(r/b))}{2\log(a/b)}\end{aligned} \tag{17}$$

Now, we define non-dimensional components as

$$\begin{aligned}&R = r/b,\ R_o = a/b,\ S_{rr} = \frac{T_{rr}}{Y},\ S_{\theta\theta} = \frac{T_{\theta\theta}}{Y},\\ &P_i = \frac{p_1 - p_2}{Y},\ P_f = \frac{p_1 - p_2}{Y_1}\end{aligned}$$

In case of initial yielding, we have expressions in non-dimensional form as following

$$P_i = \left|\frac{2\big((R_o)^m - 1\big)}{m(R_o)^m}\right| \tag{18}$$

$$\begin{aligned}S_{rr} &= -P_{i2} + \frac{P_i}{(R_o)^{-m} - 1}\left[1 - (R)^{-m}\right],\\ S_{\theta\theta} &= -P_{i2} + \frac{P_i}{(R_o)^{-m} - 1}\left[1 - (1 - \frac{m}{2})(R)^{-m}\right]\end{aligned} \tag{19}$$

Expressions in non-dimensional form for fully plastic state are

$$P_f = |2\log(1/R_o)| \tag{20}$$

$$\begin{aligned}\frac{T_{rr}}{Y_1} &= -P_{f2} - P_f\frac{\log(R)}{\log(R_o)},\\ \frac{T_{\theta\theta}}{Y_1} &= -P_{f2} - P_f\frac{(1+2\log(R))}{2\log(R_o)}\end{aligned} \tag{21}$$

For Isotropic Case: Material constants are reduced to two only, i.e., in terms of Lame's constants as

$$\begin{aligned}&C_{11} = C_{12} = C_{33},\\ &C_{12} = C_{21} = C_{13} = C_{31} = C_{23} = C_{32} = C_{11} - 2C_{66},\\ &\alpha_1 = \alpha_2 = \alpha_3 = \alpha,\\ &C_{12} = \lambda,\ \ C_{11} = \lambda + 2\mu,\ \ C_{66} = \frac{1}{2}(C_{11} - C_{12}) = \mu\end{aligned} \tag{22}$$

By using Eqs. (19) and (22), the stress components for isotropic material can be expressed as

$$S_{rr} = -P_{i2} + \frac{P_i}{(R_o)^{-2C} - 1}\left[1 - (R)^{-2C}\right],$$
$$S_{\theta\theta} = -P_{i2} + \frac{P_i}{(R_o)^{-2C} - 1}\left[1 - (1 - C)(R)^{-2C}\right] \tag{23}$$

The expressions obtained are same as Verma [20].

4 Numerical Discussion

In order to discuss the impact of initial yielding and fully plastic state on spherical shell, the numerical values of elastic coefficients for different types of materials are shown in Table 1. Pressure is computed for different types of materials in Table 2. It is observed that spherical shell made up of brass and magnesium requires high

Table 1 Elastic coefficients of the materials in units 10^{10} N/m^2

Materials	C_{11}	C_{12}	C_{13}	C_{33}
Brass (isotropic material)	3.0	1.0	1.0	3.0
Magnesium (transversely isotropic material)	5.97	2.62	2.17	6.17
Beryl (transversely isotropic material)	2.746	0.98	0.67	4.69

Table 2 Pressure required to start initial yielding and fully plastic state in spherical shell

Materials	Radii ratio R_o	Pressure required for initial yielding (P_i)	Pressure required for fully plastic state (P_f)	Percentage increase in pressure $((P_f - P_i)/P_f) * 100$
Brass (isotropic material)	0.2	1.324795989	1.397940009	5.232271719
	0.4	1.198933985	1.045757491	−14.64742023
	0.6	1.058045685	0.795880017	−32.94035057
	0.8	0.904813255	0.602059991	−50.28622864
Magnesium (transversely isotropic material)	0.2	1.351096843	1.397940009	3.350870918
	0.4	1.218708766	1.045757491	−16.5383731
	0.6	1.072325706	0.795880017	−34.73459346
	0.8	0.914569729	0.602059991	−51.90674388
Beryl (transversely isotropic material)	0.2	1.092747678	1.397940009	21.83157566
	0.4	0.924133005	0.795880017	−16.11461341
	0.6	0.680666356	0.443697499	−53.40775125
	0.8	0.370845216	0.193820026	−91.33482951

pressure for yielding as compared to spherical shell made up of beryl at $R_o = 0.2$, but percentage increase in pressure required for initial yielding to become fully plastic is maximum for beryl material. It is observed from Fig. 1 that yielding in spherical shell made up of beryl starts at low values of the pressure as compared to other cases. It is also stated that yielding takes place quickly in thinner shells as compared to the thicker shell for different types of materials. In Fig. 2, stress distribution is shown for different types of materials along R. It is seen that radial as well as hoop stresses

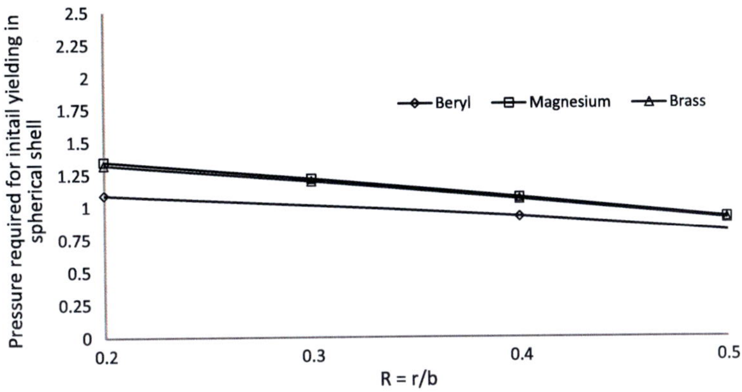

Fig. 1 Yielding in spherical shell for different pressures along radii ratio

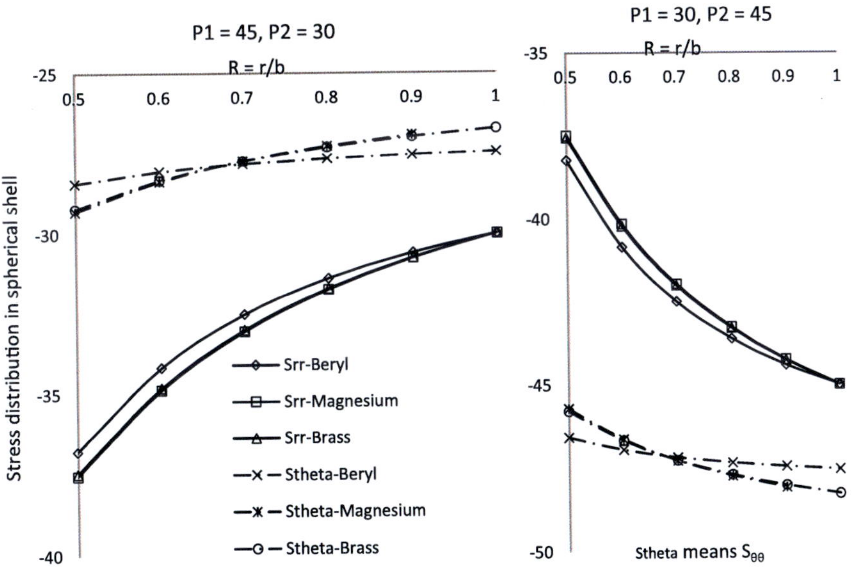

Fig. 2 Stress behavior of spherical shell when internal pressure = 45, external pressure = 30 and internal pressure = 30 and external pressure = 45

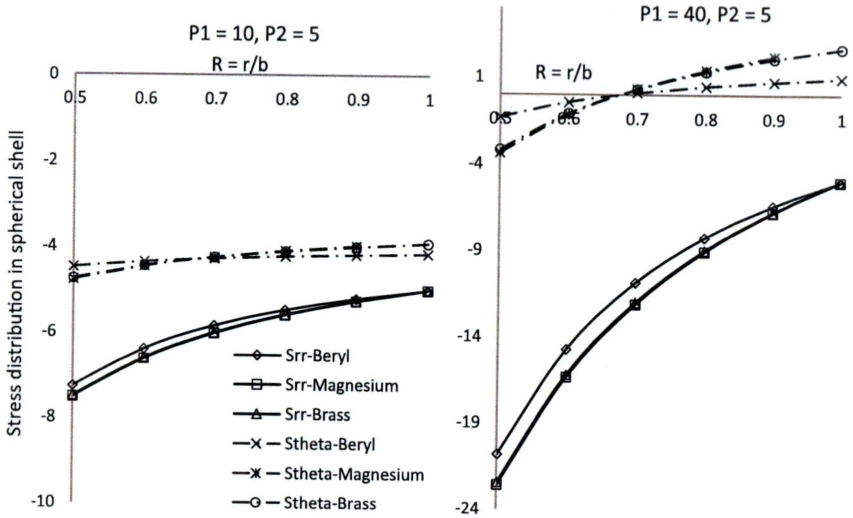

Fig. 3 Stress behavior of spherical shell when internal pressure = 10, 40 with fixed external pressure = 5

have maximum influence on the interior side of the spherical shell as compared to exterior side of the spherical shell when $P_1 > P_2$, but this effect gets reversed when $P_2 > P_1$. It is also seen that radial stresses are maximum at internal part of the spherical shell when internal pressure is more than external pressure, whereas hoop stresses are maximum on exterior part of the spherical shell when external pressure is more than internal pressure. Radial stresses as well as hoop stresses are found to be maximum for isotropic material brass as compared to the transversely isotropic materials magnesium and beryl. In Fig. 3, effect of internal pressure P_1 is studied on the spherical shell by keeping external pressure $P_2 = 5$. It is seen that there is increase in effect of radial stress on internal part of the shell with increase in internal pressure. In Fig. 4, stress behavior of spherical shell is discussed for increasing external pressure by keeping internal pressure $P_1 = 5$. More is external pressure; more is impact of hoop stresses on the external part of the spherical shell. In Figs. 5 and 6, the independent effect of external and internal pressure is shown on the spherical shell, respectively. It can be observed that stresses obtained are compressive in nature in Fig. 5, whereas hoop stresses are tensile in nature in Fig. 6. In Fig. 7, stress distribution in spherical shell is shown for fully plastic state. It is seen that effect of radial stresses is very high on the internal part of shell and leads to damage the internal part of shell for fully plastic state when $P_1 = 45, P_2 = 0$. In similar way, hoop stresses lead to damage to external part of the spherical shell for fully plastic state when $P_1 = 0, P_2 = 45$.

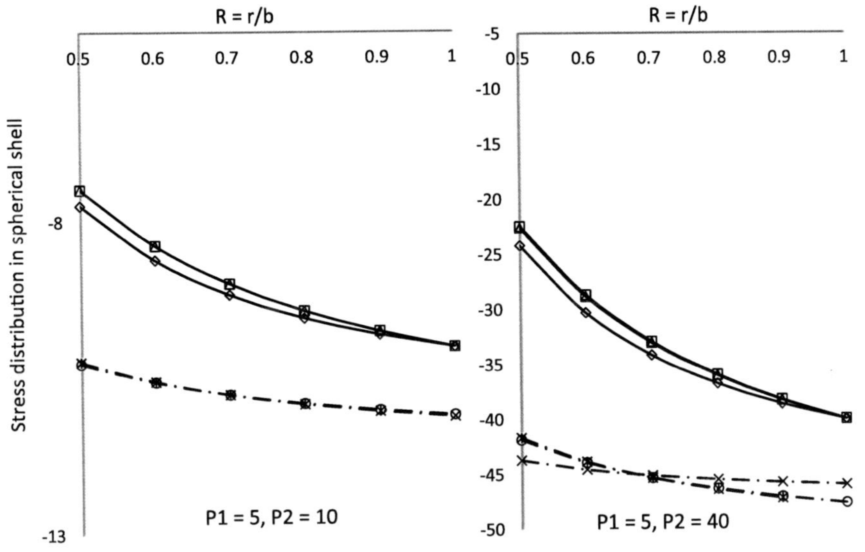

Fig. 4 Stress behavior of spherical shell when external pressure = 10, 40 with fixed internal pressure = 5

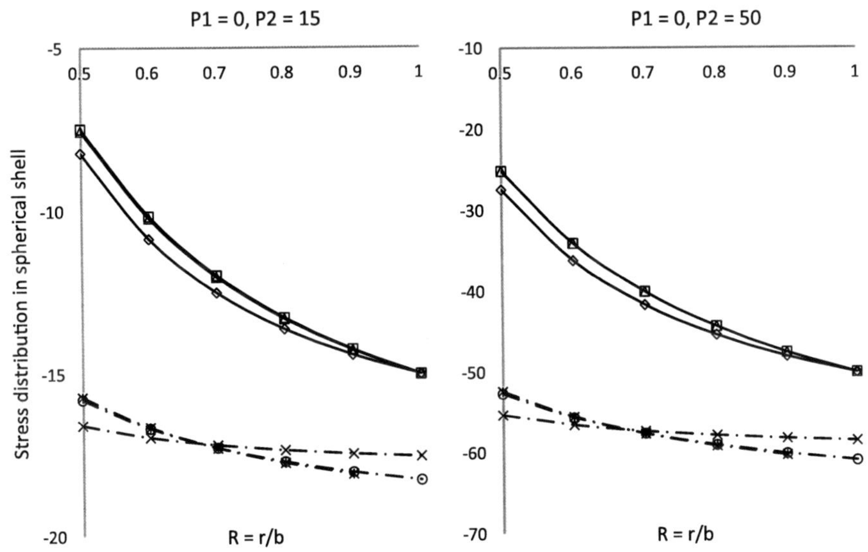

Fig. 5 Stress behavior of spherical shell when external pressure = 15, 50 without internal pressure

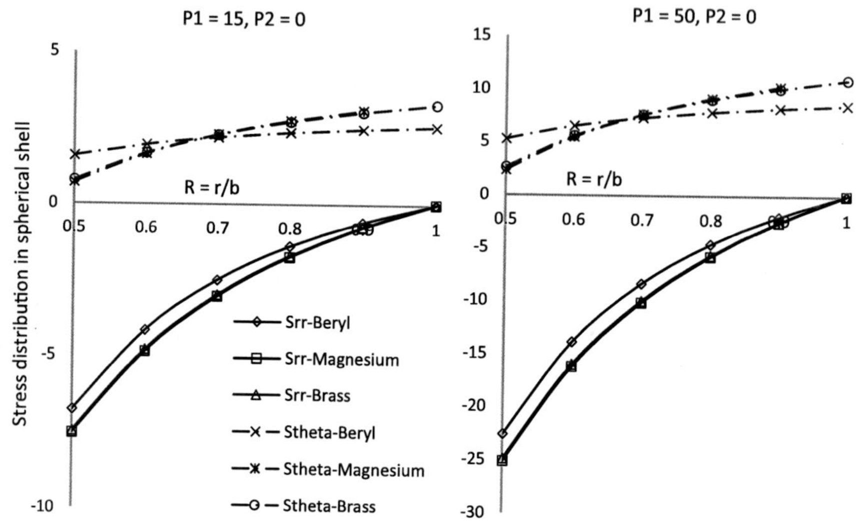

Fig. 6 Stress behavior of spherical shell when internal pressure = 15, 50 without external pressure

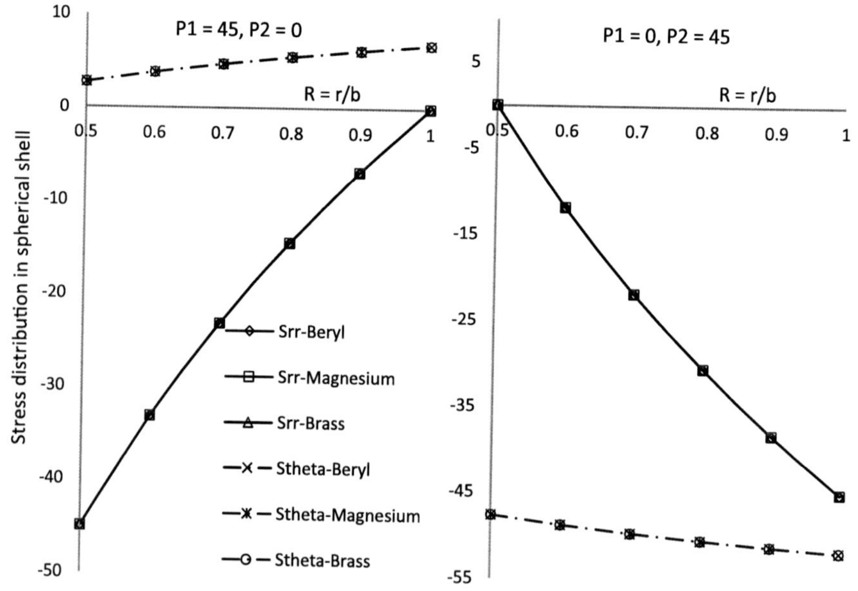

Fig. 7 Stress behavior of spherical shell for fully plastic sate when internal pressure = 45, external pressure = 0 and internal pressure = 0 and external pressure = 45

5 Conclusion

This problem concludes that the spherical shell made up of beryl is better for long life of spherical structure as compared to other materials. The main reason is that effective pressure required to attain fully plastic state from elastic state is maximum in case of beryl. Radial stresses and hoop stresses cause maximum damage of the spherical shell in case of isotropic material brass and transversely isotropic material magnesium.

References

1. Brandon M, Brock JS, Williams TO, Smith BM (2013) Benchmark analytic solution of the dynamic response of a spherical composed of a transverse isotropic elastic material. Int J Solids Struct 50:4089–4097
2. Żak M, Krawczuk A (2018) A higher order transversely deformable shell type spectral finite element for dynamic analysis of isotropic structures. Finite Elem Anal Des 142:17–29
3. Akhmedov NK, Sofiyev AH (2019) Asymptotic analysis of three-dimensional problem of elasticity theory for radially inhomogeneous transversely -isotropic thin hollow spheres. Thin-Walled Struct 139:232–241
4. Komijani M, Mahbadi H, Eslami MR (2013) Thermal and mechanical cyclic loading of thick spherical vessels made of transversely isotropic materials. Int J Press Vessels Pip 107:1–11
5. Mantari JL, Soares CG (2012) Analysis of isotropic and multilayered plates and shells shells by using a generalized higher-order shear deformation theory. Compos Struct 94(8):2640–2656
6. Thakur P (2009) Elastic-plastic transition stresses in a transversely isotropic thick-walled cylinder subjected to internal pressure and steady-state temperature. Therm Sci 13(4):107–118
7. Thakur P, Singh SB (2015) Elastic-plastic transitional stresses distribution and displacement for transversely isotropic circular disc with inclusion subject to mechanical load. Kragujevac J Sci 37:25–36
8. Gupta SK, Kumari P (2005) Elastic-plastic transition in a transversely isotropic disc with variable thickness under internal pressure. Indian J Pure Appl Math 36(6):329–344
9. Sharma S, Sahni M (2009) Elastic-plastic transition of transversely isotropic thin rotating disc. Contemp Eng Sci 2:433–440
10. Sharma S, Sahni M, Kumar R (2009) Elastic-plastic transition of transversely isotropic thick-walled rotating cylinder under internal pressure. Def Sci J 59:260–264
11. Sharma S, Sahni M, Kumar R (2010) Thermo creep transition of transversely isotropic thick—walled rotating cylinder under internal pressure. Int J Contemp Math Sci 5:517–527
12. Seth BR (1962) Transition theory of elastic- plastic deformation, creep and relaxation. Nature 195:896–897. https://doi.org/10.1038/195896a0
13. Seth BR (1966) Measure concept in mechanics. Int J Non-Linear Mech 1(1):35–40
14. Thakur P, Chand S, Sood S, Sethi M, Kaur J (2020) Density parameter in a transversely and isotropic disc material with rigid inclusion. Struct Integrity Life Serbia 20(2):159–164
15. Thakur P, Singh SB, Kumar S (2016) Steady thermal stresses in solid disk under heat generation subjected to variable density. Kragujevac J Sci 38:5–14
16. Thakur P, Singh SB, Kumar S (2015) Mechanical load in a circular rotating disk with shaft for different materials under steady-state temperature. Sci Tech Rev Mil Tech Inst Ratka Resanovića, Belgrade, Serbia 65(1):36–42
17. Thakur P, Verma G, Pathania DS (2017) Elastic-plastic stress analysis in a Spherical shell under internal pressure and steady state temperature. Struct Integr Life Serbia 17(1):39–43

18. Thakur P, Verma G, Pathania DS (2017) Elastic-plastic transition on rotating spherical shells in dependence of compressibility. Kragujevac J Sci 39:5–16
19. Verma G, Gupta S (2014) Elastic-plastic transition in shells under internal pressure. Int J Emerg Technol Adv Eng 4(8):126–129
20. Verma G, Pathania DS (2019) Elastic-plastic stress analysis of spherical shell under internal and external pressure. Struct Integr Life Serbia 19(1):3–7
21. Verma G (2020) Creep modelling of spherical shell under influence of internal and external pressure. Struct Integr Life Serbia 20(2):93–97
22. Verma G, Pathania DS (2020) Thermal creep stress analysis of functionally graded spherical shell under internal and external pressure. Struct Integr Life Serbia 20(3):297–302

Single Image Defogging Algorithm-Based on Modified Haze Line Prior

Pooja Pandey, Nidhi Goel, and Rashmi Gupta

Abstract In this paper, an effective single image defogging algorithm is proposed for the natural and synthetic datasets. The proposed algorithm is based on modified haze line prior technique for calculation of transmission map and atmospheric light parameter. Transmission map calculated using haze line prior is not refined because of some halos and artifacts. Due to this reason, the output obtained using haze line prior is not sharp and visually not appealing. In the proposed algorithm, guided filter has been introduced to overcome this problem. Both qualitative and quantitative comparisons have been done with some of the existing defogging methods to establish effectiveness of proposed method. The key benefit of the proposed algorithm is its capability to restore the sharpness of a foggy image without any post-processing and pre-processing of the image.

Keywords Defogging · Transmission map · Guided filter · Airlight

1 Introduction

Large particles existing in the atmosphere during bad weather conditions like fog causes several physical phenomena like scattering and absorption of light coming from the object. Due to these effects, visibility of the object reduces to a great extent and becomes a major concern in many computer and vision algorithms [1]. Thus, removal of fog from a given image is highly required for achieving satisfactory output in many vision algorithms. Addressing such problems are termed as image defogging algorithm.

Several defogging algorithms are based on the physical model of atmospheric scattering [2]. These methods are termed as restoration-based defogging methods. Restoration-based defogging methods calculate defog image using equation (1) [3].

P. Pandey · N. Goel (✉)
IGDTUW, Delhi, India

R. Gupta
NSUT, Delhi, India

G. Gupta et al. (eds.), *Proceedings of Academia-Industry Consortium for Data Science*, Advances in Intelligent Systems and Computing 1411,
https://doi.org/10.1007/978-981-16-6887-6_19

$$\hat{X}(p) = X(p)t(p) + A(1 - t(p)) \tag{1}$$

where $\hat{X}$ is foggy image; X is fog free image; t is transmission map, p is the location of pixel and A is atmospheric light. Equation (1) shows the relation between foggy image and clear image. Two parameters t, and A are required to calculate a clear image from a given foggy image. Thus, in a restoration-based defogging algorithm, accuracy of output depends on the estimation of transmission map t and atmospheric light A. Once these parameters are calculated, defog image can be calculated using Equation (2).

$$X(p) = \frac{\hat{X}(p) - A(1 - t(p))}{t(p)} \tag{2}$$

In many research works, more than one image (multiple images) has been used to calculate different parameters. But in recent years, the defogging algorithm proposed by different researchers is based on a single image only. The advantage of a single image defogging algorithm over multiple image defogging algorithms is its independency on external factors like change in environmental conditions. Single image defogging algorithms are simple to implement. Calculation of different parameters from single foggy image is a difficult task, and to overcome this, some assumptions or priors are used.

The proposed algorithm is a single image defogging algorithm based on modified haze line priors. In this, guided filter has been used at the intermediate stage to further refine the transmission map. Proposed algorithm restores fine features of the image like edge, texture, and contrast. The output shows that defog image is much enhanced as compared to the input foggy image and visually more appealing. The proposed algorithm is implemented on both synthetic as well as natural datasets, and output is compared with existing methods. Comparison results show that the proposed algorithm outperforms many existing defogging techniques.

This paper has been organized in different sections. Related works of the existing defogging methods are discussed in Sect. 2. Proposed work is explained in different steps in Sect. 3. Result analysis has been covered in Sect. 4. In Sect. 5, the paper is concluded with future work.

2 Related Work

Removal of fog from a given image has always been a demanding work. Different algorithms are proposed by several researchers to get a high-quality image from the degraded foggy images. Some researchers applied enhancement techniques to get clear image. In some defogging algorithms, restoration approach based on the physical model has been used. Fusion-based approaches have also been used by some researchers to calculate fog-free images.

The defogging algorithm based on enhancement methodology employs traditional image processing techniques such as histogram equalization, retinex method, frequency transform methods for contrast enhancement, edge restoration, and visibility improvement [4]. In histogram equalization-based method, different features of a foggy image are enhanced by expanding the dynamic range of a foggy image. Contrast enhancement using histogram stretching is also used to improve the contrast and visibility of the image [5, 6]. Histogram equalization-based defogging algorithms give good results for the dense foggy image as compared to retinex-based algorithm. But for natural datasets, enhancement of contrast using histogram equalization gives unrealistic appearance.

The defogging algorithm based on restoration approach uses atmospheric scattering model. In these methods, two parameters, transmission map and atmospheric light, are estimated. Fog-free images are then calculated using these parameters. In multiple images defog algorithm, more than one image is used to calculate different parameters [7]. For real-time application like driving assistance, multiple images based defogging algorithm cannot be used because of its dependency on environmental factors. In recent years, many single image defogging algorithms have been proposed by different researchers. Defogging method based on linear operation requires less processing time and suitable for real-time application [8]. Prior based defogging methods estimate transmission map and atmospheric light using some assumptions. He et al. proposed dark channel prior (DCP) [9] based on the assumptions that outdoor image captured in fog-free environment contains some pixels whose intensity is very less in at least one of the color channel. This algorithm is simple to implement, but it fails when the brightness of the scene is equivalent to atmospheric light. Similarly, Zhu et al. defogging method is based on color attenuation prior [10]. This prior is based on the assumptions that difference between brightness and saturation of pixel in foggy weather is more as compared to clear weather conditions.

In fusion-based defogging algorithms, two or more images are calculated from a single foggy image and calculated images are fused together to get fog-free image. In Galdron method [11] and Zhu et al. method [12], multiple images are generated from a single foggy image using Gamma operations, and fog-free image is calculated through fusion technique.

3 Proposed Algorithm

The proposed method is based on modified haze line prior technique. In the initial stage, haze line prior (HLP) [13] has been used for the calculation of transmission map and atmospheric light. Guided filter [14] further refines the transmission map and preserves different features of a foggy image. Figure 1 shows the block diagram of the proposed method.

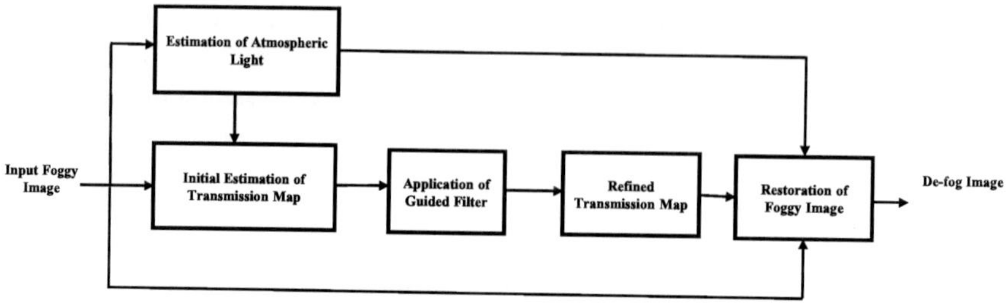

Fig. 1 Block diagram of the proposed method.

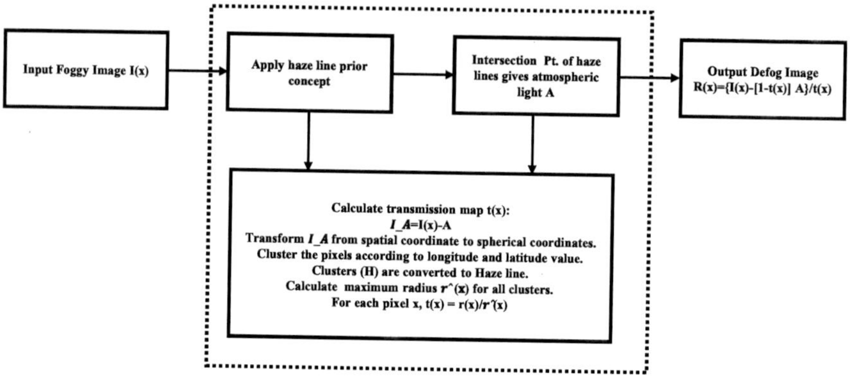

Fig. 2 Different steps of HLP method.

3.1 *Haze Line Prior (HLP)*

HLP is based on the observation that the colors of clear (fog-free) image are well estimated by a few hundred distinct colors. These colors are in the form of clusters in fog-free image. But if the same image is taken in foggy weather, then these clusters are converted into a line called a haze line. In clear weather conditions, pixels of the clusters are not local and distributed in the entire RGB space. In foggy conditions, the transmission map of a pixel depends exponentially on the distance of pixels. Due to this reason, pixels of the same clusters have different transmission map and thus converted into haze line. Different steps of HLP method are shown in Fig. 2.

3.2 *Estimation of Atmospheric Light*

Estimation of atmospheric light is based on dark channel [9]. Local region is selected in a dark channel of the captured scene which contains upper 0.1 percent high intense pixels, and then, the brightest pixel is selected from the original image corresponding to this local region. This brightest pixel is considered as atmospheric light A.

3.3 Initial Estimation of Transmission Map

Equation (1) shows that restoration of foggy image is based on transmission map parameter. Different steps which are followed for the estimation of transmission map can be enumerated as:

Step 1: Equation (1) is transformed, so that airlight becomes the origin.

$$\hat{X}_A(p) = \hat{X}(p) - A \tag{3}$$

Using Equation (1) and Equation (3), it can be written as:

$$\hat{X}_A(p) = [X(p)t(p) + A(1 - t(p))] - A \tag{4}$$

$$\hat{X}_A(p) = [X(p) - A]t(p) \tag{5}$$

Step 2: $\hat{X}_A(p)$ is transformed into spherical coordinates as represented in equation (6).

$$\hat{X}_A(p) = [r(p), \theta(p), \phi(p)] \tag{6}$$

Here, r represents distance from the origin, θ represents longitude, and ϕ latitude. Transmission map depends on distance of the pixels and only affects r. Thus, pixels having the similar θ and ϕ belong to the same haze line.

Step 3: For the same haze line, radius of the pixels from the origin can be defined as:

$$r(p) = \| X(p) - A \| \, t(p), 0 \leq t(p) \leq 1 \tag{7}$$

$$r_{max}(p) = \| X(p) - A \| \tag{8}$$

Using Equation (7) and Equation (8), transmission map of a pixel can be defined as [13]:

$$t(p) = \frac{r(p)}{r_{\max}(p)} \tag{9}$$

Step 4: To reduce the effect of noise, minimum value of the transmission map is maintained. Using Equation (1), minimum value of the transmission map can be written as:

$$t_{\min}(p) = 1 - \min_{c\epsilon(r,g,b)} \frac{X_c(p)}{A_c} \tag{10}$$

Thus, transmission map of a pixel after imposing minimum value can be written as:

$$\hat{t}(p) = \max[t(p), t_{\min}(p)] \tag{11}$$

3.4 Smoothening of Transmission Map

The transmission map obtained using Equation (11) contains halos and artifacts. To remove these artifacts, further smoothening (refinement) of transmission map is done using guided filter. In guided filter, detail layers are computed and then combined for better edge preservation. For any guidance image I_G and target image I_T, the filtering output I_o for pixel position 'i' and 'j' is can be written as:

$$I_{oi} = \Sigma_j W_{i,j}(I_G)I_{Tj} \tag{12}$$

Kernel of the filter W_{ij} depends only on I_G and independent of I_T. There is linear relation between input image and target image. In terms of filter's coefficient, output of the guided filter can be expressed as:

$$I_{oi} = a_k I_{Gi} + b_k, \forall i \epsilon w_k \tag{13}$$

where a_k and b_k are filter's coefficient, and it is considered constant in window w_k. Local linear relationship shows that output image I_O has an edge only if guidance image I_G has an edge. To determine filter coefficient, cost function which is minimized in the window is given as:

$$C(a_k, b_k) = \Sigma_{i\epsilon w_k}((a_k I_{Gi} + b_k - I_{Ti})^2 + \epsilon a_k^2) \tag{14}$$

ϵ is regularization parameter, and it prevents filter coefficients from very large value. Linear regression method is used to find solution of cost function mentioned in Equation (14), and it is calculated as:

$$a_k = \frac{\frac{1}{|\omega|}\Sigma_{i\epsilon\omega_k} I_{Gi} I_{Ti} - \mu_k \bar{I_{Tk}}}{\sigma_k^2 + \epsilon} \tag{15}$$

$$b_k = \bar{I_{Tk}} - a_k \mu_k \tag{16}$$

where, μ_k is mean and σ_k^2 is variance of guided image I_G in w_k, $| w |$ is total pixels number in considered window, and

$$\bar{I_{Tk}} = \frac{1}{| \omega |}\Sigma_{i\epsilon\omega_k} I_{Ti} \tag{17}$$

Local window is used to cover the entire image, and after computing filter's coefficient a_k and b_k for each window w_k, output of the guided filter is obtained as:

$$I_{Oi} = \frac{1}{| \omega |}\Sigma_{k:i\epsilon\omega_k} a_k I_{Gi} + b_k \tag{18}$$

Thus, transmission map calculated in Equation (10) is refined using guided filter. Refined transmission map ($t_{refined}(p)$) is further used for restoration of the foggy image.

3.5 *Restoration of Foggy Image*

The final defog image is recovered using Equation (19). Thus, the final restored image is based on a refined transmission map which is smooth. It does not contain halos and artifacts.

$$X(p) = \frac{\hat{X}(p) - A(1 - t_{refined}(p))}{t_{refined}(p)} \quad (19)$$

4 Result Analysis

In this section, result of the proposed work has been analyzed both qualitatively and quantitatively. Also, output of the proposed algorithm is compared with some of the existing defogging methods. For fair analysis, both natural and synthetic foggy images are considered. To see the robustness of the proposed algorithm, more than 300 foggy images have been used and analyzed. The proposed algorithm is simulated using MATLAB (2018b) software with Core i3 processor using 64-bit system and 8 GB RAM.

4.1 *Qualitative Analysis*

Poor visibility is a major problem in foggy weather conditions. Images captured in poor visibility are low in contrast, sharpness, and looks dull. Qualitative analysis deals with the visual perception of an image. Images which have strong contrast, natural color, and sharp edges are visually appealing. Defogging method which gives good visual effect is considered as an effective defogging method. In Fig. 3, step by step output of the proposed method on five different types of images have been shown. It can be seen that refined transmission map is smooth as compared to the initial transmission map. Due to this reason, defog image is also enhanced and gives visually appealing output.

Figure 4 shows defogging result of some existing methods and proposed method. Six different foggy images are taken and named as FI1, FI2, FI3, FI4, FI5, and FI6 as shown in first row of Fig. 4. Output of the proposed method has been shown in the last row. It can be clearly seen that images are sharp and visually more appealing as compared to some of the existing techniques. In Tarel et al. [8] technique and He et al. [9] technique, defogging of a image is based on patch-based analysis. It causes

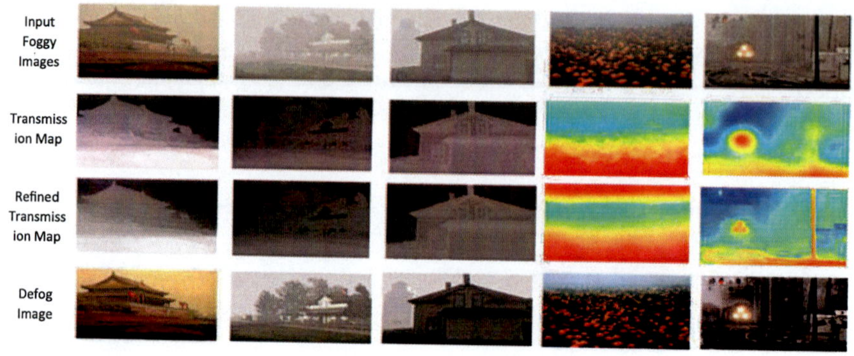

Fig. 3 Step-by-step output of proposed method.

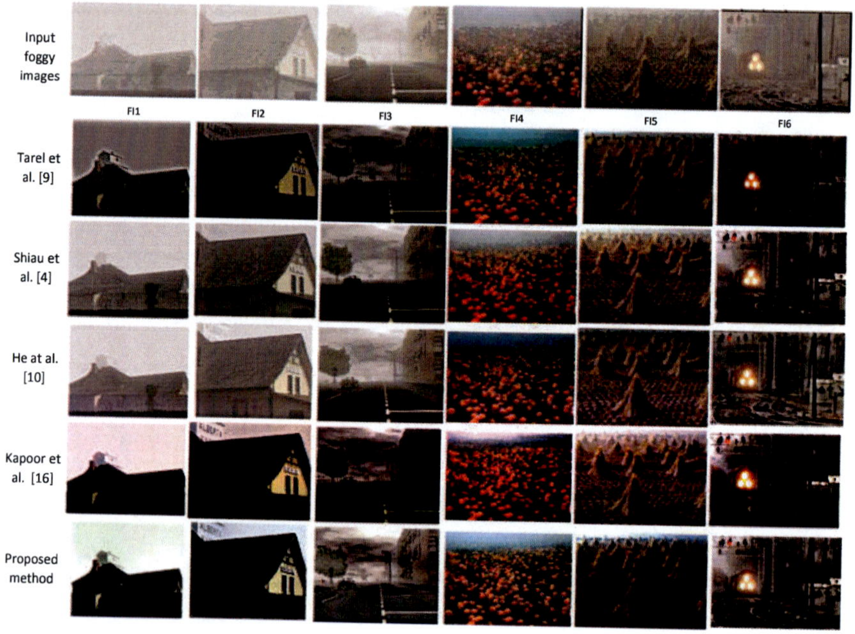

Fig. 4 Visual comparison of proposed method with some existing method.

artifacts in the estimated transmission map, and thus, defog image is not sharp. In contrast to this, the proposed method is based on global analysis and performs on a pixel-to-pixel basis. It gives output defog image enhanced in features like sharpness and contrast.

Table 1 FRF of some existing defogging methods and proposed method

Images	Tarel et al. [8]	Shiau et al. [19]	He et al. [9]	Kapoor et al. [20]	Proposed method
FI1	0.52	0.38	0.66	0.70	0.83
FI2	1.22	0.92	1.06	1.34	1.66
FI3	1.04	1.12	1.46	1.49	1.39
FI4	2.04	1.26	1.33	2.22	2.22
FI5	1.27	1.09	1.20	1.58	1.63
FI6	1.43	0.94	1.66	1.98	2.01

4.2 *Quantitative Analysis*

Parameters used for objective evaluation of proposed method are: fog reduction factor (FRF), structural similarity index measure (SSIM) and time complexity.

Fog reduction factor (FRF) measures the difference in fog density of input image and output image. Fog density of an image is calculated using fog-aware density evaluator (FADE) [15]. To evaluate fog density in a given image, different metrics are calculated like local mean, variance, saturation, entropy, colorfulness index, dark channel, and sharpness. Based on these metrics, fog density in a given image is calculated. Mathematically, FRF can be calculated as[16]:

$$FRF = D_i - D_o \quad (20)$$

where D_i and D_o represent fog density of input image and output image, respectively. In this paper, the input image is a foggy image, and an output image is defog image. High value of FRF means output image which is clear as compared to the input foggy image and defog algorithm is performing better. Table 1 shows comparison analysis of proposed method with some existing methods in terms of FRF. It can be seen that proposed method gives high value of FRF as compared to other methods.

Structural similarity index measure (SSIM) is a quantitative measure which is used for finding the quality of defog image [17, 18]. It mainly compares the similarity between two images in terms of luminosity, contrast, and structure. SSIM of each result is calculated using the following equation:

$$SSIM(I_o, I_i) = l[I_o, I_i] * c[I_o, I_i] * s[I_o, I_i] \quad (21)$$

where l is luminous which depends on average value of I_o, I_i, c is contrast which depends on variance of I_o, and I_i and s is structure which depends on variance and covariance of I_o and I_i.

SSIM comparison is given in Table 2. In this paper, input foggy image is considered as a reference image for calculation of SSIM parameter. Thus, low value of SSIM shows that defogging method is more effective. Proposed method achieves less SSIM

Table 2 SSIM of some existing defogging methods and proposed method

Images	Tarel et al. [8]	Shiau et al. [19]	He et al. [9]	Kapoor et al. [20]	Proposed method
FI1	0.56	0.42	0.43	0.42	0.35
FI2	0.46	0.50	0.36	0.39	0.28
FI3	0.61	0.50	0.52	0.41	0.41
FI4	0.58	0.55	0.41	0.38	0.30
FI5	0.51	0.44	0.40	0.28	0.26
FI6	0.49	0.41	0.37	0.29	0.25

Table 3 Time complexity analysis (in second) of some existing defogging methods and proposed method

Images	Tarel et al. [8]	Shiau et al. [19]	He et al. [9]	Kapoor et al. [20]	Proposed method
600x450	8.23	8.44	12.23	14.07	2.25
1024x768	69.29	49.54	36.89	40.02	22.37
1536x1024	218.04	46.31	73.57	80.63	22.92
1803x1080	101.10	66.39	90.72	100.08	58.06
1600x1080	94.03	117.21	102.66	112.99	55.72

value as compared to many existing methods. In the proposed technique, transmission map is calculated using guided filter. Guided filter preserves various features like edges and texture of image. Restored defog image is mainly based on transmission parameter. Better estimation of transmission parameter gives better result in terms of SSIM.

Time complexity is a parameter that measures the total time required to get final output image from an input image. Time complexity analysis for an algorithm is important because different applications require different time. For real-time applications like driving assistance and object tracking, time complexity parameter is an important criterion.

Table 3 shows time complexity analysis of proposed method and some other defogging techniques. In proposed method, post-processing has not been used, and so, it requires less processing time. In He et al. method, soft matting has been used which requires large memory and high processing time. Kapoor et al. defogging method is based on post-processing and requires high processing time.

5 Conclusion

In this paper, defogging technique using single image has been proposed for visibility improvement. For image restoration, transmission map and atmospheric light parameter are estimated and final output is calculated. In the proposed technique, guided filter is used to remove artifacts of the transmission map. Refined transmission map restores edges and texture of the image. Thus, defog image calculated using refined transmission map is sharp and visually more appealing. For analysis purpose, both natural and artificial datasets are taken and compared with existing defogging methods. Result analysis shows that the proposed method gives better results as compared to many existing methods on different datasets.

The proposed algorithm works on per pixel basis of an image, and thus, time complexity is high for a large size image. But for real-time applications, processing time required is much less. To reduce the time complexity, an embedded system can be designed to calculate different parameters required for image defogging. In future work, hardware-based defogging methods will be taken into consideration, and proposed work is extended in the direction of real- time-based applications.

References

1. Li B et al (2019) Benchmarking single-image dehazing and beyond. IEEE Trans Image Process 28(1):492–505
2. Lee S et al (2016) A review on dark channel prior based image dehazing algorithms. J Image Video Proc 2016:4
3. Narasimhan SG, Nayar SK (2001) Removing weather effects from monochrome images. In: Proceedings of IEEE conference computer vision and pattern recognition (CVPR), II-186–II-193
4. Wang W, Yuan X (2017) Recent advances in image dehazing. IEEE/CAA J Autom Sinica 4(3):410–436
5. Mathur M, Goel N (2018) Enhancement of underwater images using white balancing and Rayleigh-stretching. In: 5th international conference on signal processing and integrated networks (SPIN), pp 924–929
6. Mathur M, Goel N (2020) Enhancement of Nonuniformly Illuminated Underwater Images. World Sci IJPRAI
7. Narasimhan S, Nayar S (2003) Interactive (de) weathering of an image using physical models. In: Proceedings of IEEE workshop color photometric methods computer vision, vol 6, pp 1–8
8. Tarel JP, Hautiere N (2009) Fast visibility restoration from a single color or gray level image. In: Proceedings of 12th IEEE international conference on computer vision, pp 2201–2208
9. He K et al (2011) Single image haze removal using dark channel prior. IEEE Trans Pattern Anal Mach Intell 33:12
10. Zhu Q et al (2015) A fast single image haze removal algorithm using color attenuation prior. IEEE Trans Image Process 24(11):3522–3533
11. Galdran A (2018) Image dehazing by artificial multiple-exposure image fusion. Signal Process 149 (Elseveir)
12. Zhu Z et al (2021) A novel fast single image dehazing algorithm based on artificial multiexposure image fusion. IEEE Trans Instrum Meas 70:1–23
13. Berman D et al (2016) Non-local image dehazing. In: IEEE conference on computer vision and pattern recognition (CVPR), Las Vegas, pp 1674–1682

14. He K et al (2010) Guided image filtering. In: European conference on computer vision
15. Choi LK et al (2015) Reference less prediction of perceptual fog density and perceptual image defogging. Image Process 24(11):388–3901
16. Liu Y et al (2019) A unified variational model for single image dehazing. IEEE Access 7:15722–15736. https://doi.org/10.1109/ACCESS.2019.2894525
17. Salazar-Colores S et al (2020) Fast single image defogging with robust sky detection. IEEE Access 8:149176–149189
18. Salazar-Colores S et al (2019) A fast image dehazing algorithm using morphological reconstruction. IEEE Trans Image Process 28(5):2357–2366
19. Shiau YH et al (2014) Weighted haze removal method with halo prevention. J Visl Commun Image Represent 25(2):445–453
20. Kapoor R et al (2019) Fog removal in images using improved dark channel prior and contrast limited adaptive histogram equalization. Multimed Tools Appl 78:23281–23307

A Reversible and Rotational-Invariant Watermarking Scheme Using Polar Harmonic Transforms

Manoj K. Singh, Sanoj Kumar, Deepika Saini, and Gaurav Bhatnagar

Abstract Polar harmonic transformation (PHT) is a well-known family of rotation and scaling image-invariant descriptors. The PHT is frequently used in the watermarking applications. The watermark bits of the image are mixed into the image moments. The modified moments are then used to reconstruct the watermarked image. The ability of the image embedded with the watermark bits to resist an attack depends on the strength of the image moments to resist the attack. Therefore, the accuracy of the moments determines the robustness of the watermarking algorithm. An improved method based on the combination of analytical and numerical techniques is used for computing the image moments. The more accurate moments can be then used for the watermarking uses. The experimental results show that the improved moments have the ability to resist rotational attacks and are completely reversible. A comparative study also suggest that the improved moments have better robustness for rotational attacks than some of the existing image moment-based watermarking techniques.

Keywords Image moments · Rotational invariance · Polar harmonic transformation · Image watermarking

M. K. Singh · S. Kumar (✉)
Department of Mathematics, University of Petroleum and Energy Studies, Dehradun, India
e-mail: sanoj.kumar@ddn.upes.ac.in

M. K. Singh
e-mail: mkumar@ddn.upes.ac.in

D. Saini
Department of Mathematics, Graphic Era (Deemed to be) University, Dehradun, India
e-mail: deepikasaini@geu.ac.in

G. Bhatnagar
Department of Mathematics, Indian Institute of Technology Jodhpur, Jodhpur, India
e-mail: goravb@iitj.ac.in

G. Gupta et al. (eds.), *Proceedings of Academia-Industry Consortium for Data Science*, Advances in Intelligent Systems and Computing 1411,
https://doi.org/10.1007/978-981-16-6887-6_20

1 Introduction

In modern world, information sharing is more easier than ever and so is the contamination of the information at the sharing channel. Multimedia is often used as a medium for the sharing of hidden information. In addition, hidden text or image in the form of a watermark is also embedded in a image used for the validation of source file and copyright protection. The channel over which images are exchanged may not be secure. Therefore, the information being intended to be shared may be compromised. Inserting watermark in the image for copyright protection of inserting watermark as hidden information, both required a robust watermarking algorithm. Over the years, various state-of-the-art watermarking techniques are developed. Many of them are very robust with many forms of transformations including rotational and resizing attacks. The ariel images are generated from various sources such as from the airborne cameras or generated from various instruments [11]. The ariel images are also shared in online mode. Therefore, their security is an important task too.

Watermarking is a very popular method for identification of the rightfulness of the images, while communicating the images can be compromised using various transformation. Then, the inserted watermark can be an useful information to test if the image is undergone a transformation. Let us consider a situation in which a watermark is inserted into an image using a robust techniques, and let the watermarked image is transformed during communication. Upon receiving the watermarked image, if the watermark can still be extracted, then the image can be verified to be correct. But for this to happen, the algorithm to insert and extract the watermark must be robust to various transformation. A large number of techniques are proposed and applied over different kind of images. Many of the algorithms are also robust against a number of image transformations. However, very few are robust against rotation. In [1], DC coefficients are used in spatial domain for watermarking, and the results shows that the watermarking scheme is robust against many attacks. The authors in [1] have also compared their algorithm with DC coefficients based methods in [8, 17]. They observed that the accuracy of their algorithm is better at rotational attack. However, the results were shown only at right angle rotational attack. Image moments based approach are also used for the watermarking. In reference [16], fractional Charlier moments were used for the image watermarking. The authors in [14] shown the applications of the fractional discrete Tchebyshev in watermarking. Tchebyshev moments are also used in [2]. A variant of the radial harmonic Fourier moments have also been used for watermarking in reference [13]. A variant of the Chebyshev moments have also been used for watermarking in reference [3]. The authors in [6] combine the DCT with Zernike moments for the image watermarking. The watermark bits can be applied over an image function which is invariant to rotational transformation. One such class of image functions is the polar harmonic transform (PHT) image moments. Theoretically, the PHT moments are known to be invariant to rotation of the image. However, since image are discrete two-dimensional functions, the invariant property of the image moments depends on accurately computed moments. In reference [5], PHT moments were computed using zeroth-order approximation, while in [4] an interpolation-based method is proposed for computing the PHT moments. In this article, a novel computational approach used in [10] is used for computing PHT moments. In reference [10], the PHFT moments are computed and applied on the medical images. The techniques used in [10] shown to improve the accuracy of moments and thereby improving the usefulness of the watermarking. The novelty of the article is the use of moment computing approach of [10] and its application for watermarking over the ariel images. The proposed techniques using PCET and PST have improved the robustness of the watermarking for rotational attacks. The proposed watermarking techniques are also reversible, and in the sense, the watermark received from the watermarked image is exactly similar to the inserted watermark. The proposed techniques for rotational invariant watermarking are compared with the PZM and ZM of [15], PHT in [4, 5] and PHFT in [7]. The comparison results show that the proposed techniques show better rotational invariance property than the recent methods discussed in this article.

This section is followed by an introduction over the PHT, their computations, and steps of watermark insertion and extraction process. The results are discussed in the Sect. 3. The article concludes with a conclusion section.

2 Polar Harmonic Transform

Polar complex exponential transform (PCET), polar sine transform (PST), and polar cosine transform (PCT) are collectively known as polar harmonic transform (PHT). PCET C_{uv} of order u in the set $\mathbb{Z}$ and repetition v in the set $\mathbb{Z}$, where $\mathbb{Z}$ is the set of integers, of a function $f(r, \phi)$ in polar domain is a complex function defined over the unit disk $|z| \leq 1$ as,

$$C_{uv} = \frac{1}{\pi} \int_0^{2\pi} \int_0^1 \overline{H_{uv}(r, \phi)} g(r, \phi) r dr d\phi \tag{1}$$

where

$$H_{uv}(r, \phi) = e^{i(2\pi u r^2 + v\phi)}, \tag{2}$$

The functions $H_{uv}(r, \phi)$ form an orthogonal system. that is,

$$\int_0^{2\pi} \int_0^1 H_{u_1 v_1}(r, \phi) \overline{H_{u_2 v_2}(r, \phi)} g(r, \phi) r dr d\phi = \pi \text{ if } (u_1, v_1) = (u_2, v_2) \tag{3}$$

$$= 0 \text{ if } (u_1, v_1) \neq (u_2, v_2) \tag{4}$$

PCT with order $u \in \mathbb{Z}_+$ where $\mathbb{Z}_+$ is the set of nonnegative integers and repetition v in the set $\mathbb{Z}$ of a function $f(r, \phi)$ in polar domain is a complex function defined over the domain $|z| \leq 1$ as,

$$C_{uv}^{c} = \Omega_u \int_0^{2\pi} \int_0^{1} \overline{H_{uv}^{c}(r,\phi)} f(r,\phi) r dr d\phi \tag{5}$$

where

$$H_{uv}^{c}(r,\phi) = \cos(\pi u r^2) e^{iv\phi}, \tag{6}$$

and

$$\Omega_u = \begin{cases} \frac{1}{\pi}, & \text{if } u = 0. \\ \frac{2}{\pi}, & \text{if } u > 0. \end{cases} \tag{7}$$

Similarly, PST with order $u \in \mathbb{Z}_+$ where $\mathbb{Z}_+$ is the set of nonnegative integers and repetition $v \in \mathbb{Z}$ of a function $f(r,\phi)$ in polar domain is a complex function defined over the domain $|z| \leq 1$ as,

$$C_{uv}^{s} = \Omega_u \int_0^{2\pi} \int_0^{1} \overline{H_{uv}^{s}(r,\phi)} g(r,\phi) r dr d\phi \tag{8}$$

where

$$H_{uv}^{s}(r,\phi) = \sin(\pi u r^2) e^{iv\phi} \tag{9}$$

Using the property given in Eq. 4, we can reconstruct the function $f(r,\phi)$ in the polar domain as:

$$f(r,\phi) = \sum_{u=-\infty}^{\infty} \sum_{v=-\infty}^{\infty} C_{uv} \tag{10}$$

2.1 PHT Computation

Let $I(i, j)$ be a square image of size $N \times N$. Let the center of image be the origin of the Cartesian coordinate system assigned to the image. The rows and the columns be parallel to the coordinate axes. Let us consider the image as a bounded region in the two- dimensional Euclidean space $\mathbb{R}^2$ with the left boundary as $x = -1$, right boundary as $x = 1$, lower boundary as $y = -1$, and the upper boundary as $y = 1$. In this coordinate system, a unit circle (a circle with equation $x^2 + y^2 = 1$) will be inscribed in the image with boundaries of the image as tangents to the circle. Both the length and width of the image will be two units, and the width of the pixel along the x-axis (δx) and along y-axis (δy) will be $\delta x = \delta y = \frac{2}{N}$. The centers of the image pixels are denoted by (x_i, y_j), $i, j = 0, 1, \ldots, N-1$. The PCET moments for an image is approximated by modifying Eq. 1 as

$$c_{uv} = \frac{2}{\pi} \sum_i \sum_j f(i, j) \overline{H_{uv}(x_i, y_j)} \delta x \delta y \tag{11}$$

where i, j runs over the pixels with centers (x_i, y_j) and the sum is restricted to run over centers lying within and on the unit circle. This approximation to the moments is known as zeroth-order approximation. The zeroth-order moment method is used by the authors of [15] and [5]. The accuracy of the moments will be more accurate if function $\overline{H_{uv}(x_i, y_j)\delta x \delta y}$ is refined as

$$\overline{H_{uv}(x_i, y_j)\delta x \delta y} = \int \int \overline{H_{uv}(r, \phi)} r dr d\phi \tag{12}$$

Here, (r, ϕ) is center (x_i, y_j) of the pixels in the polar coordinates and lying inside or on the unit circle. The region of the double integration depends on location of center of the pixel. If the center is inside the unit circle, then the boundaries of the pixels are taken as the limits of the double integration. If the center is outside the unit disk, then the integration in Eq. 12 is 0 and the pixel contributes zero to the moments. When the circular boundary passes through a given pixel, the limits of the integral will no longer be only rectangular boundaries. At least two limits of the integration are circular curves. In previous studies, such circular boundaries were not considered as per our knowledge. This implies that the complete unit circular domain was not taken into account for computing the moments. Since in this study, circular boundaries were also considered, the limits of integration for intersecting pixels in Eq. 12 are no more constants. The image function is discrete and finite in nature, and therefore, their integration is obtained numerically. Quadpack [9] which is a subroutine package for automatic integration is used often in numerical integration. In this study, two steps are used for the computation of the moments. The first step is finding analytically the limits of the integration. In the second step, these limits are passed on to the function quadpack for computing the double integration. The image is reconstructed using Eq. 13

$$f(i, j) = \sum_{u=0}^{u\max} \sum_{v=-v\min}^{v\max} C_{uv} H_{uv}(x_i, y_j) \tag{13}$$

Since these integration process are computationally intensive, to save computational time, Eq. 11 was observed carefully. It was observed that to compute moments of different images, we need to compute the more computationally intensive part is done only once for every pixel. The computed moments then are applied on all the images to evaluate the image moments.

2.2 Watermarking Algorithm

Let $W(i, j)$ be a watermark image with size $n \times n$. Let W_i, $i = 1, 2, \ldots, n^2$ represents the n^2 bits of the watermark image. Each watermark bit is embedded into single image moment. Therefore, n^2 different image moments were computed for the watermark embedding. The selection of better moments are also an integral task for the watermark inserting and extracting process. The watermark bits are inserted into the absolute values of the complex-valued moments. The absolute value of the moments at lower order has higher values, while the absolute value of the higher moments has lower value. Therefore, inserting watermark bits into higher moments might decrease the quality of the watermarked image considerably. Therefore, the order of the moments was restricted between 0–25 and repetition between -25–25. In addition, $C_{uv} = \overline{C_{u\,-v}}$, that is the moment with repetition v is the conjugate of the moment with the repetition negative v. Therefore, only repetitions with positive sign are selected as candidate for the watermark bit insertion. From these $26 \times 25 = 650$ candidate moments ($u >= 0$ and $v > 0$), n^2 moments were selected randomly. Let M_i, $i = 1, 2, \ldots, n^2$ represent the selected moments. The absolute values of the moments M_i are denoted with A_i, i.e., $A_i = |M_i|$. Let θ_i denoted the argument of the moments M_i for $i = 1, 2, \ldots, n^2$. Using the division algorithm, two real numbers Q_i and $0 \leq R_i < \Delta$ can be found, such that $A_i = Q_i\Delta + R_i$, where Δ represents a quantization step. The watermarking algorithm similar is line with the algorithm in [7]. The watermark bits are inserted into the absolute values A_i by modifying it as the following equations.

$$A'_i = \begin{cases} (Q_i + 1.5)\Delta, & \text{if } W_i + Q_i = 2k + 1 \text{ for some integer } k. \\ (Q_i + 0.5)\Delta, & \text{if } W_i + Q_i = 2k \text{ for some integer } k \end{cases} \tag{14}$$

for $i = 1, 2, \ldots, n^2$. The modified absolute moments leads to the modified image moments M_i as following equation,

$$M'_i = A'_i e^{j\theta_i}, \quad i = 1, 2, \ldots, n^2 \tag{15}$$

here $j = \sqrt{-1}$.

The watermarked image are obtained from the recomputed moments using the following equation,

$$I'(i, j) = 2\text{Re}\left(\sum_{k=0}^{n^2} M_k H_k\left(x_i, y_j\right)\right) \tag{16}$$

The 'Re' represents the real part of complex-valued number, and the sum runs over k containing the indices of u and v. Therefore, $H_k(x_i, y_j)$ is obtained from the corresponding u and v. If the moment M_k is modified, then the moment corresponding to order u and repetition $-v$ will also be modified in similar way. Therefore, the conjugate complex numbers in the right-hand-side of Eq. 13 will contribute two times the real part of any one of then, to the sum, that is,

$$C_{uv}H_{uv}(x_i, y_j) + C_{u-v}H_{u-v}(x_i, y_j) = 2\mathrm{Re}(C_{uv}H_{uv}(x_i, y_j))$$

Let $I(i, j)$ denotes the original image, and the reconstructed image with the original moments is represented by $g(i, j)$. The difference between the given image and the reconstructed image can be represented as an image $e(i, j)$, that is,

$$e(i, j) = I(i, j) - g(i, j) \tag{17}$$

$i, j = 0, 1, \ldots, N-1$. Let the reconstructed image is represented with $h(i, j)$ by changing M_k to M'_k in Eq. 17. The watermarked image should be generated at the end by adding $h(i, j)$ and $e(i, j)$, that is,

$$I'(i, j) = h(i, j) + e(i, j) \tag{18}$$

2.3 *Watermark Extraction*

Since watermark bits are mixed into the randomly selected image moments, to extract the watermarks, correct identification of the moments is essential. The moment indices are generated from the seed of the random number generator. Only the seed needs to be communicated to know the moments used for the watermark embedding. Let M'_k be the required moments of the watermarked image, and let $A'_k,\ k = 1, 2, \ldots n^2$ be their absolute values. Applying the division algorithm, Q_k and $0 \leq R_i < \Delta$ can be obtained such that, $A'_k = Q_k\Delta + R_k$ for $k = 1, 2, \ldots, n^2$. The watermarked bits W'_k are extracted from using the following equations.

$$W'_k = \begin{cases} 1, & \text{if } Q_k \text{ is odd.} \\ 0, & \text{if } Q_k \text{ is even} \end{cases} \tag{19}$$

for $k = 1, 2, \ldots, n^2$.

3 Experimental Result and Discussions

The watermarking algorithm has been applied on to the ariel images of size 512×512 obtained from the USC-SIPI image database[12]. Five of the ariel images are shown in Fig. 1. A binary image of size 16×16 is selected as of watermark image (Fig. 1). The quantization step Δ is fixed as 0.2. The quality of the watermarked image when compared with the base image is measured as peak signal-to-noise ratio (PSNR).

Fig. 1 Selected host images from the USC-SIPI image database and the selected watermark. Three images shown in the top row and first two images from left are host images, and the watermark image is shown at right-hand-side

The error present, if any, in the extracted watermark is evaluated in the form of bit-error-ratio (BER) and is defined in following equation

$$\text{BER} = \frac{b}{N^2} \tag{20}$$

where b is the total number of mismatched pixel between the watermark image and the extracted watermark image. Since, $b \geq 0$, N^2 is positive and $b \leq N^2$, the BER will be in the range 0–1. The original-extracted watermark pair will look similar if number of mismatched pixels, b will be very small and the BER in this case will be close to 0. On contrary, if the two watermark images look dissimilar, that is the number of mismatched pixels, b is high then the BER will be closer to 1. The watermark algorithm presented in this paper is compared with the ZM, PZM in [15], PHT in [4, 5], and PHFT in [7]. Many forms of attacks are applied on the watermarked images to test the robustness of the watermarking. Experiments presented in the paper is implemented using the Python 3 on an Ubuntu machine with 8GB RAM.

Table 1 shows the average results of 37 images in the ariel images database. The first column shows the proposed techniques along with the techniques compared for the watermarking. The second column represents the average BER of the extracted watermark when the techniques listed in the first column is applied over the ariel images. It is clear from the table that the proposed PCET, PCT, and PST are com-

Table 1 First column shows the techniques selected for the comparison among watermarking algorithms on the ariel images

	0°	10°	20°	30°	40°	50°	60°	70°	80°	90°
Proposed PCET	0.0000	0.0003	0.0020	0.0088	0.0230	0.0196	0.0078	0.0020	0.0003	0.0000
Proposed PCT	0.0000	0.0001	0.0045	0.0249	0.0604	0.0574	0.0216	0.0044	0.0003	0.0000
Proposed PST	0.0000	0.0000	0.0018	0.0130	0.0356	0.0355	0.0118	0.0019	0.0002	0.0000
ZM	0.4955	0.5012	0.5014	0.5000	0.4964	0.4964	0.5009	0.4925	0.5081	0.4955
PZM	0.4553	0.4901	0.5019	0.4963	0.4996	0.5122	0.4988	0.4932	0.4941	0.4553
PCET	0.0936	0.0940	0.0946	0.1010	0.1124	0.1117	0.1017	0.0951	0.0937	0.0936
PCT	0.0346	0.0346	0.0368	0.0493	0.0750	0.0708	0.0476	0.0363	0.0346	0.0346
PST	0.0291	0.0291	0.0306	0.0381	0.0580	0.0557	0.0382	0.0308	0.0291	0.0291
PCET	0.0000	0.0004	0.0022	0.0092	0.0261	0.0228	0.0088	0.0021	0.0005	0.0000
PCT	0.0000	0.0000	0.0018	0.0151	0.0404	0.0392	0.0130	0.0014	0.0000	0.0000
PST	0.0250	0.0246	0.0264	0.0374	0.0609	0.0597	0.0374	0.0265	0.0241	0.0252
PHFT	0.0000	0.0015	0.0045	0.0165	0.0427	0.0423	0.0170	0.0055	0.0012	0.0000

The second column is for the BER of the extracted watermark. The last nine columns represent the BER of the watermark with rotational attacks at angles 10° – 90° at an interval of 10°. The rows 6–8 represent PCET, PCT, and PST in Li et al.(2012) and 9–11 represent PCET, PCT, and PST in Hosny and Darwish (2017)

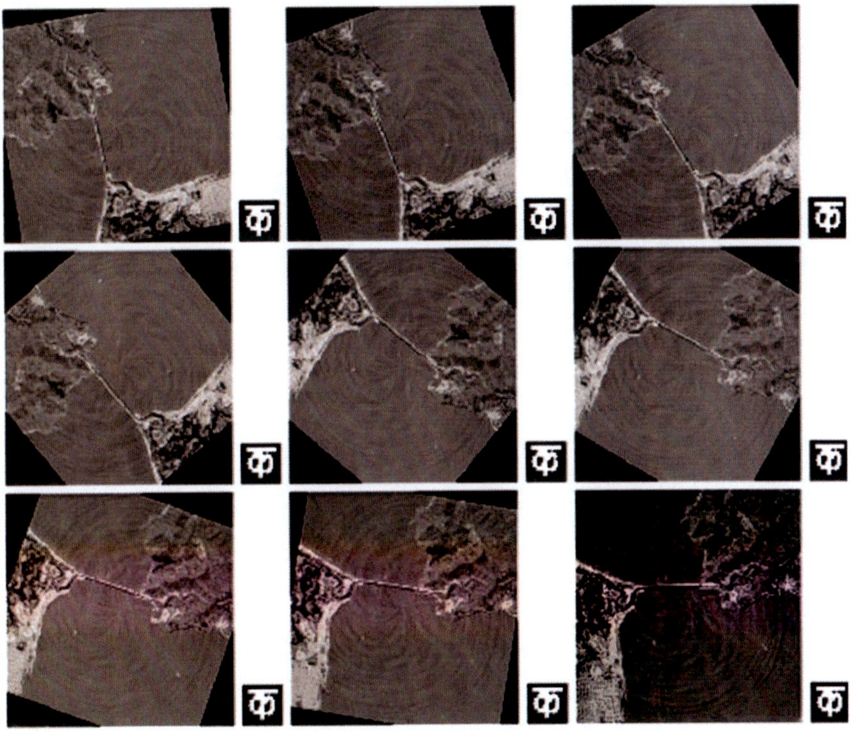

Fig. 2 Watermarked images using PCET and rotated counter-clockwise at angles 10°–90° at an interval of 10° along with the corresponding extracted watermarks

pletely reversible. In addition to the proposed method, the PCET, PCT, PST in [4], and PHFT in [7] are also completely reversible. Other methods shown in Table 1 do not show the reversibility. When the watermarked images are rotated by $10° - 20°$ and $70° - 80°$, the proposed PST perform better than the proposed PCET and PCET. At $30° - 60°$, the BER in the proposed PCET has smallest when compared with the proposed PCT and PST. On comparing the BER of the proposed PCET with the BER in recent PCET methods and PHFT in [7], it is observed that the BER in the proposed PCET is smallest at all rotation angles. The BER in the proposed PCT is smaller than the BER of the PCT in [5]. However, it is observed that the BER in the PCT of [4] is smaller than the proposed PCT. The BER in the proposed PST is smallest among the BER of PST in [5] and in [4].

Figure 2 shows the watermarked images (one of images shown in the Fig. 1) using the proposed PCET after rotating in counter clockwise directions at angles 0°, 10°, 20°, 30°, 40°, 50°, 60°, 70°, 80°, and 90°. The corresponding extracted watermarked is also shown along with the rotated watermarked images. The BER of the watermark is 0 for all angles in this case except at 50° at which the BER is 0.0039. In addition, it is also apparent from Fig. 2 that the proposed PCET is closely followed the rotational invariance.

Fig. 3 shows the watermarked images (one of images shown in Fig. 1) using the proposed PCT after rotating in counter-clockwise directions at angles 0°, 10°, 20°, 30°, 40°, 50°, 60°, 70°, 80°, and 90°. The corresponding extracted watermarked is also shown along with the rotated watermarked images. The BER of the watermark is 0.0039 at 40°, 0.0156 at 50°, and 0.0039 at 60° In addition, it is also apparent from Fig. 3 that the proposed PCT is closely followed the rotational invariance.

The figures in Fig. 4 show the images with watermark (one of images shown in the Fig. 1) using the proposed PST after rotating in counter-clockwise directions at angles 0°, 10°, 20°, 30°, 40°, 50°, 60°, 70°, 80°, and 90°. The corresponding extracted watermarked is also shown along with the rotated watermarked images. The BER is 0 at all the angles. In addition, it is also apparent from Fig. 4 that the proposed PST is closely followed the rotational invariance.

4 Conclusion

In many fields, image moments are used. Their accurate computation is an important task. The PHT moments are known as rotational invariant. However, due to the computational error associated with the moments, the image moments do not show the rotational invariance closely. Therefore, an improved method is required for computing the PHT moments at better accuracy. In this paper, analytic integration method combined with numerical computations is used for computing PCET, PCT, and PST

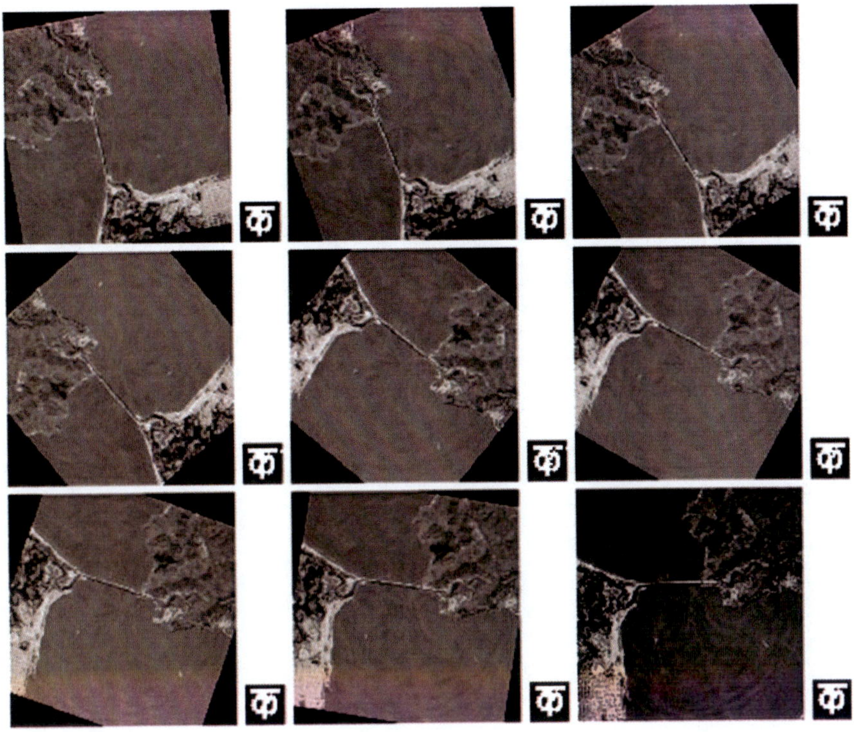

Fig. 3 Watermarked images using PCT and rotated counter-clockwise at angles 10° – 90° at an interval of 10° along with the corresponding extracted watermarks

moments at better accuracy compared to recent techniques used in the literature. The moments are used in watermarking applications over the ariel images obtained from the USC-SIPI image database. The accuracy of the extracted watermark is compared with the ZM, PZM, and PHFT. The accuracy is also compared with the PCET, PCT, and PST used in [5] and [4]. On comparison, the proposed PCET and PST show better rotational invariance than the above-discussed techniques. The computation of the moments are computationally intensive. Therefore, it can be time consuming if applied over a database. To overcome this difficulty, a template for the moments has been created and the template was applied over each image to compute the image moments.

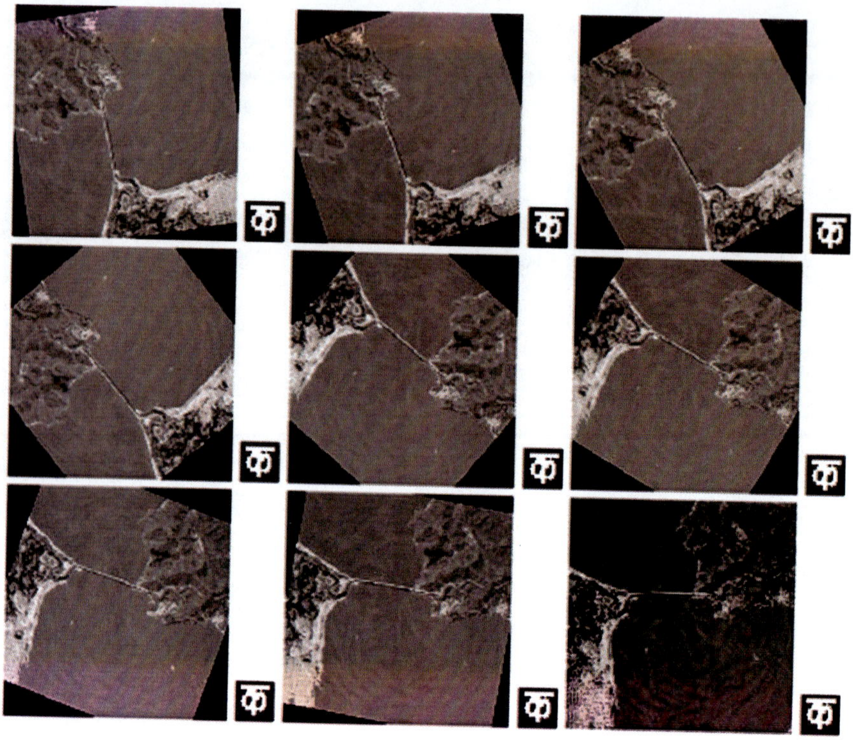

Fig. 4 Watermarked images using PST and rotated counter-clockwise at angles $10^\circ - 90^\circ$ at an interval of 10° along with the corresponding extracted watermarks

References

1. Ali M. Ahn CW, Pant M, Kumar S, Singh MK, Saini D (2020) An optimized digital watermarking scheme based on invariant dc coefficients in spatial domain. Electronics 9(9)
2. Ernawan F, Kabir MN (2019) An improved watermarking technique for copyright protection based on tchebichef moments. IEEE Access 7:151985–152003
3. Hosny KM, Darwish MM (2019) Resilient color image watermarking using accurate quaternion radial substituted chebyshev moments. ACM Trans Multimedia Comput Commun Appl 15(2)
4. Hosny KM, Darwish MM (2017) Invariant image watermarking using accurate polar harmonic transforms. Comput Electr Eng 62:429–447
5. Li L, Li S, Abraham A, Pan JS (2012) Geometrically invariant image watermarking using polar harmonic transforms. Inf Sci 199:1–19
6. Lutovac B, Daković M, Stanković S, Orović I (2017) An algorithm for robust image watermarking based on the dct and zernike moments. Multimedia Tools Appl 76(22):23333–23352
7. Ma B, Chang L, Wang C, Li J, Wang X, Shi YQ (2020) Robust image watermarking using invariant accurate polar harmonic Fourier moments and chaotic mapping. Signal Process 172:107544
8. Parah SA, Loan NA, Shah AA, Sheikh JA, Bhat GM (2018) A new secure and robust watermarking technique based on logistic map and modification of dc coefficient. Nonlinear Dyn 93(4):1933–1951
9. Piessens R, de Doncker-Kapenga E, Überhuber CW, Kahaner DK (2012) Quadpack: a subroutine package for automatic integration, vol 1. Springer Science & Business Media (2012)

10. Singh MK, Kumar, S., Ali, M., Saini, D.: Application of a novel image moment computation in x-ray and mri image watermarking. IET Image Processing pp. 1–17 (2020)
11. Singh MK, Gautam R, Gatebe CK, Poudyal R (2016) Polarbrdf: a general purpose python package for visualization and quantitative analysis of multi-angular remote sensing measurements. Comput Geosci 96:173–180
12. USC-SIPI image database (2020). http://engineering.purdue.edu/~mark/puthesis
13. Wang C, Wang X, Xia Z, Zhang C (2019) Ternary radial harmonic fourier moments based robust stereo image zero-watermarking algorithm. Information Sciences 470:109–120
14. Xiao B, Luo J, Bi X, Li W, Chen B (2020) Fractional discrete tchebyshev moments and their applications in image encryption and watermarking. Information Sciences 516:545–559
15. Xin Y, Liao S, Pawlak M (2007) Circularly orthogonal moments for geometrically robust image watermarking. Pattern Recognition 40(12):3740–3752
16. Yamni M, Daoui A, El ogri O, Karmouni H., Sayyouri, M., Qjidaa, H., Flusser, J.: Fractional charlier moments for image reconstruction and image watermarking. Signal Processing **171**, 107509 (2020)
17. Zeng G, Qiu Z (2008) Image watermarking based on dc component in DCT. In: 2008 international symposium on intelligent information technology application workshops, pp 573–576 (2008)

Discrete Wavelet Transform-Based Reversible Data Hiding in Encrypted Images

Sara Ahmed, Ruchi Agarwal, and Manoj Kumar

Abstract In the era of multimedia technology, not only reinstatement of the original media without any loss of information is required but also the security of the content is of utmost importance. In this paper, a novel discrete wavelet transforms (DWT)-based RDH-EI scheme is proposed. The approximation band of DWT has been utilized for data embedding. Also, disarrangement of DWT sub-bands is practiced for enhancing the security. Proposed scheme is simple to implement and takes less time to execute. Standard grayscale images are tested and compared with existing techniques. The results reflect supremacy of proposed scheme in terms of correlation, PSNR, and computational efficiency.

Keywords Discrete wavelet transforms · Encryption · Reversible data hiding · Security · PSNR

1 Introduction

Urbanization has led to the development in assorted sectors. Huge amount of data is being exchanged over the Internet via multimedia such as images, audio, video, and animation. Sometimes, this multimedia contains information which should only be accessible to the sender and its intended receiver. For instance, in hospitals, it is preferred that the patient's personal information such as medical history should only be available to the concerned specialists. In such cases, there is a need to hide this information from the unintended recipients. Content hiding and maintaining security are now the top priorities. Here, the concept of data hiding comes into play. Data hiding techniques help to avoid any intended or unintended manipulations. This is implemented by embedding secret data into the cover media. Later, this embedded data can be retrieved only by the sender and the intended receiver. During the embedding process or exchange between sender and receiver, the image is sometimes distorted which can result in loss of information. Thus, data hiding techniques are referred to as lossy approaches.

S. Ahmed · R. Agarwal (✉) · M. Kumar
Babasaheb Bhimrao Ambedkar University, Lucknow, Uttar Pradesh, India

G. Gupta et al. (eds.), *Proceedings of Academia-Industry Consortium for Data Science*, Advances in Intelligent Systems and Computing 1411,
https://doi.org/10.1007/978-981-16-6887-6_21

Some critical scenarios strictly prohibit this type of distortion. Reversible data hiding (RDH) was proposed to resolve this issue. RDH is also referred to as lossless/invertible data hiding. This proposed technique helped to recover both the embedded data and the cover image losslessly. The first RDH algorithm was put forward in 1997 in a US patent by Barton [1]. He suggested embedding the data into a digital media and then permit only authenticated individual to retrieve the information from the received data by verifying their authenticity. RDH is implemented in diverse fields such as image authentication [2], data coloring within cloud [3], and processing of medical image [4, 5] . The basis of RDH is that the degradation of the image quality should be very low and the embedding process should be reversible.

RDH is implemented using the following techniques: lossless compression [6–8], difference expansion (DE) [9–11], histogram shifting (HS) [12–14]. Lossless compression is easily implemented by compressing the bits of the image. This method leads to high distortion and low embedding capacity. Tian suggested a DE-based scheme [9]. In this technique, the integer average and the difference for a given set of pixel pairs are calculated. The difference is expanded, while the integer average is left unaltered. This technique utilizes a pixel pair to embed one bit which gives 0.5 bpp (bits per pixel) of embedding capacity. On comparing DE with lossless compression, we obtain lesser distortion and embedding potential. Later, various techniques were introduced to implement DE such as integer-to-integer transformation [15], predictive error expansion [16] and adaptive embedding [17]. A location map was used in these techniques to keep a track of selected expandable locations. This location map is additional information and an overhead to this technique. Implementing RDH through histogram shifting (HS) is a successful approach which creates space for higher embedding by altering the histogram generated from the grayscale image. Ni et al. [12] proposed the very first technique based on histogram shifting. In [12], zero point and peak points were determined and used in the embedding process. Higher the peak point more was the embedding capacity. A setback to this scheme was if the peak point and zero point were far apart though both helped in embedding more bits, but at the same time, distortion would increase.

Recent advancement made in RDH has aided in providing better security and information hiding facilities in encrypted domains. This creates a new area of interest for academia termed as RDHEI (RDH in encrypted images). RDHEI is carried out by encrypting the image and passing it onto the data hider. The data hider embeds secret data into the encrypted image without having any knowledge of the cover image. The secret data and the original image are completely recovered by the recipient using the encryption key. RDHEI can be implemented by two approaches: vacating room before [18–23] and after encryption [24–30].

Vacating room before encryption (VRBE) as the name suggests creates space/room for embedding before the encryption process. VRBE was introduced by Ma et al. [18]. They used the existing RDH techniques to pick the LSBs of certain pixels for embedding and then encrypted the image using image partitioning. Later, Zhang [19] proposed a new RDH technique utilizing histogram of predicted error. These predicted errors were encrypted and then used for embedding using histogram shifting. Further, various techniques utilizing prediction-error expansion

[20–23] were put forward which reserved rooms for embedding in the MSBs of the pixel values. In [23], a prediction-error method was explored to hold room for embedding in the MSBs. A location map was used to keep track of embeddable and non-embeddable pixels. The MSBs of the embeddable pixels were zero (this was possible only when the predicted errors were less than 64). The secret data was hidden in these positions. VRBE methods can be improved in terms of embedding capacity as embedding rate of 0.5 bpp can be attained, but it might be ineffectual due to requirement of extra preprocessing in the images. Overcoming this issue, VRAE methods are proven more efficient.

In vacating room after encryption (VRAE), an encryption key is used to directly encrypt the cover image and sent to the data hider for adding additional bits. The extraction process in VRAE depends on the domain: extracting data in the plaintext domain, extracting data in cipher domain, and extracting data in both domains. VRAE technique was introduced by Zhang [24]. In [24], the secret data were embedded by decomposing the encrypted image into blocks, and then, the last three least significant bits (LSBs) of the encrypted pixel in each block are flipped. In order to reduce the error during extraction, a function was introduced in [25] which measured smoothing and detected the matching edges. Still, the recovered image had large distortion. Overcoming this issue, a technique was proposed in [26] which created room for additional data by compression of LSBs in encrypted image pixels. Some of the recent works [27–30] have made modification in these ideologies. In [27, 28], data embedding in vacate rooms was done by a parameter binary label. Ge et al. [29] proposed block-based encryption where each block of histogram in both encrypted and original image were identical. In [30], prediction and encryption algorithm were used to reserve rooms for embedding secret information. Since the quality of image was not taken into consideration, a high embedding capacity was attained. Still, there is scope of enhancing security and embedding capacity.

In addition to these techniques, there are some other approaches such as pixel value ordering (PVO) [31] and cryptographic techniques [32–36] which are explored in RDH-EI. Until now, DWT, in the best of our knowledge, has been used for other image processing applications, but not for embedding. In the proposed technique, RDH in encrypted images is implemented by utilizing DWT. Encryption is done through permutation, while embedding is implemented using DWT. Double-layer encryption is used to amplify the security of the proposed scheme. The results of the proposed scheme, when compared with recent published works, outplay on the ground of computational time, encryption, and embedding capacity.

Further sections in paper are sorted as: In Sect. 2, the basic terms are explained; Sect. 3 is an elaboration of the proposed work, whereas experimental results along with security analysis are covered in Sect. 4. Section 5 compares the proposed work with resent existing work. Section 6 concludes our proposed study.

2 Related Definitions

Discrete Wavelet Transform (DWT)
Wavelets allow simultaneous analysis of both time and frequency in localized sense. This helps in synchronous evaluation of spatial and frequency aspects. DWT is a signal processing transformation which can decompose wavelets into multiple sub-bands and shows the decomposed sub-images in multiple resolutions.

2D images are divided into approximation and detailed parts. Low-frequency sub-band LL is named as approximation part, whereas LH, HL, and HH (high-frequency sub-bands) are named as the detailed parts (Fig. 2a). The approximation part can be further decomposed into four sub-bands and can be continued up to any level (e.g., second round of encryption, which can be seen in Fig. 2b). For recovering the original image, inverse DWT is applied on the decomposed sub-bands.

For an image $f(x, y)$ with M × N size, DWT is calculated as [37],

$$W_{\emptyset}(j_0, m, n) = \frac{1}{\sqrt{MN}} \sum_{x=0}^{M-1} \sum_{y=0}^{N-1} f(x, y) \emptyset_{j_0,m,n}(x, y) \tag{1}$$

$$W_{\varphi}^{i}(j, m, n) = \frac{1}{\sqrt{MN}} \sum_{x=0}^{M-1} \sum_{y=0}^{N-1} f(x, y) \varphi_{j,m,n}^{i}(x, y) \tag{2}$$

where $W_{\emptyset}(j_0, m, n)$ indicates the approximation part for the image $f(x, y)$ and $W_{\varphi}(j, m, n)$ represents the horizontal, vertical, and the diagonal parts.

Inverse DWT is calculated as

$$\begin{aligned} f(x, y) = {} & \frac{1}{\sqrt{MN}} \sum_{m} \sum_{n} W_{\emptyset}(j_0, m, n) \emptyset_{j_0,m,n}(x, y) \\ & + \frac{1}{\sqrt{MN}} \sum_{i=H,V,D} \sum_{j=j_0}^{\infty} \sum_{m} \sum_{n} W_{\varphi}^{i}(j, m, n) \varphi_{j,m,n}^{i}(x, y) \end{aligned} \tag{3}$$

3 Proposed Scheme

The proposed scheme as shown in Fig. 1 is an attempt to explore DWT in RDH in encrypted images. It involves three components: content owner, data hider, and receiver. These three components independently process by encrypting the image, embedding secret information, extracting data, and recovering the original image. Initially, the cover image is decomposed into four sub-bands using DWT as depicted in Fig. 2b. Further each band of the image is encrypted using permutation-based

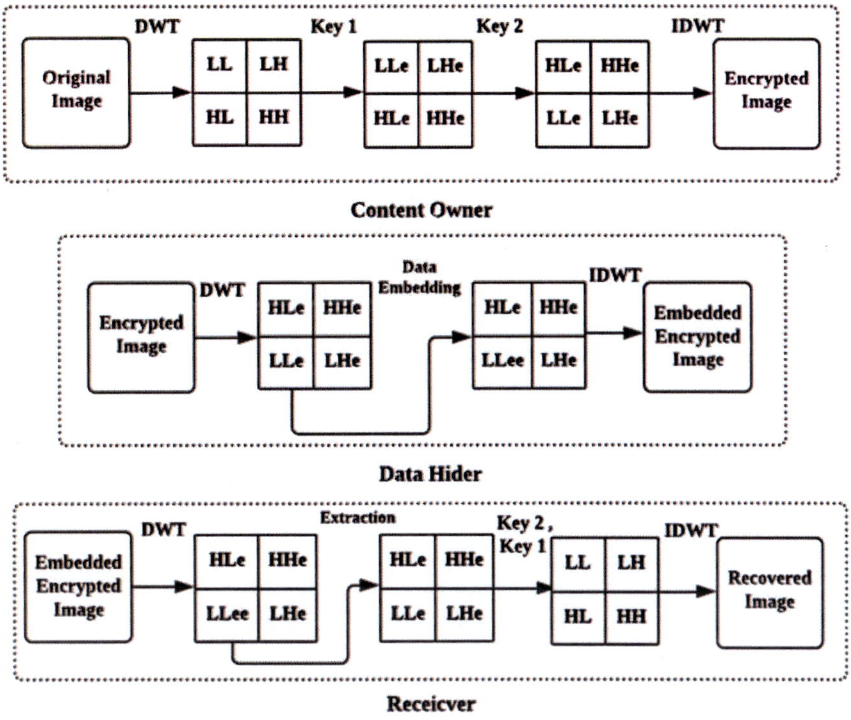

Fig. 1 Flow diagram of proposed scheme

Fig. 2 **a** Lena image, **b** first level of DWT, **c** second level of DWT

encryption. After encryption, all four sub-bands are permuted with a seed value generated with the help of pseudorandom number generator (PRNG) to inflate the level of security. Then, data hider uses DWT to embed secret information in the approximation part. Finally, the intended receiver performs data extraction and recovers the original image by reverse data embedding and encryption technique.

3.1 Encryption

Encryption in our proposed scheme is implemented in two levels. Initially, DWT is used to decompose the image into four bands. For encryption, permutation is used to interchange the pixels of the cover image which give a scrambled image. PRNG is used to generate random numbers. This generator is controlled by a seed value for image encryption. The seed value (*key 1*) scrambles the pixels in each band. To furnish it with better security, second round encryption is also imposed. In second round encryption, the four sub-bands are permuted using the same approach mentioned above, but with a different seed value (*key 2*). Random permutation method is used to permute a given set of numbers which makes sure that the original and encrypted image are completely different.

Encryption Algorithm

1. Apply DWT on the cover image *(I);* consequently, it divided into four sub-bands: LL, LH, HL, and HH (as in Fig. 2(b)).
2. Generate a seed value (*key 1*) using PRNG.
3. Use this seed value to permute the pixel positions of each sub-band.
4. Generate *key 2* using PRNG for double-layer security.
5. Permute the sub-bands using *key 2* and apply inverse DWT to reconstruct the encrypted image *(E).*

3.2 Data Embedding

After successful execution of the encryption process at the content owner's end, the data hider receives the encrypted image for embedding secret data. The content owner also marks the sub-band in which embedding has to be done. The data hider runs the following algorithm.

Data Hiding Algorithm

1. Decompose the encrypted image *(E)* into four sub-bands *LLe, LHe, HLe,* and *HHe* using DWT.
2. Prior information provided by the content owner is used to select a specific sub-band, for instance *LLe* sub-band, and perform the following steps.

   a .Check the MSB of each pixel; if it is *0,* precede further; otherwise, mark it as a non-embeddable pixel. Record the embeddable and non-embeddable pixels in a location map *(LM).*

   b .Let the secret data to be embedded is *(e).* Check the bit to be embedded; if it is *0,* leave the pixel; otherwise, replace the MSB of pixel with *1.*

3. Perform the above step for all the pixels in a sub-band *LLe.*
4. Apply inverse DWT to reconstruct the embedded encrypted image *E'* and send it to the receiver.

3.3 Data Retrieval and Recovery

At the receiver's end, the secret data are mined from the embedded encrypted image, followed by retrieval of original image. The receiver runs the following algorithm for data extraction and original image recovery.

Extraction Algorithm

1. Decompose the marked encrypted image E' into four sub-bands using DWT.
2. Extract the secret data from the specific sub-band using the following norms:
(a) Check the location map to identify the embeddable and non-embeddable pixels.
(b) In the embeddable pixels, check the MSB of each pixel; if the MSB found to be 1, retrieve secret data (e') as 1, and if MSB is *0*, extract the secret data (e') as 0.
3. Recover the original pixel by replacing the MSB with 0 if it is 1, otherwise, the pixel remains the same.
4. Decrypt the processed image using *key1* and *key2* and retrieve the original image *(I)*.

The above steps result in lossless recovery of secret data and original image (having PSNR value infinity between the original and the recovered images) in frequency domain. On changing frequency domain to spatial domain, recovered and original image are same along with lossless recovery of secret data.

4 Experiment Results and Analysis

MATLAB is used in our proposed scheme using several test images of size 512 × 512 × 8 (as in Fig. 3). These grayscale images are encrypted using permutation, and then, secret data are embedded using a sub-band of DWT. Encryption is done at two levels to boost the security. The corresponding embedded encrypted images are shown in Fig. 4. Focus of the proposed scheme is not only increasing the embedding capacity, but also enhancing the encryption potential.

Various experiments are implemented to test the efficiency of the proposed scheme. Results of the first and second levels of encryption are shown in Table 1 and Table 2, respectively. As the embedding is performed on only approximation sub-band of DWT, and thus, the maximum embedding capacity can reach upto $\frac{s}{4}$ bpp, where $s = m * n$ and *m, n* are the number of rows and columns of an image. To enhance the embedding capacity, multiple rounds of embedding can be performed using the same criteria. In that case, the maximum capacity (MEC) would be

$$\text{MEC} = \frac{s}{3}[1 - \left(1/4\right)^{l}]$$

Here, l is the level of round of embedding.

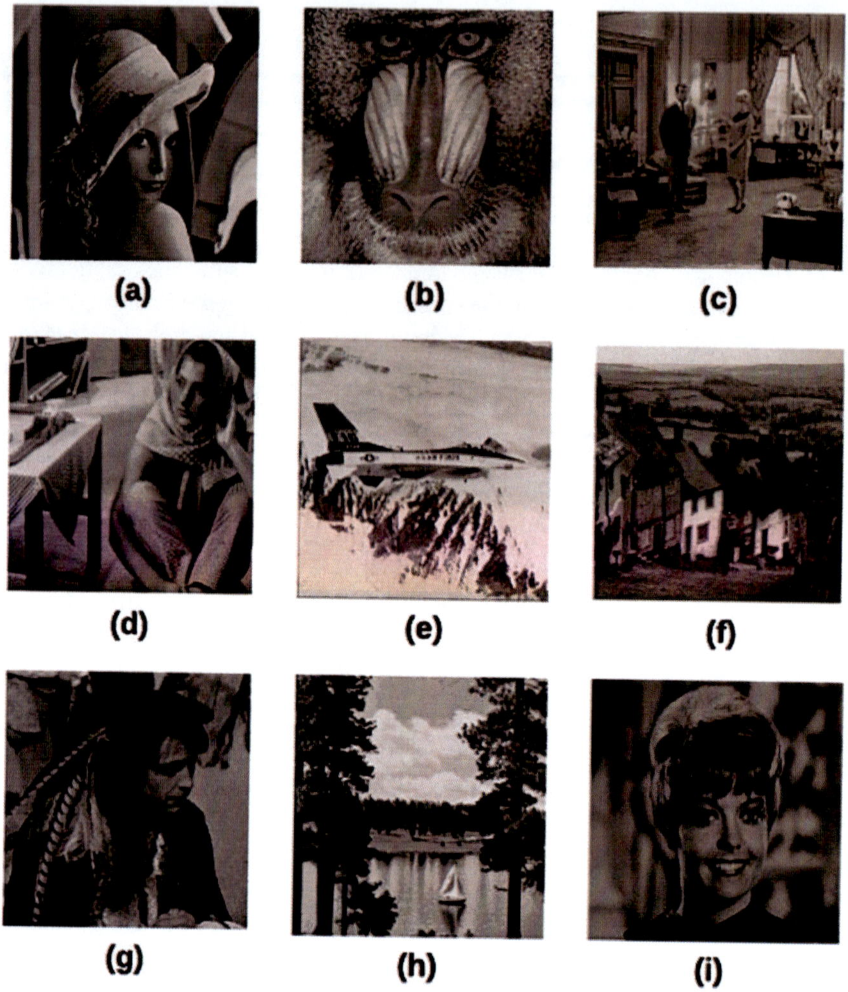

Fig. 3 Test images: **a** Lena, **b** Baboon, **c** House, **d** Babara, **e** Airplane, **f** Goldhill, **g** Man, **h** Sailboat, **i** Zelda

4.1 Security Analysis

Statistical metrics are evaluated using various parameters such as PSNR, SSIM, NPCR, UACI, CORR, and entropy to inspect the efficiency of encryption in our proposed scheme.

PSNR

To estimate the distortion and visual quality between original and marked image, PSNR is used. Higher PSNR value reflects better visual quality, whereas lower PSNR value reflects poor visual quality.

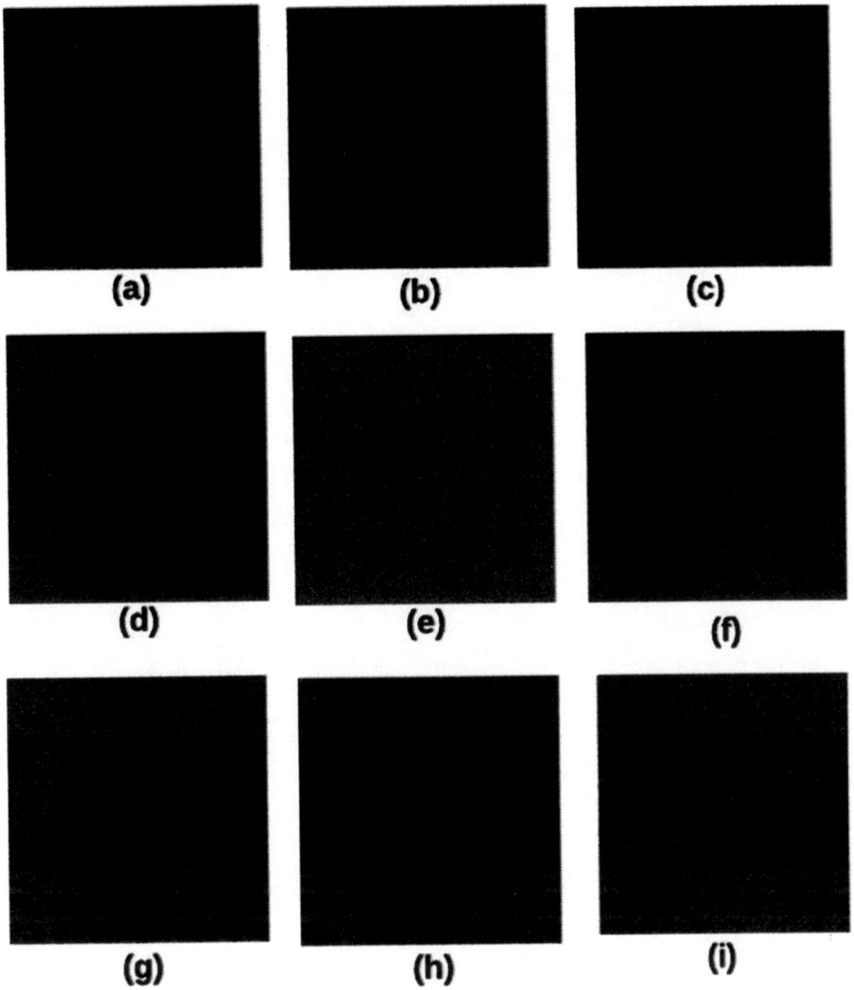

Fig. 4 Encrypted images: **a** Lena, **b** Baboon, **c** House, **d** Babara, **e** Airplane, **f** Goldhill, **g** Man, **h** Sailboat, **i** Zelda

$$\text{PSNR} = 10 \times \log_{10}\left(\frac{255 \times 255}{\text{MSE}}\right) \tag{4}$$

where mean square error (MSE) is evaluated as

$$\text{MSE} = \frac{1}{pq}\sum_{a=0}^{p-1}\sum_{b=0}^{q-1}[I(a, b) - E(a, b)]^2 \tag{5}$$

Table 1 Statistical metrics between original and encrypted images (first round encryption)

Test images	PSNR	SSIM	NPCR	UACI	CORR	Entropy
Lena	10.6536	0.0209	0.9942	0.2374	– 0.0012	7.5983
Baboon	11.6690	0.0251	0.9934	0.2108	– 0.0058	7.3481
House	12.1958	0.0315	0.9911	0.1940	– 0.0033	7.0581
Babara	10.3596	0.0192	0.9946	0.2473	– 0.0036	7.6321
Airplane	12.0052	0.0414	0.9857	0.1794	– 0.0065	6.6776
Goldhill	11.3019	0.0293	0.9934	0.2143	– 0.0058	7.4777
Man	9.7951	0.0145	0.9920	0.2627	– 0.0044	7.3573
Sailboat	8.7981	0.0150	0.9933	0.2921	– 0.0018	7.4752
Zelda	12.9536	0.0359	0.9930	0.1832	– 0.0028	7.2668

Table 2 Statistical metrics between original and encrypted images (second round encryption)

Test images	PSNR	SSIM	NPCR	UACI	CORR
Lena	8.5421	0.0118	0.9970	0.3128	– 0.0001
Baboon	6.8493	0.0102	0.9967	0.3840	0.00439
House	7.9702	0.0092	0.9951	0.3359	– 0.0004
Babara	7.5738	0.0092	0.9972	0.3537	– 0.0006
Airplane	5.3238	0.0070	0.9931	0.4403	0.00205
Goldhill	8.171	0.0113	0.9966	0.3274	0.00444
Man	9.0279	0.0118	0.9751	0.2839	– 0.0025
Sailboat	6.5799	0.0074	0.9966	0.3918	– 0.0004
Zelda	9.9202	0.0150	0.9965	0.2703	– 0.0010

where I, E are the original and the encrypted images, respectively; p, q are the number of rows and columns of the respective images; a, b are the index variables.

Structural Similarity Index Metric (SSIM)

To detect any degradation in the quality of the recovered image, SSIM is calculated whose value ranges between 1 and -1. 1 suggests that the reconstructed and the original images are identical. This is calculated by:

$$\text{SSIM} = \frac{(2\mu_i\mu_e + C_o)(2\sigma_v + C_1)}{\left(\mu_i^2 + \mu_e^2 + C_0\right)\left(\sigma_i^2 + \sigma_e^2 + C_1\right)} \quad (6)$$

where C_0 and C_1 are predefined constants; μ_i and μ_e are the averages corresponding to original and encrypted images; σ_v is covariance between the original and encrypted image, and σ_i and σ_e are the variances corresponding to original and encrypted images.

NPCR (Number of changing pixel rate) and UACI (Unified average changed intensity)

NPCR and UACI are used in case of differential attacks for security analysis in encrypted images. They test the strength of encryption. The NPCR of a secure encrypted image should be close to its own upper limit, i.e., close to 100%.

$$\mathrm{NPCR}(I, E) = \sum_{a,b} \frac{B(a, b)}{P} \times 100\% \tag{7}$$

$$\mathrm{UACI}(I, E) = \sum_{a,b} \frac{|I(a, b) - E(a, b)|}{S.P} \times 100\% \tag{8}$$

where I, E are original and encrypted images; B is bipolar array as in [38], P represents the amount of pixels in the cipher text, and S is the largest supported pixel. |.| denotes the absolute function; a, b are the index variables.

Correlation Coefficient (CORR)

To check the linear relationship between two images, correlation coefficient is used which indicates that the two images are similar. Its value varies from 1 to -1. In case of ideal encryption method, the value of this coefficient should be theoretically zero. It can be defined as:

$$\rho_{AB} = \frac{Cov(I, E)}{\sqrt{V(I)}\sqrt{V(E)}} \tag{9}$$

I, E denotes the original and the marked image, respectively. Cov is the covariance between the original and the marked image, and V is the variance of the respective images.

Entropy

Entropy is the measure of randomness. Minimum entropy (which is zero) is achieved at any location where the pixel values are constant. Maximum entropy relies on the number of gray scales. This value is found when all histogram bins are constant or the image intensity is uniformly distributed.

$$H(I) = -\sum_{i} p(x_i) \log_2 p(x_i) \tag{10}$$

$H(I)$ is information entropy, $p(x_i)$ is the probability of the pixel having intensity x_i.

In Table 1, the PSNR for Lena image and encrypted image is 10.6536 which indicate that the original and encrypted images are not similar. Therefore, the encrypted image does not reveal any clue about the original image; SSIM indicates structural similarity between two images; here, SSIM values are below zero indicating no similarity between the encrypted and original image; NPCR indicates the efficiency of

Table 3 PSNR between original and directly decrypted images for various embedding capacities (in bits)

Images	EC				
	256	512	786	1024	4096
Lena	39.1339	36.0055	34.1840	33.0455	27.0039
Baboon	39.1336	36.0052	34.1841	33.0452	27.0031
House	39.1334	36.0054	34.1839	33.0453	27.0035
Babara	39.1338	36.0053	34.1842	33.0455	27.0034
Airplane	39.1331	36.0057	34.1841	33.0457	27.0037
Goldhill	39.1332	36.0058	34.1843	33.0453	27.0039
Man	39.1335	36.0054	34.1842	33.0456	27.0032
Sailboat	39.1337	36.0056	34.1841	33.0457	27.0031
Zelda	39.1332	36.0057	34.1839	33.0453	27.0040

encryption, in the given Table 1; NPCR is nearly 99% indicating a high level of security; the correlation between original and encrypted image is negative indicating no similarity between the two images.

We perform second-level encryption (reflected in Table 2) for increasing the efficiency of encryption. It shows lesser values of PSNR, SSIM, UACI, and CORR when compared with Table 1 indicating the increase in encryption level. NPCR is close to 100% depicting high level of security.

In Table 3, the average of PSNR values is more than 34 specifying that the original image and the directly decrypted image are identical as a result no traces of embedded data can be seen. The difference between the two images is not visible to the human eye. For instance, in Lena image, the average PSNR is 33.8745.

4.2 Comparison

The proposed scheme is compared on the basis of PSNR, CORR, and encryption capacity with Ren et al. [31] in Table 4. It indicates an improvement as all the factors

Table 4 Comparison with Ren et al. [31] on the ground of encryption

	[31]			Proposed scheme		
Test images	CORR	PSNR	Encryption time	CORR	PSNR	Encryption time
Lena	0.0032	9.5294	0.7173	– 0.0001	8.5421	0.1309
Baboon	– 0.0005	9.8092	0.6150	0.0041	6.8493	0.0748
Airplane	0.0016	8.6297	0.5395	0.0020	5.3238	0.1495
Man	0.0069	9.1213	0.6041	– 0.0025	9.02799	0.1541

Table 5 Comparison with existing works based on embedding capacity (bpp)

Images	Lena	Baboon	Airplane	Man	Sailboat
Zhang [24]	0.0039	0.0009	0.0039	0.0015	0.0039
Hong [25]	0.0039	0.0100	0.0039	0.0200	0.0300
Zhang [26]	0.0350	0.0100	0.0300	0.0200	0.0300
Chen [32]	0.0020	0.0020	0.0020	0.0020	0.0020
Shiu [33]	0.0040	0.0040	0.0040	0.0040	0.0040
Zhang [34]	0.0080	0.0080	0.0080	0.0080	0.0080
Li [35]	0.0080	0.0080	0.0080	0.0080	0.0080
Proposed scheme	0.0695	0.0110	0.0048	0.1134	0.0400

are comparatively less (for instance, in Lena image the CORR is 0.0032 in [31] where as in the proposed scheme it is − 0.0001). Table 5 reflects that proposed scheme has better embedding capacity when compared with other existing schemes [24–26, 32–35].

5 Conclusion

A new approach for RDH in encrypted images is suggested which successfully extracts the embedded data and retrieves the original image. The proposed scheme utilizes DWT for embedding as well as encryption. DWT decomposes the original image into four bands. To enhance the security, double-layer encryption is applied. The first round of encryption shuffles the pixels in each band, whereas in the second round the bands are shuffled. As a result, any person other than the content owner and the intended recipient cannot identify the original image when they see the encrypted version. This encrypted image is passed on to the data hider who manipulates the MSBs of pixels of a particular band to embed the secret data. The embedded encrypted image is decrypted by the intended receiver who extracts the secret data and retrieves the original image applying the reverse mechanism of the proposed encryption and embedding techniques. Proposed work is examined using various standard test images, and the experimental outcomes clearly depict potency of proposed scheme. For analyzing the security and the efficiency of encryption, various statistical image quality metrics are used to compare encrypted image with the original one. Since, only a single band of DWT is modified for data embedding, there is still more to explore where other bands can also be used providing more space for embedding.

References

1. James B (1997, July 8) Method and apparatus for embedding authentication information within digital data, U.S. Patent 5,646,997
2. Chris H., Paul, J., Majid, R., and James, S., Lossless recovery of an original image containing embedded data, U.S. Patent 6,278,791 (2001, August 21)
3. Kai H, Deyi L (2010) Trusted cloud computing with secure resources and data coloring. IEEE Internet Comput 14(5):14–22
4. Feng B, Robert D, Beng O, Yanjiang Y (2005) Tailored reversible watermarking schemes for authentication of electronic clinical atlas. IEEE Trans Inf Technol Biomed 9(4):554–563
5. Gouenou C, Clare G, Jean C, Christian R (2009) Reversible watermarking for knowledge digest embedding and reliability control in medical images. IEEE Trans Inf Technol Biomed 13(2):158–165
6. Fridrich J, Goljan M, Du R (2002) Lossless data embedding—new paradigm in digital watermarking. EURASIP J Adv Signal Process 986842
7. Mehmet C, Gaurav S, Tekalp A (2006) Lossless watermarking for image authentication: a new framework and an implementation. IEEE Trans Image Process 15(4):1042–1049
8. Mehmet C, Gaurav S, Tekalp A, Eli S (2005) Lossless generalized-LSB data embedding. IEEE Trans Image Process 14(2):253–266
9. Jun T (2003) Reversible data embedding using a difference expansion. IEEE Trans Circ Syst Video Technol 13(8):890–896
10. Chih L, Hsien W, Chwei T, Yen C (2008) Adaptive lossless steganographic scheme with centralized difference expansion. Pattern Recogn 41(6):2097–2106
11. Ju H, Ke C, Morris C (2009) Block-based reversible data embedding. Signal Process 89(4):556–569
12. Zhicheng N, Yunqing S, Nirwan A, Wei S (2006) Reversible data hiding. IEEE Trans Circ Syst Video Technol 16(3):354–362
13. Gouenou C, Wei P, Nora B, Frédéric C, Christian R (2013) Reversible watermarking based on invariant image classification and dynamic histogram shifting. IEEE Trans Inf Forensics Secur 8(1):111–120
14. Suah K, Xiaochao Q, Vasily S, Hyoung K (2019) Skewed histogram shifting for reversible data hiding using a pair of extreme predictions. IEEE Trans Circ Syst Video Technol 29(11):3236–3246
15. Jessica F (2009) Steganography in digital media: principles, algorithms, and applications. Cambridge Univ. Press, Cambrid
16. Mehdi F (2008) Reversible image data hiding based on gradient adjusted prediction. IEICE Electron 5(20):870–876
17. Lute K, Henk H (2005) Reversible data embedding into images using wavelet techniques and sorting. IEEE Trans Image Process 14(12):2082–2090
18. Kede M, Weiming Z, Xianfeng Z, Nenghai Y, Fenghua L (2013) Reversible data hiding in encrypted images by reserving room before encryption. IEEE Trans Inf Forensics Secur 8(3):553–562
19. Weiming Z, Kede M, Nenghai Y (2014) Reversibility improved data hiding in encrypted images. Signal Process 94:118–127
20. Smita A, Manoj K (2017) Mean value based reversible data hiding in encrypted images. Optik 130:922–934
21. Pauline P, William P (2018) An efficient MSB prediction-based method for high-capacity reversible data hiding in encrypted images. IEEE Trans Inf Forensics Secur 13(7):1670–1681
22. Yi P, Zhaoxia Y, Zhenxing Q (2008, December 1–7) Reversible data hiding in encrypted images with two-MSB prediction. In: Proceedings of IEEE international workshop information forensics security, Hong Kong
23. Asad M, Hong W, Yanli C, Ahmad K (2020) A reversible data hiding in encrypted image based on prediction-error estimation and location map. Multimedia Tools Appl. 79:1–24

24. Xinpeng Z (2011) Reversible data hiding in encrypted image. IEEE Signal Process Lett 18(4):255–258
25. Wien H, Tung C, Han W (2012) An improved reversible data hiding in encrypted images using side match. IEEE Signal Process Lett 19(4):199–202
26. Xinpeng Z (2012) Separable reversible data hiding in encrypted Image. IEEE Signal Process Lett 7(2):826–832
27. Shuang Y, Yicong Z (2019) Separable and reversible data hiding in encrypted images using parametric binary tree labeling. IEEE Trans Multimedia 21(1):51–64
28. Youqing W, Youzhi X, Yutang G, Jutang T, Zhaoxia Y (2020) An improved reversible data hiding in encrypted images using parametric binary tree labelling. IEEE Trans Multimedia 22(8):1929–1938
29. Haoli G, Yan C, Zhenxing Q, Jianjun W (2019) A high capacity multi-level approach for reversible data hiding in encrypted images. IEEE Trans Circ Syst Video Technol 29(8):2285–2295
30. Delu H, Jianjun W (2020) High-capacity reversible data hiding in encrypted image based on specific encryption process. Signal Process Image Commun 80:115632
31. Ren H, Shaozhang N, Xinyi W (2019) Reversible data hiding in encrypted images using pob number system. IEEE Access 7:149527–149541
32. Chen Y-C, Shiu C-W, Horng G (2014) Encrypted signal-based reversible data hiding with public key cryptosystem. J Vis Commun Image Represent 25(5):1164–1170
33. Chih S, Yu C, Wien H (2015) Encrypted image-based reversible data hiding with public key cryptography from difference expansion. Signal Process Image Commun 39:226–233
34. Xinpeng Z, Jing L, Zichi W, Hang C (2016) Lossless and reversible data hiding in encrypted images with public-key cryptography. IEEE Trans Circ Syst Video Technol 26(9):1622–1631
35. Ming L, Yang L (2017) Histogram shifting in encrypted images with public key cryptosystem for reversible data hiding. Signal Process 130:190–196
36. Yu C, Tsung H, Sung H, Chih S (2019) A New Reversible data hiding in encrypted image based on multi-secret sharing and lightweight cryptographic algorithms. IEEE Trans Inf Forensics Secur 14(12):3332–3343
37. Rafael G, Richard W (2013) Digital image processing, 3rd edn. Pearson Education, New Delhi
38. Wu Y (2011) NPCR and UACI randomness tests for image encryption. Cyber J J Sel Areas Telecommun

A Partial Outsourcing Decryption of Attribute-Based Encryption for Internet of Things

Dilip Kumar and Manoj Kumar

Abstract Internet of things (IoT) is a fast-growing technology in the modern digital era that attracts people from industries as well as academic fields. IoT provides the platform to connect all the devices to share data over the Internet. This emerging technology suffers from many security issues due to the heterogeneity of the environment and resource-constrained IoT devices. Attribute-based encryption (ABE) is a modern encryption scheme that provides security along with an access control mechanism. This technique is applied for secure communication among IoT devices over unsecured communication channels. The bilinear pairing operation is generally used for encryption and established sharing between parties. In this paper, outsourcing-based decryption of ABE using a three-party Diffie–Hellman key exchange protocol is proposed for the IoT environment where devices have limited computational and battery power. The main aim of the proposed scheme is to lessen the complexity of computation for the decryption operation of ABE for the resource-constrained devices of IoT.

Keywords IoT · Attribute-based encryption · CP-ABE · Bilinear pairing · Secret sharing

1 Introduction

An emerging technology called the Internet of things is affecting our daily life adversely. IoT was firstly introduced by Ashton [1]. IoT mainly consists of actuators, sensors, and physical objects embedded with chips. In the IoT, things are connected to share information among them. It provides the platform where everything is connected through the Internet. IoT is applicable in various domains like smart devices, smart cities, smart vehicles, smart buildings, smart healthcare, smart farming, smart grids, smart industry, smart home, smart transport, and many more. IoT network is a collection of various devices that are purposely connected. Information sharing among these devices can be vulnerable to various security attacks. Security in IoT

D. Kumar (✉) · M. Kumar
Babasaheb Bhimaro Ambedkar University, Lucknow, Uttar Pradesh, India

G. Gupta et al. (eds.), *Proceedings of Academia-Industry Consortium for Data Science*, Advances in Intelligent Systems and Computing 1411,
https://doi.org/10.1007/978-981-16-6887-6_22

is the most challenging task in today's world where a huge number of devices are interconnected. The connecting devices come from different embedded technologies including different protocols and distributed environments. Confidentiality, integrity, and availability are the basis of information security. Besides, privacy, authorization, non-repudiation, key management, and access policy must be considered to improve the security of the IoT systems.

Traditional cryptographic solutions are not appropriate in the IoT environment. Traditional cryptographic techniques (symmetric or asymmetric) are used for enciphering and deciphering. If the same key is used for both enciphering and deciphering, then it is called symmetric key encryption. If one key is used for enciphering and another key for deciphering, then it is called asymmetric key encryption. In both encryption techniques, one-to-one communication takes place between intended users. Symmetric and asymmetric encryptions provide better security in the IoT environment but fail to provide access control. ABE is a modern encryption scheme with an access control mechanism that is used to secure data in the IoT.

ABE is an encryption scheme that provides an access control mechanism of private data by using attributes set [2]. In ABE, the key belonging to the user and ciphertext both depend on attributes. Encryption of data can be performed using the attributes but ciphertext decryption is done if the user's attributes match with the attributes of the ciphertext. Achieve collusion resistance in the ABE scheme is a big challenge. Collusion resistance means if multiple parties having different keys put together their attributes to decipher the ciphertext, they should be successful if and only if any one of them could have decrypted individually. Key Policy ABE (KP-ABE) [3] and Ciphertext Policy ABE (CP-ABE) [4] are two types of ABE scheme. In the KP-ABE, the ciphertext is embedded with attributes set and the private key of the user is embedded with the access structure [3, 5]. But in the case of CP-ABE, the conditions are reversed. Shamir secret sharing scheme [6] is used in ABE schemes [3, 4].

In the literature, major work is found on security solutions for IoT devices. The security challenges of devices in the IoT networks need new technical solutions to beat traditional solutions. ABE is the best-known public key encryption technique that provides access control besides privacy for users within the IoT system. The process of decryption in ABE is required comparatively much time to perform operations, and it is not feasible in the case of IoT networks where most of the connected devices are resource-constrained. This gives rise to the need of outsourcing the computation of ABE over the cloud server or other party. Outsourcing is an alternative solution to reduce the computational complexity but it is a challenging task in the IoT environment due to the availability of various intercepter in the networks that are always ready to attack sensitive information.

Here, we propose an outsourcing partial decryption of ABE for resources-constrained devices connected in the IoT networks. The CP-ABE scheme [4] is used for outsourcing the decryption over the cloud server. In the scheme, the encryptor intelligently decides who should or should not have the access to the ciphertext. In the proposed system, Joux's three-party Diffie–Hellman key exchange algorithm is applied to share the secret key. The data is enciphered by the data owner and then

Table 1 Notations and meaning

Notation	Meaning
G_{r0}, G_{r1}	Bilinear groups
$\hat{e}$	Bilinear map
g	Generator of group G_{r0}
$\hat{Q}_x, \hat{Q}$	Random polynomials
τ	Tree access structure
$a, b, c, r, s, \alpha, \beta$	Random elements from Z_p
CT'	Partially decrypted ciphertext
ATT	Attribute
$\mathcal{SA}$	Set of attributes
$\mathcal{SL}$	Set of leaf nodes
CS	Cloud server

forward to the data user. Outsourcing the process of decryption to the cloud server is done by the data user. After performing partial decryption, the data user gets the ciphertext from the cloud server. The data user makes use of the shared secret key to retrieve the original message. Notations and their meaning are given in Table 1.

2 Related Work

The concept of the ABE scheme is introduced by Sahai and Waters [2] in which secret keys and ciphertext are embedded with an attribute set. A threshold access structure is used in the ABE scheme. Goyal et al. [3] have put forward the KP-ABE scheme and applied this scheme in secure forensic analysis and broadcast encryption. The audit log problem can also be solved by using the KP-ABE scheme. To make the ABE scheme more appealing and better access control, the CP-ABE scheme [4] is introduced in which *AND* and *OR* gates are used in the access structure. A scheme [7] is proposed to deploy on mobile devices that have limited power resources. A CP-ABE scheme [8] is presented for storing keys of short length compared to existing schemes. An ABE scheme [9] is proposed that supports a non-monotonic access structure. In this scheme, the user's key is represented as an access formula over attributes. A policy compacting method [10] is proposed to compact ciphertext. An extended CP-ABE scheme [11] has been presented using pre-computation techniques to compute energy-saving potential.

The majority of the ABE schemes have a single authority that works as a central authority. The central authority performs the various tasks in the ABE scheme. In this scenario, if the security is compromised by the central authority, then the system security will be also compromised. If the central authority is breakdown, then the whole network will be broken down. To make a reliable ABE system, Lewko et al.

[12] have introduced a multi-authority ABE (MA-ABE) system where the role of central attribute authority is replaced by decentralized multi-authorities. In MA-ABE, any party can become an authority and generates secret and public keys for users in the system. The MA-ABE scheme creates a system more reliable because there are multiple authorities and even if one authority of the system is failed despite the system will remain reliable. A decentralized CP-ABE scheme [13] is proposed to satisfy properties of access control, fast decryption, optimal ciphertext size, and constant-size secret key. This decentralized CP-ABE scheme is applicable for lightweight device applications. An encryption system [14] is presented to ensure access control on data with the help of the ABE scheme. Authors in [15] have used the access control technique for use of data in the IoT, and this scheme has a key that is constant in size and the ciphertext.

Generally, bilinear pairing [3, 4] is an expensive operation used in various ABE schemes. Bilinear pairing increases the computational complexity of the ABE system. The point scalar multiplication of elliptic curve cryptography (ECC) is an alternative solution to replace the bilinear pairing. Yao et al. [16] have presented an ABE scheme which is based on ECC to outline privacy and security issues in the IoT. A hierarchical KP-ABE scheme is presented to enhance the security of the scheme in [16] by the authors in [17]. Ding et al. [18] put forward a novel CP-ABE scheme based on ECC for the IoT. In this scheme, simple multiplication on an elliptic curve replaces the bilinear pairing to reduce overall computation. A new key distribution technique is designed to revocate a user or an attribute during the attribute revocation phase without altering other user's keys. An access policy is represented by a linear secret sharing scheme [5, 18].

However, the use of ECC reduces the computation but for lightweight devices, the computation cost is still needed to reduce. For that outsourcing, the decryption process is a better choice. Green et al. [19] firstly proposed outsourcing of ABE to minimize the users' decryption cost by storing the ciphertext into the cloud and a user provides a single transformation to the cloud for translation of ABE ciphertext into an ElGamal ciphertext. In an online/offline scheme [20], the encryption algorithm is divided into phases: a preparatory phase that does encryption and key generation and a second phase that assembles an ABE ciphertext or key. By using MA-ABE, an ABE scheme [21] is introduced for partial decryption of ABE by a trusted cloud server. A scheme [22] is proposed in which ABE encryption is outsourced to the malicious cloud server. An outsourcing scheme [23] is presented to outsource the encryption process over the cloud service provider to reduce the computational complexity of encryption from exponentiation to a constant. An outsourcing ABE scheme [24] is proposed to support ABE key generation and decryption. Based on prime order groups, authors in [25] have constructed an MA-CP-ABE system with a monotonic access structure and decryption is outsourced.

3 Preliminaries

3.1 Joux's Three-Party Diffie–Hellman Key Exchange Algorithm

The Diffie–Hellman (D-H) algorithm is used to key exchange. The sharing of the secret key is performed as follows: Suppose three parties Attribute Authority (*AA*), Data Owner (*DO*), and Data User (*DU*) want to share a secret key using the D-H algorithm. First of all, they have agreed on the cyclic groups G_{r0} and G_{r1} of prime order p. In the next step, *AA*, *DO*, and *DU* pick their secret keys and do not reveal them to anyone.

AA picks a secret integer $a \in Z_p$ and a is kept secret. *AA* computes A as

$$A = g^a \tag{1}$$

DO picks a secret integer $b \in Z_p$ and b is kept secret. *DO* computes B as

$$B = g^b \tag{2}$$

DU picks a secret integer $c \in Z_p$ and c is kept secret. *DU* computes C as

$$C = g^c \tag{3}$$

After computing the values of A, B, and C, they broadcast these values. Finally, *AA* uses his secret integer a to compute the shared secret key.

$$K_{abc} = \hat{e}(g^b, g^c)^a = \hat{e}(g, g)^{abc} \tag{4}$$

DO uses his secret integer b to compute the shared secret key

$$K_{abc} = \hat{e}(g^a, g^c)^b = \hat{e}(g, g)^{abc} \tag{5}$$

DU uses his secret integer c to compute the shared secret key

$$K_{abc} = \hat{e}(g^a, g^b)^c = \hat{e}(g, g)^{abc} \tag{6}$$

4 Proposed Scheme

The emergence of IoT technology provides a platform where various devices are interconnected to communicate with each other. Resource-constrained IoT devices are ineffective to perform heavy computations because of limited resources. Therefore, it is required to outsource the high computational tasks over the cloud server

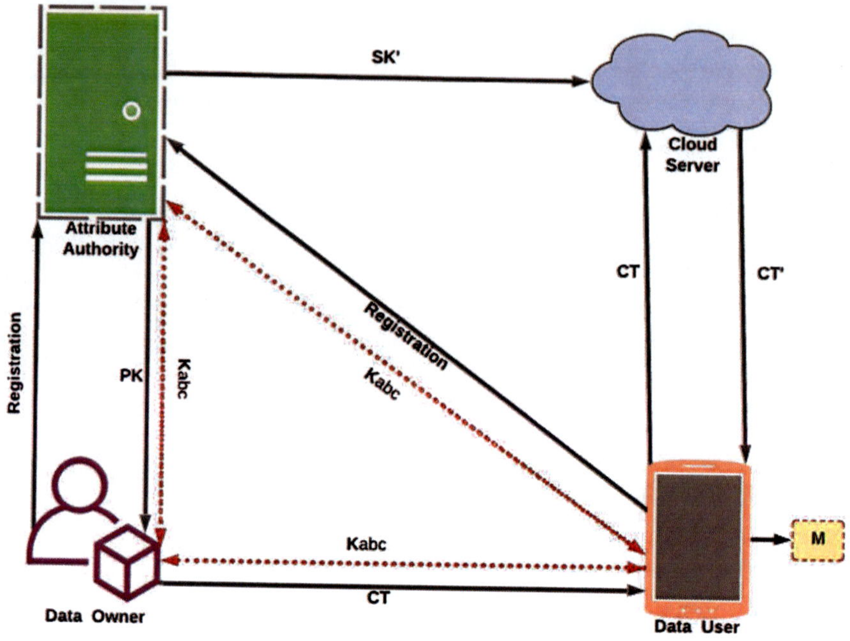

Fig. 1 Architecture of the proposed scheme

to make the system reliable for secure communication. There are several practical challenges in outsourcing storage and computation applications. Determination of computation is not known in advance so it needs a proper set of parameters. Later changes of parameters can undergo a full re-encryption or resettlement of the whole system. The size of ciphertexts varies inappropriately. Classical techniques used for the outsourcing of computation may be challenging to adapt for an encrypted domain. To overcome these issues, we propose outsourcing the decryption of the ABE scheme to perform a partial decryption process by the cloud server. The CP-ABE scheme provides better security solutions in IoT environments. The CP-ABE scheme [4] is considered as the backbone of our proposed scheme. The D-H algorithm is used to share the secret among *AA*, *DO*, and *DU*.

4.1 Definition of Outsourcing Scheme

Our outsourcing scheme comprises six algorithms:

Setup Security parameters are the input of setup algorithm, and public parameters *PK* and a master key *MK* are outputs of this algorithm.

Encryption The encryption algorithm considers a message *M*, an access structure and *PK* as inputs and produces the ciphertext *CT* as output.

KeyGeneration$_{ATT}$ The master key MK and an attribute set are inputs for this algorithm, and a secret key SK' is output.

KeyGeneration$_{SS}$ This algorithm runs to share the secret key among three parties: attribute authority, data owner, and data user by using Joux's three-party Diffie–Hellman key exchange algorithm. The data user gets its secret key *SK*.

Decryption$_{CS}$ This algorithm is run by a cloud server for partial decryption of the ciphertext by using *SK'*. The cloud server sends the ciphertext *CT'* to the data user.

Decryption$_{DU}$ The decryption algorithm considers input the secret key *SK* obtained from the Diffie–Hellman key exchange algorithm to decrypt partially decrypted ciphertext *CT'* received from the cloud server. It outputs the message *M* if his shared secret key is satisfied, otherwise produces an error.

Figure 1 shows the architecture of our outsourcing scheme of ABE decryption in the IoT environment. There are four main entities: Attribute Authority (*AA*), Data Owner (*DO*), Data User (*DU*), and Cloud Server (*CS*). The basic concept of this model is the part of decryption runs on IoT devices outsourced to the cloud server for partial decryption. *AA* generates public keys for registered users. The *DO* gets his public key *PK*, and the cloud server gets the secret key SK' from attribute authority. $KeyGeneration_{SS}$ is used to share a secret key K_{abc} among attribute authority, data owner, and data user with the help of Joux's three-party Diffie–Hellman key exchange algorithm. The *DO* enciphers the message and sent it to the *DU*. On receiving the ciphertext, the *DU* sends the *CT'* to the *CS* for partial decryption. The *CS* uses $Decryption_{SS}$ for decrypting the ciphertext partially with the help of the secret key SK' and then sends it again to the *DU*. The *DU* uses $Decryption_{SS}$ to decipher the ciphertext *CT'* with help of the shared secret key *SK*.

4.2 Proposed Methodology

In the proposed system, the cloud server is responsible for partial decryption. The random secret key is masked in the ciphertext by using an encryption algorithm. The shared secret key K_{abc} is shared among *AA*, *DO*, and *DU* by using three-party D-H algorithm.

The proposed scheme contains the following six algorithms: $\mathcal{SU} = Setup$, $\mathcal{KG}_{ATT} = KeyGeneration_{ATT}$, $\mathcal{KG}_{SS} = KeyGeneration_{SS}$, $\mathcal{EN} = Encryption$, $\mathcal{DE}_{CS} = Decryption_{CS}$ and $\mathcal{DE}_{\mathcal{DU}} = Decryption_{DU}$.

Setup Let G_{r0} and G_{r1} be the cyclic groups that have prime order p and G_{r0} contains a generator g. Let $\hat{e} : G_{r0} \times G_{r0} \rightarrow G_{r1}$ be the bilinear map. A security parameter κ calculates the size of groups. The Lagrange coefficient is defined as $\Delta_{i,\mathcal{SA}}(x) = \prod_{j \in \mathcal{SA}, j \neq i} \frac{x-j}{i-j}$, where $i \in Z_p$ and $\mathcal{SA}$ is a set of elements from Z_p. Let $H : \{0, 1\}^* \rightarrow G_{r0}$ be a hash function. The $\mathcal{SU}$ algorithm selects $\alpha, \beta \in Z_p$ and generates

$$PK = (G_{r0}, g, h = g^{\beta}, f = g^{\frac{1}{\beta}}, \hat{e}(g, g)^{\alpha}), MK = (\beta, g^{\alpha}). \tag{7}$$

KeyGeneration$_{ATT}$ This algorithm generates the SK' for every user with the help of *MK*. It chooses $r \in Z_p$ at random and then chooses $r_j \in Z_p$ randomly for $j \in \mathcal{SA}$. The secret key is calculated as

$$SK' = (D = g^{(\alpha+r)/\beta}, \forall j \in \mathcal{SA} : D_j = g^r.H(j)^{r_j}, D_j' = g^{r_j}). \tag{8}$$

KeyGeneration$_{SS}$ Attribute authority is responsible to share the secret key with users who have valid set of attributes, otherwise it selects randomly $t \in Z_p$ and g in G_{r0}. This $KeyGeneration_{SS}$ algorithm takes place among three-party as follows: *AA*, *DO* and *DU* participate in the secret key sharing process.

- *AA* select a secret integer $a \in Z_p$ and computes $A = g^a$
- *DO* select a secret integer $b \in Z_p$ and computes $B = g^b$
- *DU* select a secret integer $c \in Z_p$ and computes $C = g^c$
- Broadcast the values of *A*, *B* and *C* among them.
- *AA* again use his secret integer a to compute the secret key. $K_{abc} = \hat{e}(g^b, g^c)^a = \hat{e}(g, g)^{bca}$
- *DO* again use his secret integer b to compute the secret key $K_{abc} = \hat{e}(g^a, g^c)^b = \hat{e}(g, g)^{acb}$
- *DU* again use his secret integer c to compute the secret key $K_{abc} = \hat{e}(g^a, g^b)^c = \hat{e}(g, g)^{abc}$
- The shared secret key is $K_{abc} = \hat{e}(g^b, g^c)^a = \hat{e}(g^a, g^c)^b = \hat{e}(g^a, g^b)^c = \hat{e}(g, g)^{abc}$. The $SK = K_{abc}$ is a shared secret key.

Encryption The $\mathcal{EN}$ algorithm enciphers a message M using τ. The secret $s \in Z_p$ is selected at random. Then for root node, set the polynomial $\hat{Q}_R(0) = s$ and for remaining nodes (other than root node) set another polynomial $\hat{Q}_x(0) = \hat{Q}_{parent(x)}(index(x))$. The ciphertext is computed as

$$CT = (\tau, \tilde{\varepsilon} = M\hat{e}(g, g)^{\alpha s} K_{abe}, \varepsilon = h^s, \forall y \in \mathcal{SL} : \varepsilon_y = g^{\hat{Q}_y(0)}, \varepsilon_y' = H(att(y))^{\hat{Q}_y(0)}). \tag{9}$$

Decryption$_{CS}$ An algorithm *DecryptNode(CT, SK', x)* is proceed if x is a leaf node from τ and $i \in \mathcal{SA}$ then

$$DecryptNode(CT, SK', x) = \frac{\hat{e}(D_i, C_x)}{e(D_i', C_x'}) = \hat{e}(g, g)^{rs}. \tag{10}$$

The recursive algorithm *DecryptNode(CT, SK', x)* is proceed if x is a non-leaf node.

$$F_x = \prod_{z \in SA_x} F_z^{\Delta_{i,SA_x'}(0)} = \hat{e}(g, g)^{rs}, \ where\ i = index(z), SA_x' = \{index(z) : z \in SA_x\} \tag{11}$$

If the $\mathcal{SA}$ satisfied access tree, then set $\mathcal{A} = DecryptNode(CT, SK', r) = \hat{e}(g, g)^{r\hat{Q}_R(0)} = \hat{e}(g, g)^{rs}$. Finally, the partially decrypted ciphertext CT' is calculated as

$$CT' = \frac{\tilde{\varepsilon}}{(\hat{e}(\varepsilon, D)/\mathcal{A})} = \frac{\tilde{\varepsilon}}{(\hat{e}(h^s, g^{(\alpha+r)/\beta})/\hat{e}(g, g)^{rs})} = M\ K_{abe}. \tag{12}$$

Decryption$_{DU}$ The data user deciphers the ciphertext using *SK* and get the original message *M* as

$$\frac{CT'}{SK} = \frac{MK_{abc}}{K_{abc}} = M. \tag{13}$$

The *DO* uses the public key to generate the ciphertext. The boolean formula is used to represent access policy. The *DO* can encipher the data and then send it to the *DU*. If the received *CT* is large, then the user outsources ciphertext to the cloud server for partial decryption because decryption is an expensive operation in the ABE and it increases computational complexity if some attributes increases. The ciphertext is decrypted by the cloud server partially and sent to the *DU*. By using the decryption algorithm, the *DU* recovers the original message with the help of his shared secret key. Even if the cloud servers are compromised, security remains stagnant in the proposed system. A data user with a low-performance device can produce a feasible result by using single key decryption. An alternate solution in IoT applications is outsourcing ABE decryption to the cloud server.

5 Security Analysis

Achievement of collusion resistance in the ABE scheme is a big challenge against malicious users. As in the scheme, [4], users private key is randomized such that they can not be combined. The secret sharing is associated with the ciphertext. If an attacker wants to decrypt then he should retrieve $\hat{e}(g, g)^{\alpha s}$. For decryption, the intruder needs to pair the key component *D* of the private key of the user and *C* from the ciphertext. The attacker gets $\hat{e}(g, g)^{\alpha s+rs}$ by pairing of *C* and *D*. He did not get the $\hat{e}(g, g)^{\alpha s}$ because result of pairing includes a blinded value $\hat{e}(g, g)^{rs}$. To blind out the value, it required the right components of the key that satisfies the secret sharing scheme. In our proposed scheme, a random secret key K_{abc} is integrated with the ciphertext. The shared secret key K_{abc} is only shared with intended users. The cloud server cannot decipher the ciphertext completely. The shared secret key is not known to the cloud server because the secret key is shared using the D-H key exchange algorithm among the attribute authority, data owner, and data user. Since the security of partially decrypted ciphertext depends on Diffie–Hellman key exchange protocol whose security depends on the discrete log problem.

6 Conclusion and Future Work

Outsourcing computation of CP-ABE scheme is an alternative approach to lessen the complexity of encryption/decryption process in the IoT networks where a huge amount of devices are connected for communication. IoT devices have limited resources to perform heavy computations. The bilinear pairing is used to share the secret key among three-party. In a single transformation, the secret key is shared among three parties by using the bilinear pairing. In the ABE scheme, computational complexity increases if the bilinear operation is used. Outsourcing of computation over the third party makes the lightweight devices perform fewer computations. The CP-ABE is adopted to achieve impenetrable security in IoT networks.

In the future, applications of ABE schemes have a wider scope to analyze vulnerabilities and security in the IoT. Designing the constant key size and ciphertext, outsourcing the computational tasks are good practices to reduce the ABE overheads. Applications of lightweight cryptographic techniques in the IoT can be used to lessen the storage and computational overheads. CP-ABE scheme has some limitations of user attributes management and policy specification. The inflexibility of attribute revocation, poor generality, and scalability can be also considered for future scope. The replacement of bilinear pairing by ECC can also be considered for future work because ECC provides similar security as bilinear pairing with a key of a smaller size.

References

1. Asthon K (2010) That internet of things Thing, RFID J., p 4986
2. Sahai A, Waters B (2005) Fuzzy identity-based encryption. Theory and applications of cryptographic techniques. Springer, Berlin Heidelberg, pp 457–473
3. Goyal V, Pandey O, Sahai A, Waters B (2006) Attribute-based encryption for fine-grained access control of encrypted data. In: Proceedings of ACM conference computer communication security, pp 89–98
4. Bethencourt J, Sahai A, Waters B (2007) Ciphertext-policy attribute-based encryption. In: IEEE symposium on security and privacy, IEEE, pp 321–334
5. Beimel A (1996) Secure schemes for secret sharing and key distribution. PhD thesis, Israel Institute of Technology, Technion, Haifa, Israel
6. Shamir A (1985) Identity-based cryptosystems and signature schemes. Lecture notes computer science (including subseries lecture notes artificial intelligence lecture notes bioinformatics), vol 196 LNCS, pp 47–53
7. Odelu V, Das AK, Khan MK, Choo KR, Jo M (2017) Expressive CPABE scheme for mobile devices in IoT satisfying constant-size keys and ciphertexts. IEEE Access 5:3273–3283

8. Guo F, Mu Y, Susilo W, Wong DS, Varadharajan V (2014) CPABE With constant-size keys for lightweight devices. IEEE Trans Inf Forensics Secur 9(5):763–771
9. Ostrovsky R, Sahai A, Waters B (2007) Attribute-based encryption with non-monotonic access structures. In: Proceedings of ACM conference on computer communication security, pp 195–203
10. Wang J, Xiong NN, Wang J, Yeh WC (2018) A compact ciphertext-policy attribute-based encryption scheme for the information-centric internet of things. IEEE Access 6(2):63513–63526
11. Oualha N, Nguyen KT (2016) "Lightweight attribute-based encryption for the internet of things. In: 2016 25th international ICCCN, conference computer communication networks, p 2016
12. Lewko A, Waters B (2011) Decentralizing attribute-based encryption. Advances in cryptology—EUROCRYPT 2011—30th annual international conference theory applied cryptography techniques, vol 6632, pp 568–588
13. Malluhi QM, Shikfa A, Tran VD, Trinh VC (2019) Decentralized ciphertext-policy attribute-based encryption schemes for lightweight devices. Comput Commun 145:113–125
14. Rasori M, Perazzo P, Dini G (2020) A lightweight and scalable attribute-based encryption system for smart cities. Comput Commun 149:78–89
15. Banerjee S, Roy S, Odelu V, Das AK, Chattopadhyay S, Rodrigues JJPC, Park Y (2020) Multi-authority CP-ABE-based user access control scheme with constant-size key and ciphertext for IoT deployment. J Inf Secur Appl 53
16. Yao X, Chen Z, Tian Y (2015) A lightweight attribute-based encryption scheme for the Internet of things. Futur Gener Comput Syst 49:104–112
17. Tan SY, Yeow KW, Hwang SO (2019) Enhancement of a lightweight attribute based encryption scheme for the internet of things. IEEE Internet Things J 6(4):6384–6395
18. Ding S, Li C, Li H (2018) A novel efficient pairing-free CP-ABE based on elliptic curve cryptography for IoT. IEEE Access 6:27336–27345
19. Green M, Hohenberger S, Waters B (2011) Outsourcing the decryption of ABE ciphertexts. In: 20th USENIX conference security, p 34
20. Hohenberger S, Waters B (2014) Online/offline attribute-based encryption. Lecture notes computer science (including subseries lecture notes artificial intelligence lecture notes bioinformatics), vol 8383 LNCS, pp 293–310
21. Belguith S, Kaaniche N, Laurent M, Jemai A, Attia R (2018) PHOABE: securely outsourcing multi-authority attribute based encryption with policy hidden for cloud assisted IoT. Comput Netw 133:141–156
22. Ohtake G, Safavi-Naini R, Zhang LF (2019) Outsourcing scheme of ABE encryption secure against malicious adversary. Comput Secur 86:437–452
23. Li J, Jia C, Li J, Chen X (2012) Outsourcing encryption of attribute-based encryption with MapReduce. Lecture notes computer science (including subseries lecture notes artificial intelligence lecture notes bioinformatics), vol 7618 LNCS, pp 191–201
24. Li J, Huang X, Li J, Chen X, Xiang Y (2014) Securely outsourcing attribute-based encryption with checkability. IEEE Trans Parallel Distrib Syst 25(8):2201–2210
25. Sethi K, Pradhan A, Bera P (2020) Practical traceable multi-authority CP-ABE with outsourcing decryption and access policy updation. J Inf Secur Appl 51:102435

Linear Derivative-Based Approach for Data Classification

Amit Verma, Ruchi Agarwal, and Manoj Kumar

Abstract In computer vision, object classification is the most essential stage for recognizing the class of the image using its features. Many models have been presented in the last few years for the classification of still images. A simple linear classification technique with moderate accuracy and low run-time complexity for still images is proposed in this paper. The proposed scheme is compared with the other state-of-the-art techniques and experimental results show the effectiveness of the proposed scheme in terms of time complexity. Thus, the proposed work can be proved beneficial for real-time applications where low computational time is required.

Keywords Artificial neural network (ANN) · Support vector machine (SVM) · Logistic regression (LR) · Run-time complexity · Data classification

1 Introduction

Nowadays, the use of the internet and multimedia technologies are growing day by day; as a result, millions of data (audios, digital images, videos, etc.) are generated with the help of several digital equipment such as high-resolution cameras, mobile phones, and other portable devices. Classification of massive amount of redundant data is a challenging task, especially in situations where unstructured and redundant data is to be reduced. In computer vision and pattern recognition, image classification [1, 2] emerged as an important research component to promote the development of artificial intelligence on theoretical as well as technical basis. Image classification includes image preprocessing [1, 2], image sensing [3], object detection [3, 4], object segmentation [3, 5], feature extraction [6], object classification [5, 7, 8], and many more with the help of machine learning (ML) which is a sub-field of artificial intelligence. It is required that computers should properly understand the content or information for accurate retrieval, management, and proper organization of data.

Image classification is an important part of digital image analysis. Its objective is to identify an object that reflects the features of an image. Autonomous driving

A. Verma · R. Agarwal (✉) · M. Kumar
Babasaheb Bhimrao Ambedkar University, Lucknow, Uttar Pradesh, India

G. Gupta et al. (eds.), *Proceedings of Academia-Industry Consortium for Data Science*, Advances in Intelligent Systems and Computing 1411,
https://doi.org/10.1007/978-981-16-6887-6_23

is one such real-world example of image classification. Various other classifications includes automatic organization of images, websites which provide licensed photos and videos, huge visual database, social networks recognizing images and faces [9, 10], etc. Classifiers are needed to achieve accuracy. From the last few decades, many notable efforts for the calculation and prediction of classes have been performed. In ML, there are mainly two types of classification techniques: supervised classification (SVM, LR, decision tree, naive Bayes classification, etc.) and unsupervised classification (clustering, principal component analysis, etc.). In supervised classification, a sample feature of an image is selected which represents specific classes. Features of an image are classified and used as reference in training sites which are used in image processing software. These training sets group similar features together. The user feeds this information keeping in mind the training area. Once information regarding each class is achieved, the image is categorized by observing each pixel and checking which signature it resembles the most. To implement this, some classification algorithms and regression techniques including linear regression [9], logistic regression [11], neural networks(NN) [12], decision tree [13], support vector machine [14], random forest [15, 16], naive Bayes [14–16], and k-nearest neighbour [17–20] have been proposed. Unsupervised learning algorithms include clustering, detection of anomaly, techniques to learn latent variable models [21–23]. Related pixels/features are grouped into a class, and the user can specify the number of output classes as per requirement. Some of the common algorithms used in unsupervised learning include cluster analysis, anomaly detection, neural networks, and approaches for learning latent variable models [23–26]. The main difference between the two classification algorithm is that in supervised classification, labeled sample data is needed to train the system whereas, in unsupervised classification, labeled sample data are not required for training the system. The main purpose of these classification techniques [19, 20, 27] is to predict the classes of given data points and on the basis of these data points, specifically, every classifier such as NN, SVM and logistic regression achieves an accurate result in a specific area with some limitations. These classifiers need to train the network and work in three phases [27]. Each phase has an important task and achieve the capability to reduce the size of the trained network. So, the main idea for these supervised classification techniques is to train the network using all aspects and extract the features with adequate set of rules.

At present, the existing classification methods aim to improve the performance by depending on the local machine environment; also their performance may decrease for unstructured data. Additionally, the computational time is one of the issues which is noticed in existing algorithms [7, 21–23]. In this paper, an analytical approach for supervised classification is proposed. The proposed work is a simple classification technique which utilizes the linear derivatives for data classification. The main aim of the proposed work is to reduce the time complexity with desirable prediction along with consistent performance on structured as well as unstructured data in a network environment.

The following sections include: Sect. 2: Related definitions, Sect. 3: the proposed methodology. Section 4 Experimental results and comparisons. Section 5 conclusion of the work.

2 Existing Classification Algorithms

Before proceeding to the proposed methodology, this section gives a brief overview of some of the existing supervised classification techniques used for comparison with the proposed work along with the basic concept of the dataset and its processing.

2.1 Support Vector Machine

SVM is a supervised model for data classification. The method was first introduced by Vapnik et al. [19] in 1963. This technique is implemented for the classification and regression of the data (image, video, etc.). If the target class is discrete, then classification is performed and if the target class is continuous, regression is performed. SVM is also known as a maximum margin classifier because it maximizes the geometric margin and simultaneously minimizes empirical classification error [4, 19]. The idea of SVM is to find a hypothesis that generates a minimum error. Originally, SVM was developed for the classification of two distinct classes, but now it can be implemented in more than two classes. It works by finding the hyperplane[1] in N-dimension space (N-features) which is utilized for differentiating the classes. This technique has more than one hyperplane for classification, but it selects a hyperplane that has maximum margin between the support vectors. SVM is applicable in many areas [14, 20] including object recognition, speech identification, text categorization, handwritten digit recognition, face detection in images, etc.

2.2 Artificial Neural Network

ANN is a classification technique that can be implemented using both supervised and unsupervised learning methods [12, 28]. The neural networks are influenced by the human brain. The human brain consists of millions of neurons that are connected with each other and form a special structure known as synapses [29]. These neurons send and process signals further synapses allow neurons to pass the signals.

[1] This Hyperplane can be understood as a decision boundary that separates our data according to their respective features. SVM uses a separating hyperplane for which it is known as a discriminative classifier. In supervised learning, with the labeled training data, the algorithm outputs an optimal hyperplane that categorize the test data. A plane is divided into two parts by a line called hyperplane where each class lays on either side in two-dimensional space.

A neural network is framed using the same foundation. It consists of a large number of simulated neurons that are trained through the neural network algorithms. Neural network is the most successful tool for the classification of a dataset with a suitable combination of learning, training, and transfer function which works on following step:

- **Error calculation**: Calculates the difference between the predicted and actual output.
- **Minimum error**: Check whether the risk is minimized or not.
- **Update parameter**: If the error is not minimized, then parameters (i.e., weight and bias) are changed.
- **Check the error**: After updating values, again check the error. Repeat these steps until the error is minimized.

The model is ready for prediction, i.e., it can predict output by taking random input after error minimization. It is one of the methods of neural networks which utilizes supervised learning techniques. In order to obtain minimum error between the predicted output and actual output, it uses weight adjustment while training.

2.3 Logistic Regression

Logistic regression is a classifier which describe the relationship between the dependent (output) variable and one or more independent (input variables). It predicts values in the form of either 'true' or 'false,' '0' or '1,' or 'yes' or 'no.' In general, it predicts the probability of occurrence of any event. The outcome is discrete in logistic regression. Binary or binomial logistic regression is the simplest form of LR in which the output can have only two possible values, i.e., either 1 or 0. So in LR, for binary output variables (dependent variable), a distribution function called 'sigmoid function' [10] is used. In this function, the value of distribution must lie between the range 0 to 1 [10, 11]. Three types of classification can be performed using LR, namely binary classification (logistic regression), multiple classifications (Polytomous Logistic Regression), and classification by hierarchy [11].

2.4 Dataset

The computational performance of the system is evaluated on Fisher's Iris dataset that is taken from the Dataset Repository of UCI [30]. The dataset consists of three classes (Iris Setosa, Iris Virginica, and Iris Versicolor) under the attribute named species. Each class contains 50 records with four attributes: sepal length, sepal width, petal length, and petal width.

Fig. 1 Iris flower species

2.5 *Data Processing*

The raw data from the dataset collected is in the form of comma-separated value file. The CSV file is preprocessed until no text data resides in it. The next step assigns the numerical values 0 and 1 to the corresponding classes. The value 0 is assigned for the Iris Setosa and 1 for the Iris Versicolor as shown in Fig. 1. So the final input will be the CSV file which has numerical values for all the responses made in the form of a target class column. This file is used for supervised learning in the proposed approach. The Iris Flower dataset is divided with an 80:20 ratio, where 80% data is training data and 20% is for testing the data.

3 Proposed Methodology

The proposed methodology has a simple linear approach for the classification of data into classes. These classes are different from each other based on their features. The proposed classification scheme works by taking into consideration the main key difference between the features of two different classes. The key feature of the proposed work is that it does not need any type of training function for the classification of classes like other existing methods. The flow of proposed model is drawn in Fig. 2.

The detailed steps of the classification algorithms are explained below:

1. Consider the input dataset $S(m, n)$ where m, n are the number of rows and columns of the dataset.
2. Divide the dataset S in 80:20 ratio into two sub-datasets $X(m1, n)$ and $Y(m2, n)$ for training and testing respectively.
3. Calculate the rowwise linear derivative $R(m1, n)$ of training set $X(m1, n)$ as:

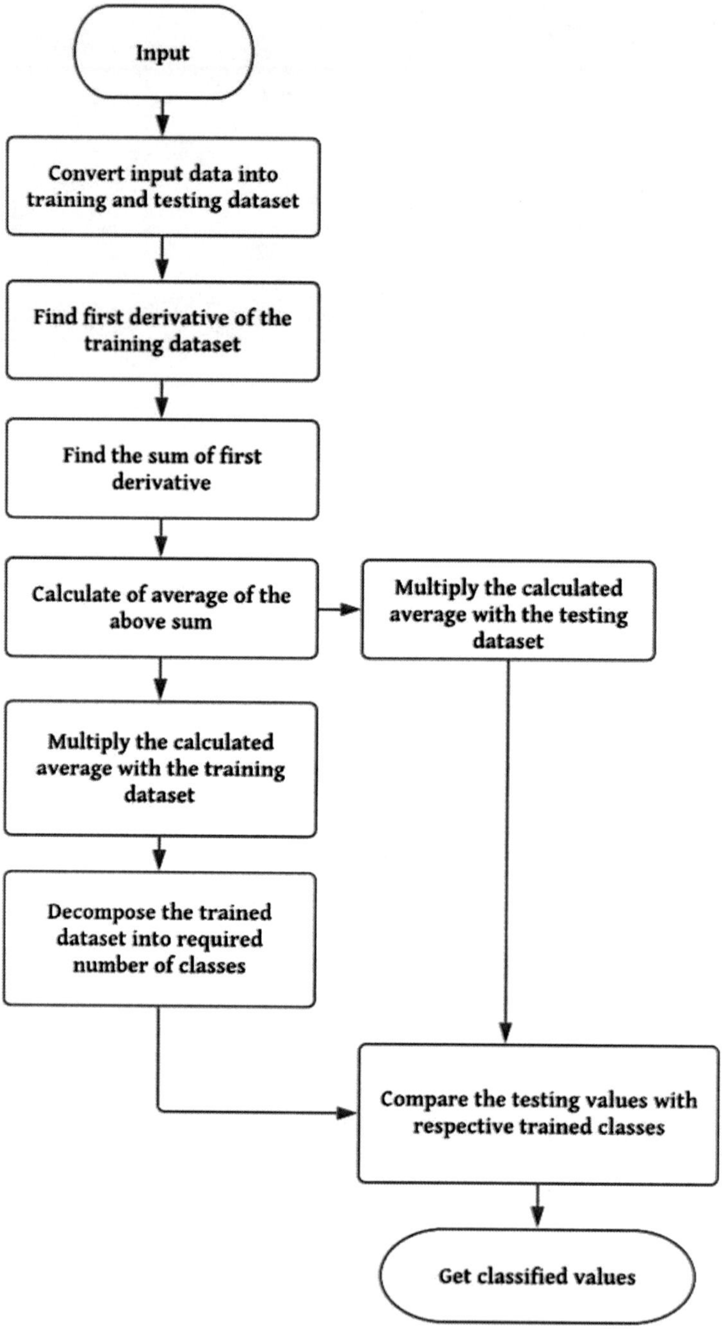

Fig. 2 Flow diagram of proposed model

$$R(m1, n) = X_{i+1,j} - X_{i,j} \tag{1}$$

where, $1 \leq i \leq m, 1 \leq j \leq n$

4. Calculate weighted averages w as:

$$w = \frac{\sum_j \sum_i R_{i,j}}{m1} \tag{2}$$

where, $m1$ is 80% size of original data, $1 \leq i \leq m, 1 \leq j \leq n$

5. Obtain the trained set z by multiplying above calculated weights w with training data X as:

$$z = w * X^T \tag{3}$$

where X^T is the transpose of X.

6. The different classes are obtained by dividing z equally into required number of classes. For instance, if there are two classes; divide z into $z1$ and $z2$.
7. Now for testing data perform following operation:

$$\text{result} = w * Y^T \tag{4}$$

8. If the *result* obtained in the previous step belongs to $z1$ then the element belongs to class first otherwise if result comes in range of $z2$, the element belongs to class second.

4 Results and Discussions

4.1 *Confusion Matrix*

After training the data, testing is performed with remaining 20% data by multiplying with the weight and finding the matrix that compares their output with the training output for each class. The accuracy of output is checked by confusion matrix which measures the performance for any classification technique. The efficiency of the proposed model shows on Table 1. Various ratios have been considered as shown in Fig. 3a–c for testing the accuracy of proposed model via confusion matrix. The terms used in Fig. 3 are briefly described below:

- **True Positive (TP)** It means that the model predicted a positive value and it is true. For the proposed model, the true Positive came out to be 10 indicating that the model predicted a true value for 10 positive values out of 10 for first class. When the model takes 30% test data, then the model predicts 15 values out of 15 for first class. Accordingly, for 40% test data in our model, 20 true positive values come out of 20 for the same class.

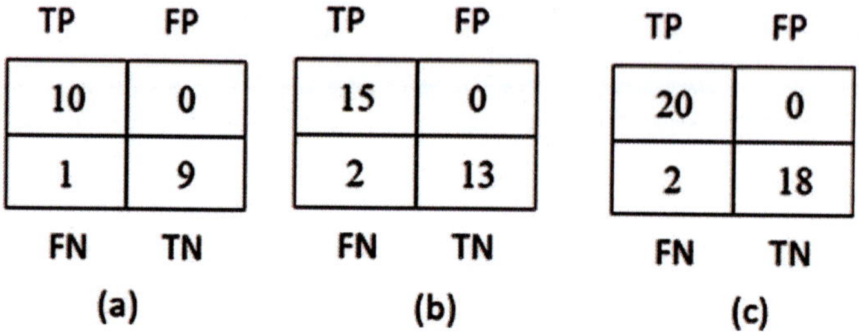

Fig. 3 Confusion matrix with respect to dataset (80/20%, 70/30%, 60/40%)

Table 1 Accuracy of the classifier for trained and tested data

Test set	TP	TN	FP	FN	Sensitivity (%)	Specificity (%)	Accuracy (%)
Test set 1	10	9	0	1	90.9	100	95
Test set 2	15	13	0	2	88.2	100	93.33
Test set 3	20	18	0	2	90.9	100	95

- **True Negative (TN)** It means that the model predicted a negative value and it is true. For our model, the true negative came out to be 9 for first class. For the sane class, when 30% test samples are considered, then 13 true negative values come out of 15 values. If our model uses 40% test samples, then the true negative value comes to be 18 out of 20 values for the same class.
- **False Positive (FP)** (Type 1 Error) It means that the model predicted a positive value and it is false. For our model, the false positive came out to be 0, i.e., the model predicted a false value 0 out of 10 when predictions are considered for class first. When the size of test data is 30%, the result comes to be 0 false positive out of 15 values.
- **False Negative (FN)** (Type 2 Error) It means that the model predicted a negative value and it is false. For our model, the false negative is 1 for second class. The model predicted a false value 1 out of 10 for the second class.

The fidelity of any classification model is measured by evaluating some regularity parameters such as sensitivity, specificity, and accuracy. The precise definition of mentioned terms is formulated in Eqs. 5, 6, and 7, respectively.

$$\text{Sensitivity} = \frac{\text{TP}}{\text{TP} + \text{FN}} \tag{5}$$

$$\text{Specificity} = \frac{\text{TN}}{\text{TN} + \text{FP}} \tag{6}$$

$$\text{Accuracy of the model} = \frac{\text{correct records find by model}}{\text{total records}} \tag{7}$$

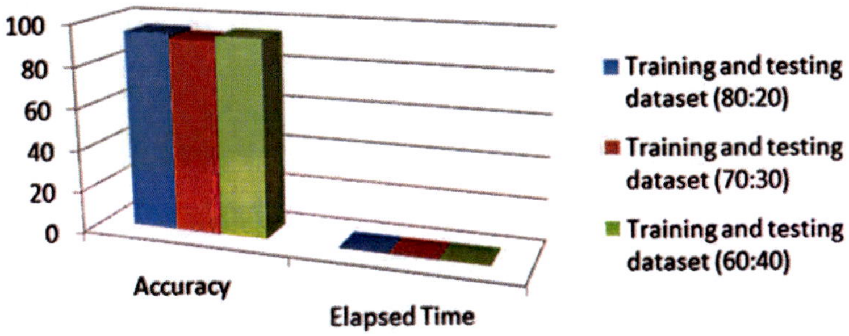

Dataset Ratio (%)	Accuracy (%)	Elapsed Time
Training and testing dataset (80:20)	95	0.114053
Training and testing dataset (70:30)	93	0.100032
Training and testing dataset (60:40)	95	0.161423

Fig. 4 Accuracy and elapsed time with respect to dataset

4.2 Experimental Results

The proposed model has been done on *Intel core i5, 2.40 GHz, 8GB RAM* in *Windows 10* operating system. The computational programming has been implemented on *MATLAB R2018a* version using the above-mentioned dataset. To verify the effectiveness of the proposed scheme, many experiments are done with different ratios of training and testing data as 80:20, 70:30, and 60:40. This training and testing data is used for testing the accuracy and time complexity which checks the feasibility of the proposed scheme. Figure 4 shows the bar graph for accuracy and elapsed time (the execution time of model). The output of the proposed model is graphically represented in Fig. 5 for different ratios.

Experimental results shown in Table 2 are better for the computational time when the average is calculated of all the different ratios taken for the testing and training of the dataset when compared with other existing classifiers.

4.3 Comparison

Comparison of the proposed model has been done with standard existing classifiers. The diagrammatically representation of existing models SVM, NN, and LR on the same dataset are shown in Figs. 6, 7, and 8, respectively. The effectiveness of the proposed scheme can be seen from Table 2 which describes that the proposed scheme provides a better computational time for almost all the different ratios of the respective dataset. For instance, the computational time of SVM on the given dataset is 62.1448 s while for the proposed technique, the computational time comes to be 0.1140, which is approx 0.18% of the time taken by SVM for prediction. This clearly highlights the efficacy of the proposed model.

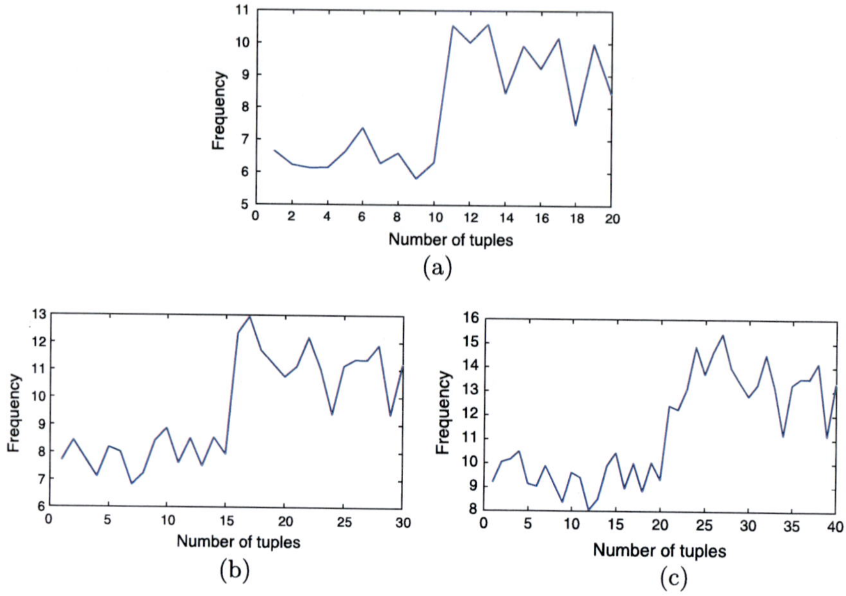

Fig. 5 Frequency plot with respect to dataset (**a** 80/20%, **b** 70/30%, **c** 60/40%)

Table 2 Comparison with existing classification approaches

Approaches	Accuracy (%)	Computational time (s)
SVM	100	62.1447565
LR	100	1.299225
NN	98.67	16.006017
Proposed scheme	95	0.114053

5 Conclusion and Future Scope

This paper provides a single derivative-based classification approach. The proposed classification is based on the features of classes to find the higher dissimilarity between two or more classes. The method is implemented using simple mathematical calculations to find the appropriate result. Easy implementation and computational time are key features of the proposed method. Experimental results with respect to the existing techniques justify the statement. In addition, the consistent results for structured as well as unstructured data with moderate accuracy prove the effective-

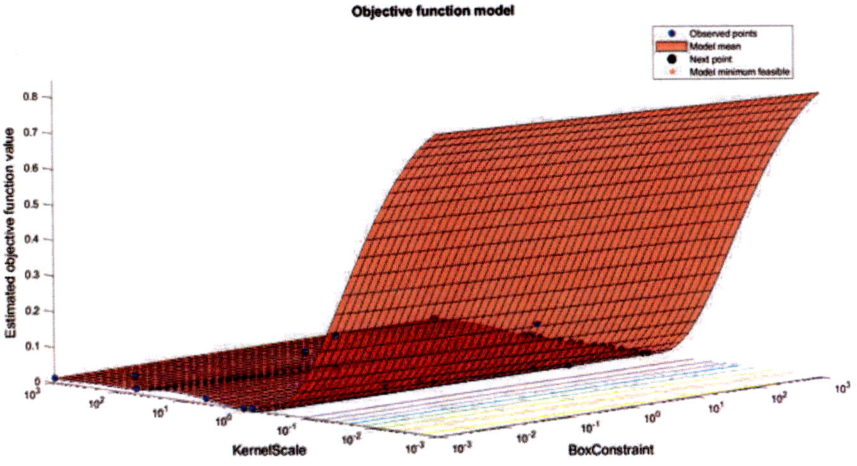

(a)

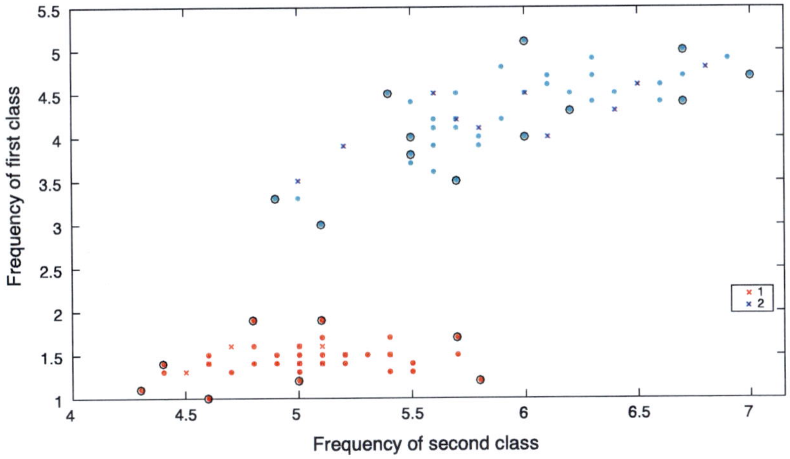

(b)

Fig. 6 **a** SVM object function model, **b** frequency object class

ness of the proposed. There is still more to explore in this novel approach where the dynamic analysis results can be augmented by updating weights of the training sample. The proposed approach can reduce the system requirement as it works on local machine instead of network environment, making it cost effective and robust.

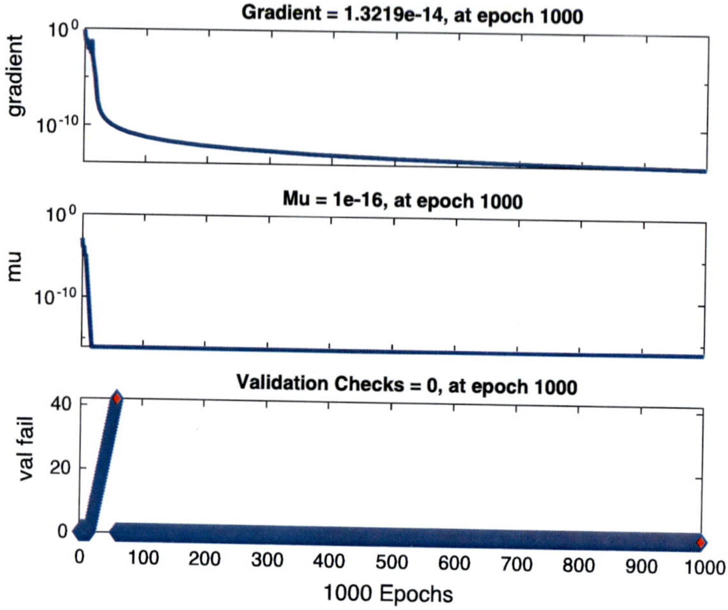

(a)

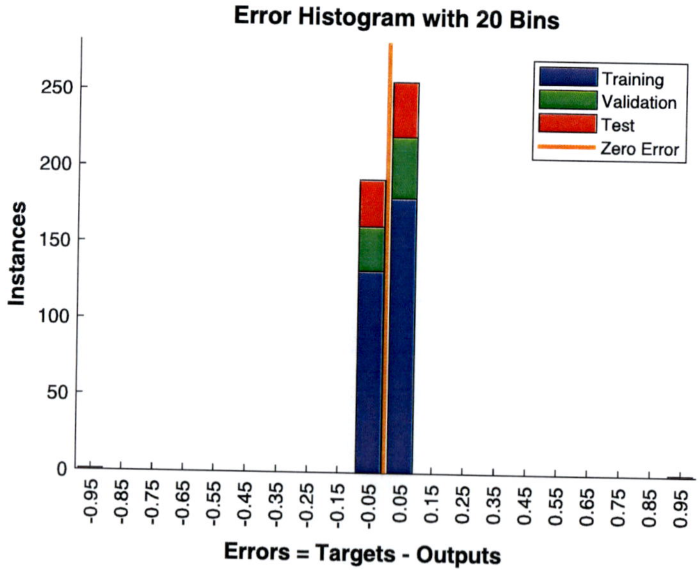

(b)

Fig. 7 **a** NN validation check, **b** histogram

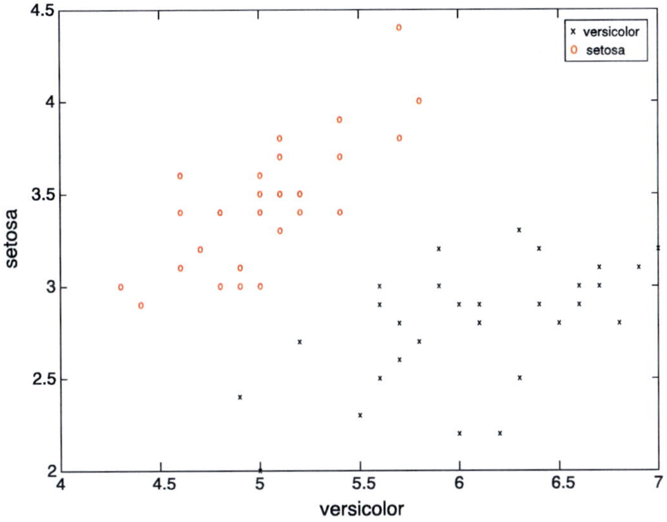

Fig. 8 Using LR plot on fisher dataset

References

1. Lu D, Weng Q (2007) A survey of image classification methods and techniques for improving classification performance. Int J Remote Sens 28(5), 823–870
2. Ying G, Miao L, Yuan Y, Zelin H (2009) A study on the method of image pre-processing for recognition of crop diseases. In: 2009 international conference on advanced computer control, Singapore, pp 202–206
3. Lin TY, Dollár P, Girshick R, He B, Hariharan K, Belongie B (2017) Feature pyramid networks for object detection. In: Proceedings of the IEEE conference on computer vision and pattern recognition. Honolulu, pp 2117–2125
4. Lin TY, Goyal P, Girshick R (2018) Focal loss for dense object detection. In: Proceedings of the 2017 IEEE international conference on computer vision (ICCV), Venice, Italy
5. Wei X, Phung SL, Bouzerdoum A (2014) Object segmentation and classification using 3-D range camera. J Vis Commun Image Representation 25(1), 74–85
6. Lin Y, Lv F, Zhu S, Yang M, Cour T, Yu K, Cao L, Huang T (2011) Large-scale image classification: fast feature extraction and SVM training, CVPR 2011, Providence, RI, 1689–1696
7. Romero A, Gatta C, Camps-Valls G (2016) Unsupervised deep feature extraction for remote sensing image classification. IEEE Trans Geosci Remote Sens 54(3), 1349–1362
8. Wang G, Hoiem D, Forsyth D (2009) Building text features for object image classification. In: IEEE conference on computer vision and pattern recognition, Miami, FL, pp 1367–1374
9. Naseem R, Togneri Bennamoun M (2010) Linear regression for face recognition. IEEE Trans Pattern Anal Mach Intell 32(11), 2106–2112
10. Cheng Q, Varshney PK, Arora MK (2006) Logistic regression for feature selection and soft classification of remote sensing data. IEEE Geosci Remote Sens Lett 3(4), 491–494
11. Li J, Bioucas-Dias JM, Plaza A (2010) Semisupervised hyperspectral image segmentation using multinomial logistic regression with active learning. IEEE Trans Geosci Remote Sens 48(11), 4085–4098

12. Giacinto G, Roli F (2001) Design of effective neural network ensembles for image classification purposes. Image Vis Comput 19(9–10), 699–707
13. Liu H, Cocea M, Ding W (2017) Decision tree learning based feature evaluation and selection for image classification. In: 2017 international conference on machine learning and cybernetics (ICMLC), Ningbo, pp 569–574
14. Maulik U, Chakraborty D (2017) Remote sensing image classification: a survey of support-vector-machine-based advanced techniques. IEEE Geosci Remote Sens Mag 5(1), 33–52
15. Thanh Noi P, Kappas M (2018) Comparison of random forest, k-nearest neighbor, and support vector machine classifiers for land cover classification using sentinel-2 imagery. Sensors 18(18)
16. Xia J, Ghamisi P, Yokoya N, Iwasaki A (2018) Random forest ensembles and extended multiextinction profiles for hyperspectral image classification. IEEE Trans Geosci Remote Sens 56(1), 202–216
17. Chunhui Z, Bing G, Lejun Z, Xiaoqing W (2018) Classification of hyperspectral imagery based on spectral gradient. SVM and spatial random forest. Infrared Phys Technol 95:61–69
18. Alfarrarjeh S, Kim SH, Agrawal M, Ashok S, Kim Y, Shahabi C () Image classification to determine the level of street cleanliness: a case study. In: IEEE fourth international conference on multimedia big data (BigMM), Xi'an, 1–5
19. Chapelle O, Haffner P, Vapnik VN (1999) Support vector machines for histogram-based image classification. IEEE Trans Neural Netw 10(5), 1055–1064
20. Karthick G, Harikumar R (2017) Comparative performance analysis of Naive Bayes and SVM classifier for oral X-ray images. In: 2017 4th international conference on electronics and communication systems (ICECS), Coimbatore, 88–92
21. Hurtik P, Molek V, Perfilieva I (2020) Novel dimensionality reduction approach for unsupervised learning on small datasets. Pattern Recogn 103:107291
22. Wickramasinghe CS, Amarasinghe K, Manic M (2019) Deep self-organizing maps for unsupervised image classification. IEEE Trans Ind Inf 15(11), 5837–5845
23. Shivakumar BR, Rajashekararadhya SV (2018) Investigation on land cover mapping capability of maximum likelihood classifier: a case study on North Canara, India, Procedia Computer Science, 143, 579–586
24. Zhou J, Zeng S, Zhang B (2020) Linear Representation-based methods for image classification: a survey. IEEE Access 8:216645–216670
25. Ying G, Miao L, Yuan Y, Zelin H (2009) A study on the method of image pre-processing for recognition of crop diseases. In: International conference on advanced computer control, Singapore, 202–206
26. Miles W, James B, Bravo S, Campos F, Meehan M, Peacock J, Ruggles T, Schneider C, Simons AL, Vandenbroucke J (219) Particle identification in camera image sensors using computer vision. Astroparticle Phys 104:42–53
27. Ren S, He K, Girshick R, Sun J (2017) Faster R-CNN: towards real-time object detection with region proposal networks. IEEE Trans Pattern Anal Mach Intell 39(6), 1137–1149
28. Sharma N, Jain V, Mishra A (2018) An analysis of convolutional neural networks for image classification. Procedia Comput Sci 132:377–384
29. Arboleda ER, Fajardo AC, Medina RP (2018) Classification of coffee bean species using image processing, artificial neural network and K nearest neighbors. In: 2018 IEEE international conference on innovative research and development (ICIRD), Bangkok, 1–5
30. Fisher RA, Iris data set, Michael Marshall (MARSHALL, PLU '@' io.arc.nasa.gov). https://archive.ics.uci.edu/ml/datasets/Iris

A Novel Logo-Inspired Chaotic Random Number Generator

Apoorva Lakshman, Parkala Vishnu Bharadwaj Bayari, and Gaurav Bhatnagar

Abstract This paper proposes a novel yet simple pseudo-random number generation technique using a sequence of one-dimensional discrete chaotic maps and a logo image. The core idea is to generate a sequence of real numbers between zero and one by combining the sequences generated by chaotic maps and a logo image, defining a relation between the two. This combination has generated numbers showing good sensitivity even to slight changes in the underlying keys. To evaluate the proposed generator, a set of statistical and security analyses are considered to demonstrate the efficiency of the proposed generator.

Keywords Pseudo-random number generation · Chaotic maps · Logo image

1 Introduction

The substantial proliferation in the development of communication technologies, network and information security, and the digital data is readily accessible, and this scenario poses a serious problem of maintaining integrity. Due to this, it has become essential to incorporate cryptographic protocols to ensure the integrity of the data [1]. One of the essential aspects of the cryptographic protocols is the generation of the sequence of random numbers. For this purpose, the underlying process is termed as random number generator (RNG), and there are some essential attributes of RNG regardless of their diverse applications, which include [2]: (1) The generated random sequence should be statistically strong, (2) no intruder can predict predecessors or

A. Lakshman (✉) · P. V. B. Bayari · G. Bhatnagar
Department of Mathematics, Indian Institute of Technology Jodhpur, Jodhpur, India
e-mail: lakshman.1@iitj.ac.in

P. V. B. Bayari
e-mail: bayari.1@iitj.ac.in

G. Bhatnagar
e-mail: goravb.1@iitj.ac.in

G. Gupta et al. (eds.), *Proceedings of Academia-Industry Consortium for Data Science*, Advances in Intelligent Systems and Computing 1411,
https://doi.org/10.1007/978-981-16-6887-6_24

successors or complete sequence from a sub-sequence of it, and (3) the prediction of the sequence must comply only with the initial state value.

Owing these attributes, the RNG are broadly categorized into two groups, namely true random number generators (TRNGs) and pseudo-random number fenerators (PRNGs). The former are in accordance with the real-world phenomena and are frequently used in the cryptographic protocols because of unique statistical properties, unpredictability, and dearth of re-production [3]. The TRNGs can be used to produce PRNGs but have some drawbacks like they are often slow, expensive, and highly dependent on the hardware. In contrast, PRNGs always generate deterministic random sequences depending on the initial seed value. This initial seed value is crucial and acts as the key for the PRNG. The prime feature of the PRNG is in the selection of the initial seed to enhance the security and evade the possibility of generating correlated sequences [4]. Usually, PRNGs are having successful predictions in a certain time bound, but the unpredictability of the sequences may be ensured by enhancing the time bound.

Recently, PRNGs with specific statistical attributes attracted the interest of research communities specifically working in the development of network-based communication systems and Internet technologies due to the fact that these produce significant amounts of data which eventually leads to the requirement of securing the same. It is thus required by the PRNGs to meet a set of requirements such as fairly long period, optimal memory consumption, unpredictability, fast implementation, and execution. Chaos theory has been attracting the attention of many researchers for the realization of these requirements in real life. The important characteristics of chaotic maps include high sensitivity to variations of initial conditions, periodicity, ergodicity, and mixing property which are extracted as a source of randomness which are in alignment with the two main operators of cryptographic systems, confusion, and diffusion [5].

Here, some of the prominent works done in the PRNGs will be highlighted. In [6], authors have proposed the use of chaotic oscillations emerged from a couple map lattice to construct a new pseudo-random number generator. In [7], authors have proposed a new PRNG to generate the pseudo-numbers for video cryptography and is inspired by the chaos theory. In [8], a novel 3D chaotic map is proposed by integrating a logistic map and piecewise linear map toward a novel PRNG. This technique shows to have high key sensitivity and an acceptable security for image encryption. The authors have proposed a new PRNG based on the discrete space chaotic map in [9] and proved to have infinite key space virtually that resulted into the increased security and high-cycle sequences. This technique requires small memory and is very useful for low memory devices. In [10], authors have presented a new PRNG algorithm that employed a piecewise linear map to exhibit asymptotic deterministic randomness. This technique shows amiable features for encryption and provides better resistance to entropy attacks. In [11], authors have proposed a PRNG, for image encryption, based on Chen's chaotic system. This technique essentially solves the non-uniformity probability distribution issue of generated random sequences. In [12], a new secure PRNG is devised by integrating three distinct chaotic maps to achieve better unpredictability, large key space, and sensitivity to initial conditions.

In [13], a PRNG is proposed in which a saw-tooth chaotic map is utilized to generate random sequences with an improved period. In [14], authors have proposed a chaotic PRNG based on Lorenz chaotic map. In this technique, authors have generated two pseudo-random sequences integrating the coordinates of the Lorenz chaotic circuits. In [15], authors have developed an improved logistic map to define an efficient and secure PRNG for hardware systems to generate pseudo-random numbers having certain statistical characteristics.

The objective of this work is to generate numbers with maximum randomness which is ensured through various experiments. The concept of the novel technique employed in this research deals with combining the sequences generated by two sources. One being a sequence of one-dimensional discrete chaotic maps and the other from a logo image. The combination of these two sequences is defined by a relation discussed in the scope of this paper. The resultant sequence lies in the real domain. Experimental analyses such as key sensitivity, histogram, and entropy have been carried out. Three one-dimensional discrete-time chaotic maps, namely logistic, sine, and tent map are arranged in a sequence for pseudo-random number generation. The parameter values are set as per the requirement of their chaotic behavior. A logo image is selected and converted into a sequence by concatenating the rows. These two sequences are combined to produce a sequence with increased randomness. The efficiency of the proposed PRNG is carried out by the different analyses, and it is observed that the proposed technique significantly increases the randomness and security.

This rest of paper is organized as follows: Sect. 2 provides the brief idea about the chaotic maps, followed by illustrating the proposed technique in Sect. 3. Section 4 presents the experimental results and discussions, while concluding remarks are finally given in Sect. 5.

2 Chaotic Maps

Chaos theory focuses on the study of systems where the disoder/irregularities are controlled by underlying patterns and highly sensitive deterministic laws [5]. There are mainly two types of chaotic maps, continuous, and discrete. However, this research is based on one-dimensional discrete-time maps, generally defined as $x_{k+1} = \tau(x_k)$ where $k = 0, 1, 2\ldots$ are state variables and τ is a mapping from current x_k to x_{k+1}. Starting with an initial value x_0 and applying τ repeatedly, a sequence (trajectory) is obtained. A sequence of such maps are used in this work for generating a chaotic stream of pseudo-random numbers.

2.1 Logistic Map

The logistic map (population growth model) is an iterative relation of degree two, often mentioned as a prototype of complexity. Chaotic behavior can be observed from very simple nonlinear dynamical equations, mathematically expressed as:

$$x_{n+1} = rx_n(1 - x_n) \tag{1}$$

where x_n lies between zero and one, representing the ratio of the existing population to the maximum possible population (dimensionless measure of the population in the nth generation). The values of interest for the parameter $r \geq 0$, the intrinsic growth rate, are those in the interval [0, 4]. The chaotic behavior can be seen by setting the initial value of the variable, $x_0 \in (0, 1)$ and the parameter r in the interval $(3.57, 4)$. This increases the randomness in number generation as the values approach and cross the chaotic threshold.

2.2 *Sine Map*

One-dimensional sine map which is similar to the logistic map with one parameter has the following form:

$$y_{n+1} = \alpha \sin(y_n) \tag{2}$$

where initial value $y_0 \in [0, 1]$ and $\alpha \in [0, 1]$ is the control parameter. For these values, chaotic behavior is observed.

2.3 *Tent Map*

Similar to the logistic map, but easier to analyze. It also depends on only one parameter and is expressed as the following mathematical equation:

$$z_{n+1} = \begin{cases} \mu z_n; & 0 \leq z < \frac{1}{2} \\ \mu(1 - z_n); & \frac{1}{2} \leq z < 1 \end{cases} \tag{3}$$

The chaotic behavior is observed when $\mu > 1$ and setting the initial value $z_0 \in (0, 1)$. The arrangement of these maps in a sequence and generating random numbers using their chaotic behavior is described further.

3 Proposed RNG Process

In this section, some of the impelling aspects of the proposed PRNG are discussed. The process is initiated by iterating the chaotic maps in the sequential manner followed by combining of generated sequences and the logo image gives the resultant

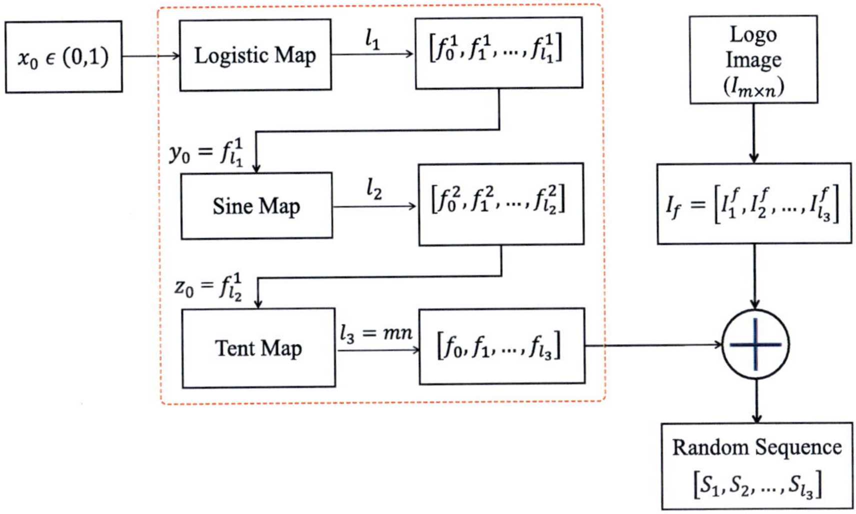

Fig. 1 Block diagram of the proposed random number generation

sequence of random numbers. The process flow of generating sequences of random numbers is illustrated in Fig. 1 and can be summarized in the following stages:

1. **Initialization**: Each map generates the sequence with the logic:

$$\text{Initialization} \rightarrow \text{Iteration} \rightarrow \text{Number Generation}$$

The process of initialization involves choosing initial values of variables from the interval (0,1). The length of the sequence (number of iterations) is pre-determined as per the given image. The pseudo-random number generator requires this due to its demand for initial seeding. Next, iteration involves iterating the values of these variables till the iteration limit is reached. Finally, obtaining the real sequence of numbers in the interval (0, 1). Different sequences are generated, varying the parameters responsible for the chaotic behavior of the maps.

2. **Chaotic Maps Synchronization**: Maps are arranged in the following sequence:

$$\text{Logistic Map} \rightarrow \text{Sine Map} \rightarrow \text{Tent Map}$$

The initial value x_0 of the logistic map (1), is chosen randomly from the interval (0,1) and the parameter r is chosen from $(3.57, 4)$ for chaotic behavior. The number of iterations (l_1) should be set to at least 1000 to maximize the randomness. The final iterated value x_{l_1} of this map is used as the initial value for the variable of the map, next in the sequence, i.e., $y_0 = x_{l_1}$. Since, the numbers generated by all these maps lie in the range of (0, 1); thus, we are abiding by the requirement of seed parameters in all the maps. The iteration limit l_2 for this map can be the same as in the previous map or increased. The parameter value α of the sine map (2), is set in accordance with the chaotic threshold ($\alpha \in (0, 1)$). Now, $z_0 = y_{l_2}$ and the iteration limit l_3 of the tent map (3) is set equal to the size of the logo image.

3. **Image Sequence**: A logo image I of size $m \times n$ is considered. The logo is first transformed into the frequency domain (say represented by I_f) using the Fourier transform. Now, the phase component ($\theta_{\rm I}$) is obtained, and the image feature matrix is constructed as per the following equation:

$$I^{\rm F} = \tanh |\pi \theta_{\rm I}| \tag{4}$$

The rows of $I^{\rm F}$ are concatenated to form the image sequence ($I_i^f \mid i = 1, 2, 3, \ldots, m \times n$).

4. **Combined Sequence**: Let the generated chaotic sequence from Step 2 be $\{f_0, f_1, \ldots, f_l\}$ where $l = m \times n$ is the pre-determined length set according to the size of the image considered in Step 3. Then, the combined sequence is given by:

$$S_i = \bmod \left((f_i)^{I_i^f}, 1 \right) \tag{5}$$

It is clear that the length of the final generated random sequence (S_i) is $l = m \times n$ and the modulus function will ensure that every element of sequence must be in the interval [0, 1].

4 Experimental Results

In this section, the efficiency of the proposed random number generation scheme is accessed. For the same, different analyses have been involving different logos in the MATLAB environment on a PC with Intel Core i7 CPU and a 4GB RAM. The logos considered in the experiments are depicted in Fig. 2. Further, the analyses to validate the performance of the proposed PRNG include key space analysis, key sensitivity analysis, histogram analysis, and numerical analysis.

4.1 Key Space Analysis

In general, to oppose the brute-force attack, the underlying keys of a PRNG should belong to a large space. The proposed PNRG utilized the initial seed of chaotic maps as the key. Although, the proposed PNRG has infinite key space theoretically as the initial seed always belongs to the interval [0, 1]. However, the key space found to be finite due to the finite precision of the computers. Owing the average calculation precision (10^{-6}), the proposed PNRG has the initial seed and other underlying control parameters would be roughly $10^{50} \approx 2^{168}$, which is fairly effective and large to oppose brute-force attacks.

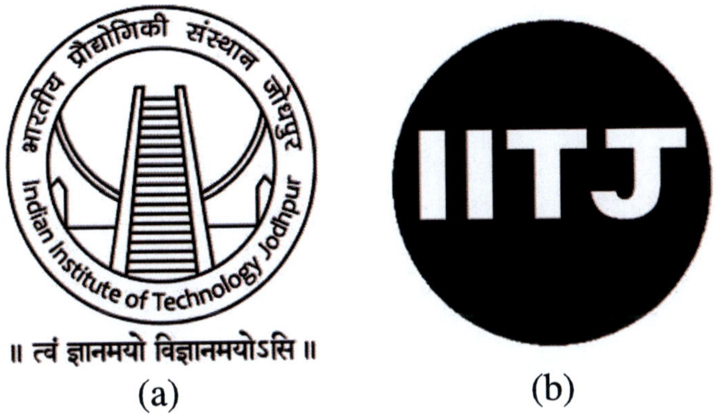

Fig. 2 Logo images used in the experiments

4.2 Key Sensitivity Analysis

Key sensitivity captures the change in the generated sequence when we perturb the key even by a small amount. The keys of proposed PNRG are the initial values of chaotic maps, their parameter values, and logo image, respectively. Among these, the effect of parameter is insignificant due to the use of multiple chaotic maps in a sequential manner. Therefore, the keys which will make proposed PNRG sensitive are initial seed and logo image. The key sensitivity of the proposed PNRG is thus assessed visually and quantitatively. For visual assessment, the distribution of the generated random sequence is considered. For the ease, only the terminal 100 entries of the generated random sequence are shown in Fig. 3, and it is clear that the proposed PNRG has good sensitivity with the initial seed and logo image.

Another aspect of the evaluation is the quantitative evaluation, and key sensitivity measure is being used for the same. Mathematically, key sensitivity measure (KSM) is given by:

$$\mathrm{KS}(S, \tilde{S}) = \frac{\sum_{j=1}^{l} \mathrm{Diff}(i)}{l} \times 100\% \tag{6}$$

where l represents the length of random sequences, S and $\tilde{S}$ be the random sequences generated with true and false key, respectively. Further, $\mathrm{Diff}(i)$ is defined as:

$$\mathrm{Diff}(i) = \begin{cases} 0; & S(i) = \tilde{S}(i) \\ 1; & S(i) \neq \tilde{S}(i) \end{cases} \tag{7}$$

The value obtained by the KSM ideally should be close to 100, which essentially signifies that the complete random sequence is changed after altering the keys associated. The KSM results of the proposed PRNG are given in Table 1. Based on Table 1,

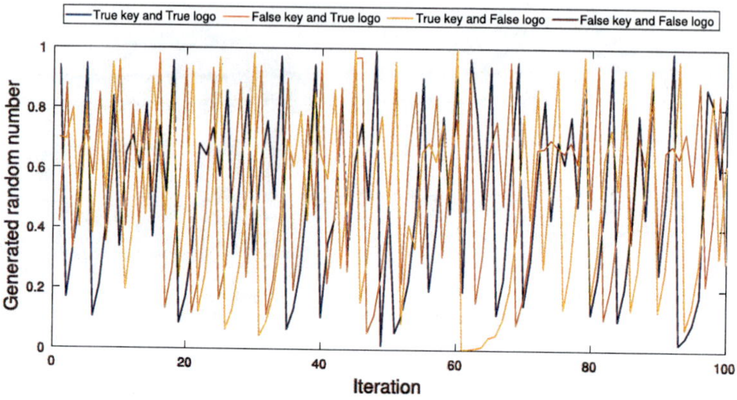

Fig. 3 Sensitivity analysis of the proposed PRNG

Table 1 Key sensitivity analysis based on KSM

Particulars	KMS (in %)
True key and false logo	99.9954
False key and true logo	99.9999
False key and false logo	99.9999

it can be observed clearly that the proposed PRNG is sensitive to small changes of keys based on KSM, which can be justified further by the obtained high values. Thus, it can be concluded that the proposed PNRG is highly sensitive to the keys.

4.3 Histogram Analysis

A histogram uses a graphical representation to show the frequency distribution of the given data. The histograms essentially provides the information of the occurrence of each element of a sequence. In general, an efficient PRNG should be as spread as possible across the range and uniform in nature, ensuring that all values are distributed equally exhibiting uniform histogram for every keys to resist against statistical attacks. The histogram of the generated random sequences using the true and false keys are depicted in Fig. 4. It can be observed that the distributions are similar to a uniform distribution, suggesting that the proposed PRNG efficiently generates uniform random sequences.

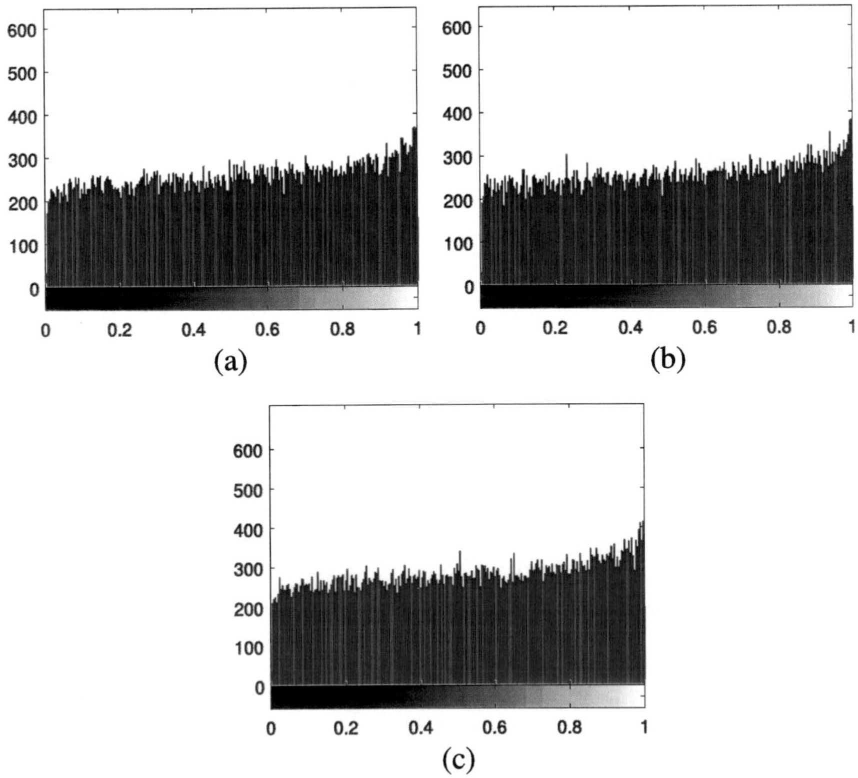

Fig. 4 Histograms analysis of the proposed PRNG: histogram of generated random sequences with **a** true key and true logo, **b** false key and true logo, and **c** false key and false logo

5 Conclusion

A simple yet efficient pseudo-random sequence generation process has been investigated in this paper. As described, a random sequence is generated by defining the relation among the sequences generated by a series of chaotic maps and a logo image. The performance and efficiency of the proposed PNRG are ensured using an exhaustive evaluation and analysis. It is revealed from the analysis that the key characteristics of the proposed PRNG are large key space, high key sensitivity, and good statistical features. Therefore, it can be deduced that the proposed PRNG is efficient and can generate random sequences which are highly suitable for diverse applications.

References

1. Bhatnagar G, Wu QMJ, Raman B (2011) A novel image encryption framework based on Markov map and singular value decomposition. In: Proceedings of international conference on image analysis and recognition, pp 286–296
2. Wold K (2011) Security properties of a class of true random number generators in programmable logic. Ph.D. thesis, Gjøvik University College, Norway
3. Zhou Q, Liao X, Wong K-W, Hu Y, Xiao D (2009) True random number generator based on mouse movement and chaotic hash function. Inf Sci 179(19):3442–3450
4. Barani MJ. Ayubi P, Valandar MY, Irani BY (2020) A new Pseudo random number generator based on generalized Newton complex map with dynamic key. J Inf Secur Appl 53:102509-1–102509-25
5. Tutueva AV, Nepomuceno EG, Karimov AI, Andreev VS, Butusov DN (2020) Adaptive chaotic maps and their application to pseudo-random numbers generation. Chaos Solitons Fractals 133(109615):1–8
6. Wang X-Y, Qin X (2012) A new pseudo-random number generator based on CML and chaotic iteration. Nonlinear Dyn 70(2):1589–1592
7. Xu H, Tong X, Meng X (2016) An efficient chaos pseudo-random number generator applied to video encryption. Optik 127(20):9305–9319
8. Sahari ML, Boukemara I (2018) A pseudo-random numbers generator based on a novel 3D chaotic map with an application to color image encryption. Nonlinear Dyn 94(1):723–744
9. Lambi CD, Nikoli CM (2017) Pseudo-random number generator based on discrete-space chaotic map. Nonlinear Dyn 90(1):223–232
10. Wang K, Pei W, Xia H, Cheung YM (2008) Pseudo-random number generator based on asymptotic deterministic randomness. Phys Lett A 372(24):4388–4394
11. Hamza R (2017) A novel pseudo random sequence generator for image-cryptographic applications. J Inf Secur Appl 35:119–127
12. François M, Grosges T, Barchiesi D, Erra R (2014) Pseudo-random number generator based on mixing of three chaotic maps. Commun Nonlinear Sci Numer Simul 19(4):887–895
13. Dastgheib MA, Farhang M (2017) A digital pseudo-random number generator based on sawtooth chaotic map with a guaranteed enhanced period. Nonlinear Dyn 89(4):2957–2966
14. Lynnyk V, Sakamoto N, Celikovsk S (2015) Pseudo random number generator based on the generalized Lorenz chaotic system. IFAC-PapersOnLine 48(18):257–261
15. Murillo-Escobar M, Cruz-Hernández C, Cardoza-Avendaño L, Méndez-Ramírez R (2017) A novel pseudorandom number generator based on pseudorandomly enhanced logistic map. Nonlinear Dyn 87(1):407–425

Applications of Deep Learning for Vehicle Detection for Smart Transportation Systems

Poonam Sharma, Akansha Singh, and Anuradha Dhull

Abstract The most important task in designing an intelligent transportation system is the robust and efficient detection of all the vehicles presents in the frame sequence. As there is huge development going on in terms of computer vision and deep learning techniques, a lot of new techniques and algorithms have been proposed for doing the same. Also with the high accessibility of video data, the training has become easier and more accurate. In this paper, a brief review of the literature in the area of vehicle detection and classification is presented. This paper also discuss the different challenges faced in on-road vehicle detection due to weather and driving conditions, camera illuminations, and occlusion. Then, it discuss various methods proposed for overcoming all these challenges using deep learning. Future scope is also presented so that efforts may be focused toward making more robust smart systems.

Keywords Vehicle detection · Computer vision · Intelligent transportation systems · Deep learning

1 Introduction

Video cameras have been extensively used in recent years for traffic surveillance due to their low cost and high field of view [1]. Also the fast and extensive research in computer vision and deep learning techniques have motivated the researchers in designing intelligent transportation and surveillance systems which are low cost and require less human intervention to operate [2].

Computer vision and deep learning techniques in traffic surveillance have become very popular in designing ITS [3]. These systems utilize visual appearance and

P. Sharma (✉)
Department of Computer Science, Amity University, Gurgaon 122051, India

A. Singh
School of Computing Science and Engineering, Amity University, Noida 122001, India

A. Dhull
School of Computing Science and Engineering, The NorthCap University, Gurugram 122004, India

G. Gupta et al. (eds.), *Proceedings of Academia-Industry Consortium for Data Science*, Advances in Intelligent Systems and Computing 1411,
https://doi.org/10.1007/978-981-16-6887-6_25

motion analysis-based techniques for vehicle detection, recognition, and tracking. After detecting and tracking, the information can be further used for incident detection, congestion analysis, speed detection, etc. [4]. A lot of research has been done in this area, but there are various challenges which are still faced by these systems [5]. All traffic scenarios have different kind of challenges associated with them such that scale and angle variations, traffic congestion, weather and lighting conditions, camera scale and illuminations, occlusions, etc., [1]. Also, the variability in types of vehicles, size, color, and angle of detection is a crucial challenge in the detection process [6]. Traffic congestion and placement of camera highly affect the performance of ITS due to increase in the probability of occlusion [3]. Several computer vision and deep learning-based methods are proposed in the literature to resolve these challenges and improve the accuracy of the system. However, there is still lack of a method which is applicable for all kind of vehicles in all conditions.

In a recent survey, a discussion on the available vehicle surveillance architecture especially about computer vision methods is presented [1]. A detailed analysis on detection and tracking is provided in [2]. A review on computer vision techniques for traffic and congestion analysis has been done in [3]. This review is giving brief concentrating on infrastructure-side. In another review, the computer vision and pattern recognition-based techniques are discussed focusing on various challenges while designing such systems [4].

With the advancement of deep learning techniques, object detection has become a great area of interest for researchers [7]. As a result, these networks can be very affectively used in designing ITS and many researchers have started proposing technique using these networks [8, 9]. The progress in the area and extensive work done since 2010 is evident from Fig. 1. The number of publications available on Google scholar and science direct Website which provides scientific publication data clearly depict that there have been a great growth in terms of research in this area.

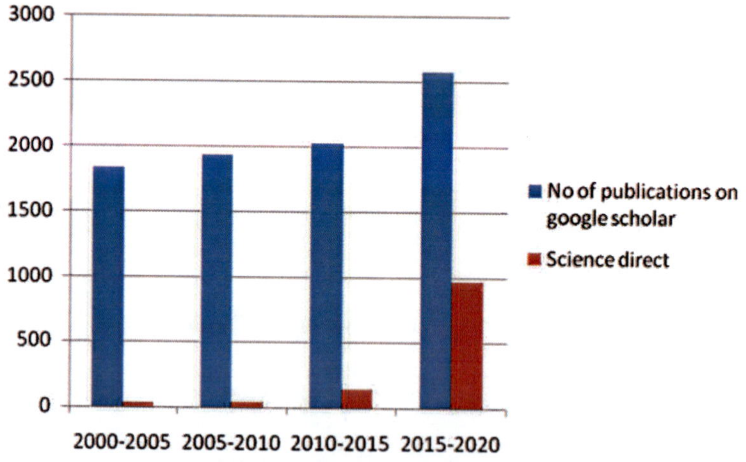

Fig. 1 Research in terms of number of publications

In this paper, a review on various detection and tracking methods is provided focusing on computer vision and deep learning. This review also highlights improvements, challenges, advantages, and disadvantages. The remaining of paper is organized as follows. In the next section, we have provided a review of the existing research done so far in the area. Section III gives a brief on deep leaning and various deep learning techniques proposed in the literature has been discussed. Finally, Section IV summarizes and concludes this paper.

2 Vehicle Detection Methods

The very first task in designing an ITS is detection and classification of vehicles. Accuracy in detection impacts further tracking and analysis process to a great extent. All the methods proposed till date for doing the detection can be categorized as shown in Fig. 2.

2.1 Motion-Based Methods

Motion-based techniques basically use motion of moving vehicles to differentiate them from the stationary background. The simplest method in this category is the frame difference method [10]. This method uses difference in consecutive frames. This difference is used to select moving vehicles by setting a threshold value. As the frames are highly dynamic in nature and much other noise occurring possibilities, the method could not work very well. Later on, the method was combined with

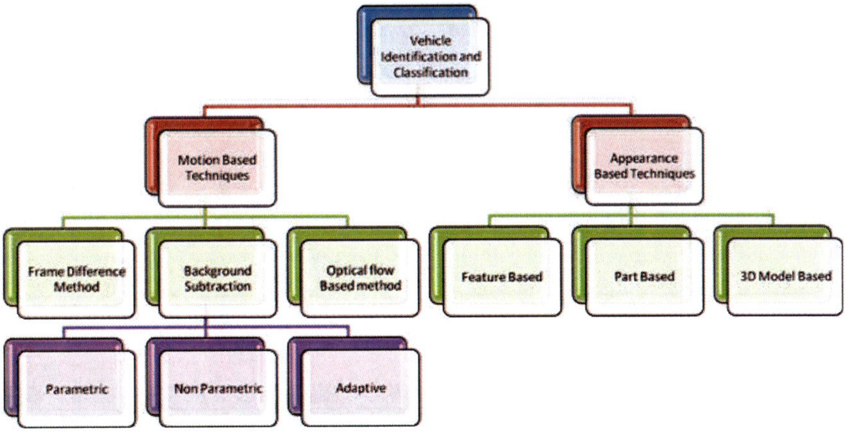

Fig. 2 Classification of vehicle detection methods

background subtraction methods [11] and Gaussian mixture models to improve the accuracy [12].

In the background detection-based methods, standard background information was first gathered. Later on, the subsequent frames were subtracted from this background to get the foreground moving objects [13]. The method can be categorized into three different techniques, namely parametric, non- parametric, and predictive techniques [14]. In parametric techniques, a probability density function is used to model every pixel of the frame. The various techniques to model every pixel include frame averaging [15], single Gaussian [16], median filter [17], sigma delta estimation [18], and Gaussian mixture model [19]. But all these methods suffer from the challenge of accurate background detection. The reason behind can be moving vehicles, dynamic illumination changes, and complex backgrounds.

In nonparametric techniques, history of the pixels is used to build a probabilistic model for coming pixels using recent values [14]. The two main techniques used are kernel density estimation (KDE) and codebook models [20]. In KDE methods, a Parzen window is used to find probability of each background pixel using recent samples [21]. This probability is then used to find the possibility of object being a vehicle. These methods are good as they are very less sensitive to quickly changing background, but these cannot be used for long durations due to high memory requirements for storing the history. In codebook-based methods, a set of codeword, which is dynamically created and handled, is used to design the probabilistic function for modeling every pixel in background. After that, clustering technique is used so that each codebook occupies a variable number which constructs a part of background.

Another popular category is optical-flow-based motion segmentation. It uses moving objects features in the form of flow vectors to detect moving regions in video frames. The fields which are generated in this way represents the speed and direction of every pixel in the form of a motion vector [22]. Also, the methods are good in providing accurate motion details but involve large iterative calculations which increase the complexity of the system.

2.2 *Appearance-Based Techniques*

These methods use visual features like color, shape, texture, dimensions, etc., to detect vehicles on road. The motion-based methods were only able to detect objects which are moving as compared to background. Stationary vehicles can be easily identified by appearance-based techniques. Appearance-based techniques can be further classified into three main categories. First one of them is feature-based techniques. A large number of features has been identified and used in the literature to identify vehicles. One of them is edge-based methods which used histogram [23] or local symmetry-based method [24]. But these were not giving very good results in case of large size and illumination variations. Some more robust features like scale invariant feature transformation (SIFT) [25], speeded up robust features (SURF) [26], histogram of oriented gradient (HOG) [27], and Haar-like features [28] are extensively used in vehicle detection literature.

Another classification of feature-based techniques includes part-based detection models. These methods consider vehicles in different parts like front rear and top views. The parts are detected using edge, shape, and orientations. On these detected parts, spatial relationships and other motion-based cues are applied to detect various vehicles [29].

In 3D modeling techniques, computer generated 3D models are used for detection purpose. Appearance matching algorithms are used for detection purpose [30]. 3D modeling suffers from the problem of finding an accurate 3D model. As a result, it can be used for few number of vehicle classes only. Also, the extraction and presentation with comparison become more complicated as the number of models increase.

All these methods, whether motion-based or appearance-based, may use different types of learning methods or classifiers for further detection. Some of the methods used statistical models which are based on principal component analysis for detection [31]. Some other has used AdaBoost classifiers which use Haar-like features for detection purpose [32, 33]. Multilayer neural networks [34] and pulse coupled neural networks were also a good choice for detection purpose [35]. Out of these neural networks have shown good results but processing was still slower due to manual feature extraction and identification required. The summary of the methods is provided in Table 1.

The datasets for vehicle detection have also become constraint in developing a good transportation system as initial datasets were small and were not containing all categories for classification. The summary of different datasets is shown in Table 2.

Therefore, deep learning is now considered as a better option for solving such problems. These networks learn important features out of the images or videos by themselves. This as a result reduces the overhead of manual feature extraction. Among all the proposed deep architectures, convolutional neural networks have revealed significant performances. These networks are least impacted by geometric transformations, deformations, and illumination which are major reasons for reducing the efficiency of any vehicle detection system. With the use of high end GPUs and process capabilities, large amount of data can be generated for the training purpose which will improve the detection accuracy.

3 Applications of Deep Learning in Intelligent Transportation

Vehicle detection and classification are very initial and most important step in designing ITS. The three different types of classifiers used for doing the same include support vector machine (SVM), AdaBoost, and neural networks. SVM is a classifier which constructs a hyper plane and learns the boundaries between the classes to make a decision. AdaBoost, on the other hand, is a type of classifier which combines

Table 1 Summary on work in vehicle type classification

Features used	Classification method	Vehicle types classified	Pros	Cons
Aspect ratio, compact ratio, length, and height of vehicle silhouettes [36]	Dimension-based thresholds	Large: Bus, van truck, trailer, truck; Small: Sedan, van, pickup	Simple features used makes it less complex	Only useful for static images
Vehicle blob features transformed by Fisher's linear discriminant analysis (LDA) [2]	weighted KNN	Sedan, pickup, SUV, van, merged, bike, truck	Accuracy is better as compared to methods above	Not all moving regions other than vehicles get avoided
3D-HOG and SD-model [3]	Matching with models	Bus, sedan, van, bike	Less complex because of template uses	If training data is less accuracy and is less. Therefore, good training templates are required
Size and shape features are bounding box perimeter, elasticity, filled area, convex area, etc., [37]	SVM, random forest, model-based matching	Car, van, bus, bike (bicycle or motorcycle)	Simple feature extraction	Little slower and do not consider occlusion and other environmental conditions
Time-spatial Images (TSI); shape- and geometry-based features, shape-invariant image moments, and texture-based statistical features [38]	Two-level KNN	Motorbike, rickshaw, auto-rickshaw, car, jeep, covered van, bus	Occlusion detection is also addressed and can be worked with larger dataset	Little bit slow due to complex two stage process
Edge orientation histograms (EOHs) [12]	Hybrid Kernel auto-associator-based classifier with rejection option	Bus, light truck, car, and van	Less complex because of template uses	Not suitable for small vehicles

(continued)

Table 1 (continued)

Features used	Classification method	Vehicle types classified	Pros	Cons
Boosted binary features and spatial pyramid-based features quantization [4]	SVM	Sedan, taxi, van, truck	Accuracy is better	Feature extraction is complex
Local and global feature generated by a two stage unsupervised semi-supervised CNN (with layer skipping) [39]	Softmax regression	Minivan, SUV, sedan, truck	Features learned by the system were discriminative enough to work with complex scenes as well	Little slow due to a complex two stage process
CNN-based architecture using fine-tuned model [40]	Unsupervised and use previously trained models	All categories	Unsupervised and use previously trained models	Not performed good under occlusions and dense scenarios
R-CNN-based models for detection [41, 42]	Unsupervised and use previously trained models	AD categories	Faster and giving very good results in normal conditions	Detection in dark and angle properties of camera still creating problems for detection

Table 2 Benchmark datasets for vehicle detection

Name of dataset	Size and description	Metrics
PETS 2001	Testing set of 2867 frames	Recall and precision
LISA 2010	Three short videos of 500, 300, and 300 frames taken in different situations	Recall and precision
Caraffi 2010	20 min driving videos on Italian highways	Recall and precision
KITTI 2012	Extended videos from stereo pairs	Recall, precision, orientation, and speed
MIO TCD 2018	Lakhs of training and testing frames	Recall, precision

local feature descriptors to improve the accuracy of the local classification networks. Neural networks have been a great topic of interest in the past decades. The reason is they can learn linear boundaries to classify into different classes which were not possible with SVM. However, neural networks suffer problem that parameters and weights need to be tuned to get the desired output. The development of deep learning

Table 3 Deep learning applications in intelligent transportation systems

S. No.	Application area	Deep learning technique	Accuracy rates (%)
1	Automatic vehicle detection [37, 38]	Convolutional neural networks	Classification into seven categories with accuracy rates between 89 and 99
2	Automatic vehicle detection [39]	Deep belief networks	96
3	Night time vehicle detection [41]	FIR and deep learning	92
4	Occlusion detection [42]	Locally connected deep models	89
5	Obstacle detection [43]	Deep stacked auto encoder-based	80

technologies, convolutional neural networks (CNNs), and much other architecture has been successfully used in vision-based vehicle detection. Fast R-CNN [36] is very popular techniques for vehicle detection, where a region proposal network (RPN) significantly reduces the computation cost and manual feature extraction. The various areas of ITS where applications of deep learning have been found is summarized in Table 3. In this sections, below all such architectures have been discussed.

3.1 Convolutional Neural Networks for Detection and Classification

In recent years, object detections have gained a great attention and success due to developments in deep learning technologies. Vehicle detection and classification being an object detection problem can also be done by applying deep learning methodologies.

Initially, some semi-supervised models of convolutional networks were proposed which could solve the problem [37]. In these networks, the filter bank of the network was learnt in unsupervised way. The data taken in these methods were completely unlabelled data. The output parameter training was on the other hand supervised with labeled data. The network in these methods consists of two different stages. First stage was used to find low level features from the frames. Second stage was used to predict global features.

Every stage in this method was consisting of five different layers. In the convolutional layer, for effective filtering of the network, the sparse Laplacian filter learning (SLFL) is used. The absolute rectification layer and local contrast normalization layers are used to provide a nonlinear transformation to the output of the convolutional layer. The pooling and sub sampling layers are used to reduce the spatial resolution and dimensions of the frame. This is achieved by the use of average pooling

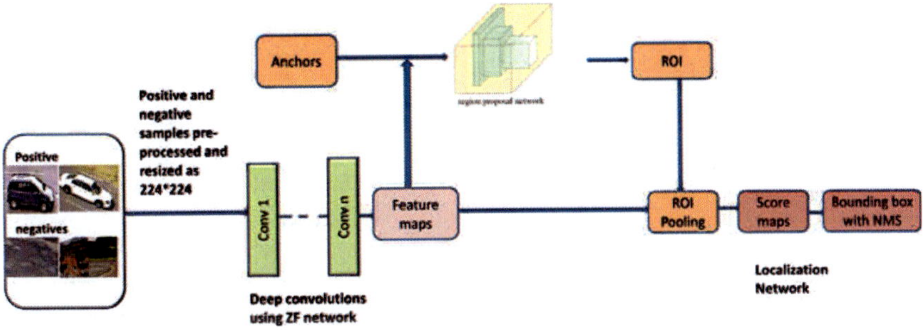

Fig. 3 RPN using ZFNet for vehicle detection and classification

operator. The representations produced in this way are robust to geometric distortions and augmentations. To compute the probability of each vehicle type, a Softmax layer is used at output. The method has shown an accuracy of 88% but was not working well for night light images and also in occlusion and other abrupt conditions.

Region proposal technology made object detection even simpler with higher accuracy rates. In 2015, Ren et al. introduced a new technique based on fast R-CNN. This method used both CNN and region proposal network (RPN) for detecting the objects of interest in the frames. This method took very less time and also increased the accuracy to a great extent.

Much different architecture has been proposed in the literature which uses RPN combined with convolutional network for vehicle detection and classification. In one of them, ZF Net is used for initial training of the network [38]. The last shareable layer features are forwarded to RPN for further detection. The detailed architecture is shown in Fig. 3.

As shown in Fig. 3, the method builds RPN on last layer (layer 5) of the ZF network. The processing function used is simple ReLU function. The features generated as result are fed to RPN. This network utilizes anchors (reference for object detections) for mapping to different kind of objects which may be present in the frames. The features obtained from RPN are used for further bounding box generation. The regression layer is used to predict the 4k coordinates which are nothing but the coordinating specifying the exact location of vehicles in the frames. The classification layer is used find probabilities of objects being a vehicle in terms of output score. The method provided very good results for classification.

3.2 *Deep Belief Networks for Vehicle Detection*

The ultimate objective in vehicle detection task using learning is to find a mapping function from training data X to the label data Y. This mapping function should be able to classify objects from frames into vehicle and non-vehicle categories. A 2D

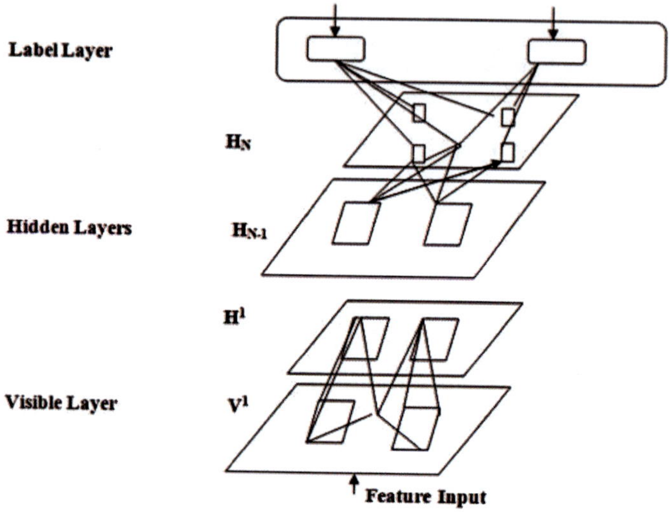

Fig. 4 DBN architecture for vehicle detection

deep belief network (2D-DBN) was proposed in [39] as shown in Fig. 4.The method learns the detection in three steps.

Step 1: The size of top most layer is determined using projection techniques. The lower layer output data are mapped to subspace and optimize to best dimension possible. This optimum dimension is used to get size of next upper layer.

Step 2: A greedy wise reconstruction method is used to refine adjacent layers of the network. Step 1 and step 2 are repeated to get all the hidden layer parameter and generate the optimum network model.

Step 3: Fine tune the network using back propagation.

The method has shown good accuracy rates of 96%.

In another DBN-based technique, a combination of DBNs having different values for pre-training epochs (PE), training epochs (TE), number of iterations, size of the batch, and size of different layers was used [40]. The architecture has shown good accuracy rates for cars.

3.3 *Vehicle Detection in Night Using Deep Learning*

Statistics show that accidents often occur at night and result in fatalities. But most of the methods developed so far are only focusing on the videos or the data taken in daylight conditions. A method based on far infrared (FIR) camera images using DBN has been introduced which works very well for night detections [41]. It is purely dependent on the heat generated in environment to produce a grayscale image. This results in a very good view as compared to other illumination methods as shown in

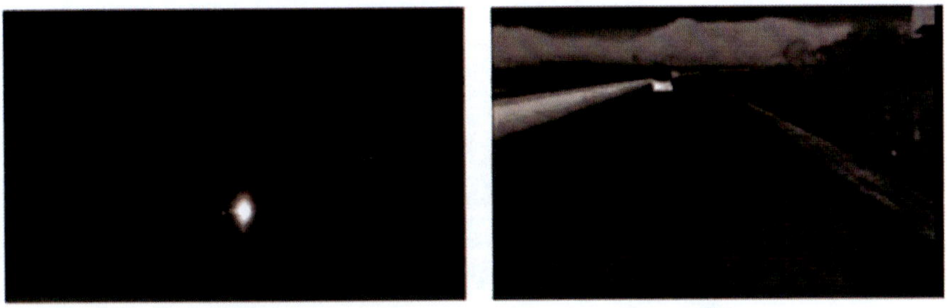

(a) (b)

Fig. 5 **a** Image captured in normal light, **b** FIR image

Fig. 5. The required heat signal can be generated by heat sources such as vehicle tire, engine, and exhaust pipe of the vehicle.

The main step of the algorithm includes

Step 1: Vehicle area extraction using image saliency.

Step 2: Vehicle candidate generation: This is done using informations like camera parameters, size of edges, contours, and texture of the vehicles.

Step 3: Final verification of the vehicles using deep belief architectures.

3.4 Occlusion Detection Using Deep Learning

The algorithms presented so far have good detection accuracies but only when the vehicles are in full view. These algorithms do not perform well on occluded vehicles. Therefore, there is a great need for detection methods to handle this problem of occlusion to improve the detection. A method based on deep learning was proposed [42]. The steps of the method are

Step 1: Suspected occluded vehicle generation: In this step, Haar features are first extracted from the frames. After that AdaBoost classifier is applied. During the classification stage, it was observed that the frames which were having very little difference from the positive samples taken and were rejected in the final stage of the classification. These samples were possible candidates of occluded vehicles. It was observed that all the samples which were classified at 14th and 15th layer of the cascade classifiers were possible candidates which were further used for learning purpose.

Step 2: Visual model establishment depicting occlusion: Through observation and analysis done in the previous step, it was found that the visual object model can be designed consisting of eight different situations and attributes.

Step 3: Training through locally connected deep models: A deep belief network is finally used to pre-train and then finally tune the network for occlusion as well as positive vehicle detections.

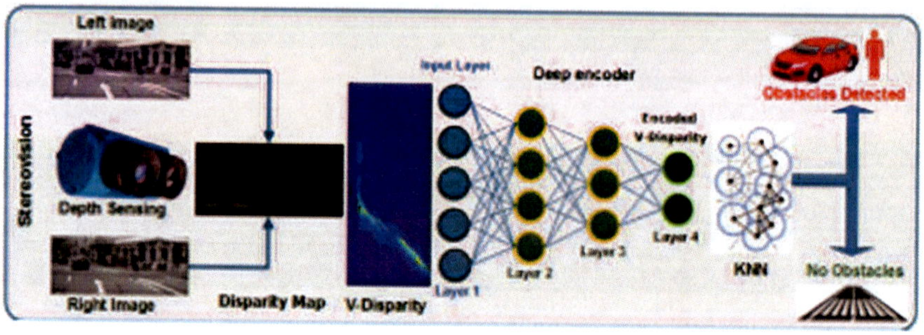

Fig. 6 Architecture of deep learning for obstacle detection [43]

It was observed that method was giving 89% accuracy results for occluded vehicles which were very good as compared to other methods.

3.5 Obstacle Detection Using Deep Learning

Detection of obstacles is again a basic requirement for designing ITS. A deep stacked auto encoder (DSA)-based obstacle detection approach was proposed in literature [43]. The detailed architecture of the method is shown in Fig. 6.

This approach consists of four layers. Each layer is used for extraction of features which are used for encoding input to output. This output is used as an input for the next layer. The input to DSA model is normalized V-disparity map. The input to the KNN as shown classifier is the output which is obtained from the last layer. The proposed deep learning architecture is trained in an unsupervised way without any labeling. The method works in two steps.

Step 1: Extraction of new features for generating a new encoded output that will be used as input for the next layer.

Step 2: The data encoded by the deep encoder model in previous step is used as the input to KNN model for obstacle detection.

As a result, we can say that deep learning has given good opportunities to the researchers working in the area of ITS. To conclude, a comparative analysis in terms of accuracy for all the work done so far has been shown in Table 4.

4 Conclusion

In this paper, first, we have introduced about importance of vehicle detection and classification in designing an intelligent transportation system. Then, a brief review is done on all the motion-based and appearance-based techniques that have been introduced in the literature so far. After that, a brief review is done on evolution and

Table 4 Comparative analysis of various methods

References	Method	Conditions	Accuracy (%)
[44]	Background subtraction	Highway frames	76
[2]	Appearance-based method using Haar-like features	Highway roads	91
[12]	Frame differencing and appearance-based method	Videos captured in urban traffic	90
[31]	HOG gradients and AdaBoost classification	Vehicle detection from all angles	96
[38]	Deep convolutional network	Videos captured from still cameras in all environmental conditions	98

developments in the area of deep neural networks. Finally, how these networks can be used for vehicle surveillance is being shown with the help of various architectures and methods proposed so far.

References

1. Tian B et al (2015) Hierarchical and networked vehicle surveillance in ITS: a survey. IEEE Trans Intell Transp Syst 16(2):557–580
2. Sivaraman S, Trivedi MM (2013) Looking at vehicles on the road: a survey of vision-based vehicle detection, tracking, and behaviour analysis. IEEE Trans Intell Transp Syst 14(4):1773–1795
3. Buch N, Velastin S, Orwell J (2011) A review of computer vision techniques for the analysis of urban traffic. IEEE Trans Intell Transp Syst 12(3):920–939
4. Wang X (2013) Intelligent multicamera video surveillance: a review. Pattern Recogn Lett 34(1):3–19
5. Kastrinaki V, Zervakis M, Kalaitzakis K (2003) A survey of video processing techniques for traffic applications. Image Vis Comput 21(4):359–381
6. Tian B et al (2014) Rear-view vehicle detection and tracking by combining multiple parts for complex urban surveillance. IEEE Trans Intell Transp Syst 15(2):597–606
7. Sharma P, Singh A (2017) Era of deep neural networks: a review. ICCCNT, New Delhi
8. Krizhevsky A, Sutskever I, Hinton G (2012) Imagenet classification with deep convolutional neural networks. In: Neural information processing systems (NIPS), pp 1097–1105
9. Ren S et al (2015) Towards real time object detection with region proposal networks. In: Proceedings of advances in neural information processing systems, pp 91–99
10. Park K, Lee D, Park Y (2007) Video-based detection of street parking violation. In Proceedings of international conference on image processing, computer vision, and pattern recognition (IPCV), pp 152–156
11. Xu L, Bu W (2011) Traffic flow detection method based on fusion of frames differencing and background differencing. In: Proceedings of IEEE 2nd international conference on mechanic automation and control engineering (MACE), pp 1847–1850
12. Zhang H, Wu K (2012) A vehicle detection algorithm based on three-frame differencing and background subtraction. In: Proceedings of IEEE 5th international symposium on computational intelligence and design (ISCID), pp 148–151

13. Zhang R et al (2013) An method for vehicle-flow detection and tracking in real time based on Gaussian mixture distribution. Adv Mech Eng 5:1–8
14. Wan Q, Wang Y (2008) Background subtraction based on adaptive non-parametric model. In: Proceedings of 7th IEEE world congress on intelligent control and automation (WCICA), pp 5960–5965
15. Kanhere NK, Birchfield ST (2008) Real-time incremental segmentation and tracking of vehicles at low camera angles using stable features. IEEE Trans Intell Transp Syst 9(1):148–160
16. Wren CR, Azarbayejani A, Darrell T, Pentland AP (1997) Pfinder: real-time tracking of the human body. IEEE Trans Pattern Anal Mach Intell 19(7):780–785
17. Cucchiara R, Grana C, Piccardi M, Prati A (2003) Detecting moving objects, ghosts, and shadows in video stream. IEEE Trans Pattern Anal Mach Intell 2(10):1337–1342
18. Manzanera A, Richefeu JC (2007) A new motion detection algorithm based on Σ-Δ background estimation. Pattern Recogn Lett 28(3):320–328
19. Stauffer C, Grimson WEL (1999) Adaptive background mixture models for real-time tracking. IEEE Comput Soc Conf Comput Vision Pattern Recogn 2:246–252
20. Mittal A, Paragios N (2004). Motion-based background subtraction using adaptive kernel density estimation. In: Proceedings of the IEEE computer society conference on computer vision and pattern recognition (CVPR), vol 2, pp II-302
21. Kim K, Chalidabhongse TH, Harwood D, Davis L (2005) Real-time foreground–background segmentation using codebook model. Real-time Imaging 11(3):172–185
22. Noh S, Jeon M (2013) A new framework for background subtraction using multiple cues. In: Computer vision (ACCV), pp 493–506
23. Agarwal S, Awan A, Roth D (2004) Learning to detect objects in images via a sparse, part-based representation. IEEE Trans Pattern Anal Mach Intell 26(11):1475–1490
24. Ma X, Grimson WEL (2005) Edge-based rich representation for vehicle classification. In: Proceedings of IEEE 10th international conference on computer vision, vol 2, pp 1185–1192
25. Lowe DG (1999) Object recognition from local scale invariant features. In: Proceedings IEEE 7th international conference on Computer vision, vol 2 pp 1150–1157
26. Bay H, Tuytelaars T, Van Gool L (2006) Surf: speeded up robust features. In: Computer vision (ECCV) pp 404–417
27. Dalal N, Triggs B (2005) Histograms of oriented gradients for human detection. On proceedings of IEEE computer society conference on computer vision and pattern recognition, vol 1, pp 886–893
28. Viola P, Jones M (2001) Rapid object detection using a boosted cascade of simple features. In: Proceedings of EEE computer society conference on computer vision and pattern recognition, vol 1, pp I–511
29. Winn J, Shotton J (2006) The layout consistent random field for recognizing and segmenting partially occluded objects. In: Proceedings of IEEE computer society conference on computer vision and pattern recognition, vol 1, pp 37–44
30. Ferryman JM, Worrall AD, Sullivan GD, Baker KD (1995) A generic deformable model for vehicle recognition. In: Proceedings of British machine vision conference (BMVC), pp 127–136
31. Yan G, Yu M, Yu Y, Fan L (2016) Real-time vehicle detection using histograms of oriented gradients and AdaBoost classification. Optik-Int J Light Electron Opt 1–34
32. Wen X, Shao L, Fang W, Xue Y (2015) Efficient feature selection and classification for vehicle detection. IEEE Trans Circ Syst Video Technol 25(3):508–517
33. Tao D, Li X, Wu X, Maybank SJ (2007) General tensor discriminant analysis and gabor features for gait recognition. IEEE Trans Pattern Anal Mach Intell 29(10):1700–1715
34. Kuo YC, Pai NS, Li YF (2011) Vision-based vehicle detection for a driver assistance system. Comput Math Appl 61(8):2096–2100
35. Ming Q, Jo KH (2011) Vehicle detection using tail light segmentation, vol 2, pp 729–732. Strategic Technology (IFOST)
36. Gu J, Wang Z, Kuen J et al (2016) Recent advances in convolutional neural networks. Pattern Recogn 77:354–377

37. Wang D, Zhang J, Cao W, Li J, Zheng Y (2018) When will you arrive? Estimating travel time based on deep neural networks. Association for the Advancement of Artificial Intelligence, pp 1–8
38. Hochreiter S, Schmidhuber J (1997) Long short-term memory. Neural Comput 9(8):1735–1780
39. Wang et al (2017) Real-time vehicle type classification with deep convolutional neural networks. J Real-Time Image Process 1–10
40. Veres et al (2019) Deep learning for intelligent transportation systems: a survey of emerging trends. IEEE Trans Intell Transp Syst 1–17
41. Wang et al (2014) A vehicle detection algorithm based on deep belief network. Sci World J 1–7
42. Wang et al (2015) Occluded vehicle detection with local connected deep model. Multimed Tools Appl 75(15):9277–9293
43. Dairi A et al (2018) Obstacle detection for intelligent transportation systems using deep stacked autoencoder and k-nearest neighbor scheme. IEEE Sens J 18(12):5152–5132
44. Sharma et al (2019) Automatic vehicle detection using spatial time frame and object based classification, J Intell Fuzzy Syst 37(6):8147–8157

A GARCH Framework Analysis of COVID-19 Impacts on SMEs Using Chinese GEM Index

Xuanyu Pan, Zeyu Guo, Zhenghan Nan, and Sangeet Srivastava

Abstract Stock market return analysis and forecasting are an important topic in econometric finance research. Since the traditional ARIMA models do not consider the variation of volatility, their prediction accuracy is not satisfactory to represent highly volatile periods of any stock market. The GARCH model family resolves the heteroskedasticity of a time series, and hence, it performs better in periods of high volatility. This paper explores the impact of the COVID-19 epidemic on Chinese small- and medium-sized enterprises (SMEs) using a GARCH model for Business as usual (BAU) simulation. We use the Chinese Growth Enterprise Market (GEM) stock index to represent the economic situation of SMEs during the COVID-19 period. Then, we extract, analyze, and predict changes in GEM stock volatility, explore the impact on and recovery status of SMEs, and predict their future trends. For BAU simulation, we first preprocess the GEM stock index between 2018 and 2020 and determine the order of autocorrelation and lags of the data to build the mean model. An ARCH effect test on the residual term of the mean equation was found to be significant and help to decide the order of the GARCH framework. Using the model, a BAU simulation was created and compared statistically with the actual GEM index during 2020. The comparison successfully demonstrated that the GEM index has increased volatility during the pandemic, which is in line with our prior hypothesis.

Keywords GARCH · BAU · COVID-19 · Chinese SMEs

1 Introduction

Beginning January 2020, the COVID-19 virus began spreading globally. According to Johns Hopkins University data [1], as of December 31, 2020, 1.8 million people have died worldwide from the COVID-19 virus. In China, the government decided to implement lockdown in cities to prevent the virus from spreading on a large-scale outside these cities. Starting February, schools were closed, factories were shut

X. Pan · Z. Guo · Z. Nan · S. Srivastava (✉)
Wenzhou-Kean University, Wenzhou, China
e-mail: srivassa@kean.edu

G. Gupta et al. (eds.), *Proceedings of Academia-Industry Consortium for Data Science*, Advances in Intelligent Systems and Computing 1411,
https://doi.org/10.1007/978-981-16-6887-6_26

down, the whole society's production came to a standstill, and many companies were hit financially. COVID-19 occurred at a time when the Chinese economy was in a tightening phase of the business and financial cycle, which would have led to more severe economic losses [2]. China is a major manufacturing country, and factory shutdowns in China have simultaneously led to disruptions in the global supply chain, affecting everything from pharmaceuticals to apparel [3]. Studies show that every additional month of the epidemic costs 2.5% of global GDP [4]. The impacts were felt everywhere, though, measuring these impacts on stock indices is still a research question. Also, different industries were hit differently. To quantify these measures, simulation is the best tool. Theoretically, small- and medium-sized enterprises will be more negatively affected during an epidemic because they are smaller, have an unstable source of orders, and are more prone to capital chain breakdowns. When we reviewed the relevant research, we did not find any research on SMEs, and even if there were individual ones, they were not the quantitative analysis that we have expected [5, 6]. In that case, our research may fill this gap. We decided to study the impact of the pandemic COVID-19 on small- and medium-sized enterprises (SMEs) in China during the year 2020.

2 Data and Methods

We used the Growth Enterprises Market (GEM) Composite Index data procured from ex-ChiNext [7] to represent the economic situation of Chinese SMEs [8]. The GEM is designed by the Chinese government for entrepreneurial companies, SMEs, and companies in high-tech industries that are temporarily unable to be listed on the main board stock market, which includes 710 companies as of 2017 [7], so the GEM Composite Index is a fair indicator of the economic situation of Chinese SMEs. The China GEM Composite Index from June 21, 2018, to December 1, 2020, was extracted using the Tushare API of Tushare Big Data Community [9]. The period was chosen due to the limited availability of the dataset and also based on the assumption that this period is sufficient to represent the changes brought about by the COVID-19. Also, we split this dataset into two segments using threshold of January 1, 2020. We assume that before January 1, 2020, the stock index was unaffected by the outbreak of COVID-19, and after that date, it is affected by the outbreak. We further assume that from March 2020 through May 2020, return-to-work season does not eliminate the impact of the pandemic and that the impact persists until December 1, 2020. This study assumes that the change of volatility in China's GEM stock index is mainly affected by the pandemic and impacts due to the other factors which are negligible, and we investigated this by modeling the returns of the GEM index and its volatility under the assumption that the stock market is heteroskedastic.

In this paper, we model the GEM index using the GARCH model and its variants. The generalized autoregressive conditional heteroskedasticity (GARCH) model family evolved from Engle's ARCH model [10], proposed by Bollerslev [11], and

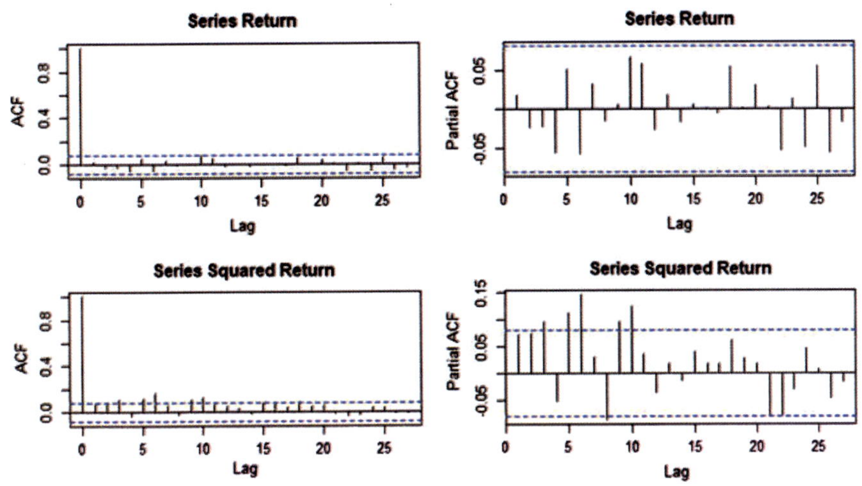

Fig. 1 ACF and PACF test

has been widely used in the financial analysis literature [12]. The model and its variants have also been used by many other studies [13–15], and they applied GARCH to detect changes in volatility due to epidemics with good performance. This paper followed these studies to apply this model to explore the impacts of COVID-19 on Chinese SMEs.

In this study, we used the classical GARCH model to represent volatility in stock indices. Before modeling the volatility, we simply used a stationary ARMA (1, 1) model to represent the mean, as ACF and PACF in Fig. 1 indicate that GEMCI has a weak autoregressive correlation. Equations. (1) and (2) represent the mean and variance equations used in the model.

$$\ln\left(\frac{P_t}{P_{t-1}}\right) = \ln(1 + r_t), \quad r_t = \mu + \text{AR} * r_{t-i} - \text{MA} * a_{t-i} \tag{1}$$

$$a_t = \sigma_t \varepsilon_t, \quad \sigma_t^2 = \omega + \sum_{i=1}^{m} \alpha_i a_{t-i}^2 + \sum_{j=1}^{s} \beta_j \sigma_{t-j}^2 \tag{2}$$

In this paper, we applied the GARCH model to two different periods, June 21, 2018, to January 1, 2019 (before the COVID-19 outbreak) to prepare the best model, and January 1, 2020, to December 1, 2020 (after the COVID-19 outbreak), to model the mean and variance of the two time periods. Then, we extended the model from the first period to the second period of January 1–December 1, 2020, as BAU simulation and compared it with the observed time series to derive the impact of the outbreak.

3 Results and Analysis

In the actual analysis stage, first, we need to fit the data using the GARCH model. We test the squared returns for the ARCH effect, and the results show that the Box–Ljung test score is 0.0003128 for an interval of 5 (Tables 1 and 2). This suggests that the returns are strongly autocorrelated and there is an ARCH effect, so the data are suitable for fitting with a GARCH model (Table 3). Next, we determined the model order to be (1, 1) using ACF and PACF tests. Then, we fit the GARCH (1, 1) model, and the results show that all parameters of the model are significant (Table 3) and the model to be well fitted (Table 4). We compared the AIC and BIC values at different orders by the information criterion and found that the GARCH performance did not improve significantly at different orders. We then performed an ARCH effect test on the residuals of the fitted model, and the results showed that the residuals no longer have an ARCH effect, suggesting to consider the residuals to be white noise.

Table 1 Box–Ljung test scores for ARCH effect of returns

	Returns	Squared returns
X-squared	4.3051	23.173
d.f.	5	5
p-value	0.5064	3.128×10^{-4}

Table 2 ARCH effect test for squared residuals

Box–Ljung test

Data: squared residuals

	Statistic	*p*-value
lag = 1	0.1314	0.71696
lag = 5	4.1452	0.23650
lag = 9	10.1207	0.04727

Table 3 GARCH model fitting result

GARCH model fit

	Estimate	Std. error	*t* value	Pr(> |*t*|)
Mu	0.001204	0.000669	1.8002	0.071836
AR	−0.869344	0.113407	−7.6657	0.000000
MA	0.904952	0.097292	9.3014	0.000000
Omega	0.000013	0.000001	11.1058	0.000000
Alpha	0.078901	0.008938	8.8274	0.000000
Beta	0.879539	0.016457	53.445	0.000000

Table 4 GARCH model adjusted Pearson goodness-of-fit test

Adjusted Pearson goodness-of-fit test		
Data: GARCH model		
Group	Statistics	p-value $(g-1)$
20	26.00	0.1302
30	28.85	0.4728
40	45.87	0.2087
50	48.69	0.4857

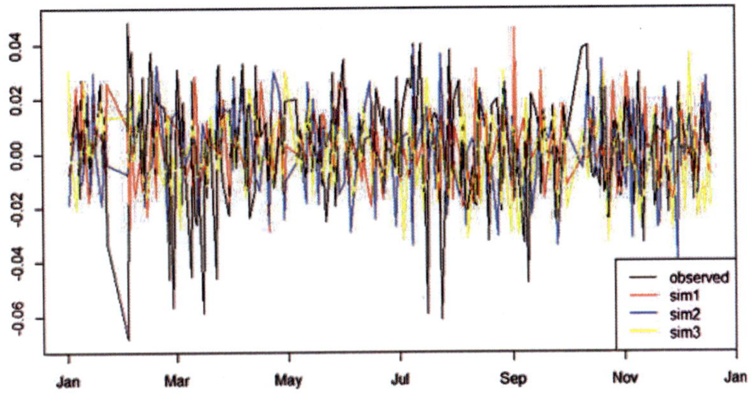

Fig. 2 2020 GEM simulated returns versus observed returns

Next, we used the fitted model to randomly simulate three different time series of the return of GEM index in 2020 and compared them with the actual GEM index in 2020, as shown in Fig. 2. We found that the observed time series (recorded data) has many downward spikes. The three simulations, on the other hand, are relatively stable and slightly above the black line. This suggests that in most cases, the observed returns are larger than the predicted returns, reflecting the increased volatility during the pandemic. However, this judgment is not straightforward, and we decided to quantify it. We calculated the mean of the three curves and obtained an averaged simulation to compare with the observed time series (Fig. 3). We calculated the mean and variance of the two time series and compared them. We can see in Table 5 the variance of the BAU is much smaller than the actual value, which proves that the pandemic did make the GEM index more volatile. And, the mean of BAU is larger than the actual value, which proves that the pandemic makes the return of the GEM index lower. We tested the significance of the differences between the means and variances of the two time series separately, and the results showed that the differences between their means and variances were significant.

In addition, several related studies have shown that pandemics lead to an increase in the RV and VIX indices, which implies an increase in market volatility and a rise

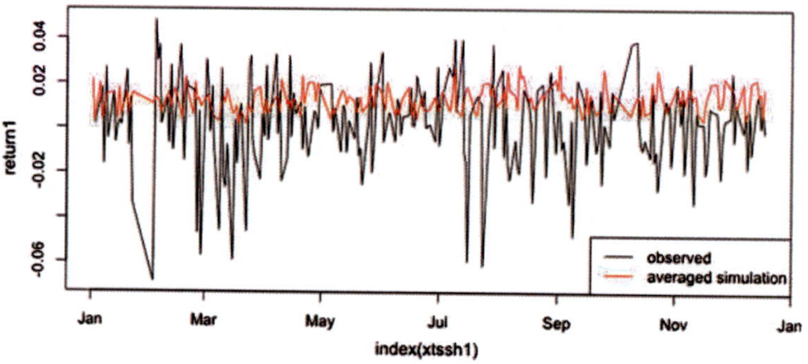

Fig. 3 2020 GEM mean simulated returns versus observed returns

Table 5 BAU versus actual returns

BAU versus actual returns		
Time series	Mean	Variance
BAU	0.0128065	$3.07 * 10^{-5}$
Actual return	0.001975034	$3.82 * 10^{-4}$
Significance of difference	t-score = 8.1555 p-value = $1.32 * 10^{-14}$	$F = 0.080285$ p-value < $2.2 * 10^{-16}$

in people's panic levels worldwide, which is consistent with our findings in China GEM [16–18]. Another paper also illustrates the trend of increasing risks in the world financial markets through volatility analysis [19].

4 Discussion and Future Work

The pandemic has impacted industries globally. As seen from the recorded data, it is found that the stock indices became more volatile during the year 2020. GARCH (1, 1) best-fit model successfully resolved the volatility that was extended further as simulated time series for the pandemic period. We compared the recorded data with that of simulated time series under business as usual volatility regime and found significant differences in the mean and variance of the two. Under the assumption of keeping other socioeconomic conditions constant for the year, the results showed that the volatility is higher for the recorded data for 2020 as compared to before the pandemic regime. Related studies show that the phenomenon of high volatility is usually associated with price level, the riskless rate of interest, the risk premium on equity, and the ratio of expected profits to expected revenues, and all four of these indicators are affected to some extent by the epidemic [20, 21]. This result could be further tested with more data for the other sectors as well as for different lockdown regimes during the pandemic event.

There are few limitations in the study. First, we can try some other variants of the GARCH model to represent the volatility of the financial market. In our study, we found that the two parameters of GARCH (1, 1) added up to close to 1. With other variants, we can increase the robustness given the condition alpha1 + beta1 = 1. We will also use other available data to confirm the robustness of our result. The data we used in this paper is Chinese GEMCI from June 21, 2018. We could try another time period, provided the data is available. Also, the China small- and medium-sized board index, and Shenzhen GEM 300 index, are good indicators for SME could be used to analyze the impacts.

Secondly, we assumed that January 1, 2020, is the start of the pandemic, and that the March 2020-May 2020 return-to-work season does not eliminate the impact of the epidemic and that the impact persists until December 1, 2020. This might not be accurate. To exclude possible error, we might choose more accurate boundaries and divide the period further, or take the consideration of the extent of COVID-19, such as numbers of cases and deaths.

Thirdly, the change in China's GEM stock index is mainly affected by the epidemic and other factors are negligible, may not be proper enough. To further confirm whether the large volatility changes we observe are linked to the novel coronavirus outbreak, we can segment to study the volatility at different periods after including significant socio-economic and political factors.

5 Conclusion

The findings suggest that GARCH (1, 1) has a better performance in portraying changes in SME volatility. This change may be associated with the outbreak of novel coronaviruses. Through segmenting the data and comparing the predicted GEM index values (BAU) in 2020 with its actual value, our model demonstrates that the COVID-19 does cause increased volatility in GEM.

References

1. Johns Hopkins Coronavirus Resource Center (2021) Retrieved 18 Jan 2021. https://coronavirus.jhu.edu/
2. Liu D, Sun W, Zhang X (2020) Is the Chinese economy well positioned to fight the COVID-19 pandemic? The financial cycle perspective. Emerg Mark Financ Trade 56(10):2259–2276
3. Gupta M, Abdelmaksoud A, Jafferany M, Lotti T, Sadoughifar R, Goldust M (2020) COVID-19 and economy. Dermatol Ther 33(4):e13329
4. Fernandes N (2020) Economic effects of coronavirus outbreak (COVID-19) on the world economy. Available at SSRN 3557504
5. Juergensen J, Guimón J, Narula R (2020) European SMEs amidst the COVID-19 crisis: assessing impact and policy responses. J Ind Bus Econ 47(3):499–510
6. Lu Y, Wu J, Peng J, Lu L (2020) The perceived impact of the Covid-19 epidemic: evidence from a sample of 4807 SMEs in Sichuan Province, China. Environ Hazards 19(4):323–340

7. En-ChiNext (2021) Retrieved 18 Jan 2021. http://www.szse.cn/English/products/equity/ChiNext/index.html
8. Raczyńska M (2019) Definition of Micro small and medium enterprise under the guidelines of the European Union. Rev Econ Bus Stud 12(2):165–190
9. Tushare Big Data Community (2021) Retrieved 18 Jan 2021. https://tushare.pro/
10. Engle R (1982) Autoregressive Conditional heteroscedasticity with estimates of the variance of United Kingdom inflation. Econometrica 50(4):987–1007. https://doi.org/10.2307/1912773
11. Bollerslev T (1986) Generalized autoregressive conditional heteroskedasticity. J Econ 31(3):307–327
12. Tsay RS (2005) Analysis of financial time-series, vol 543. Wiley
13. Ali M, Alam N, Rizvi SAR (2020). Coronavirus (COVID-19)—An epidemic or pandemic for financial markets. J Behav Exp Finance 27:100341
14. Adenomon MO, Maijamaa B, John DO (2020). On the effects of COVID-19 outbreak on the Nigerian Stock Exchange performance: evidence from GARCH models. PPR155344
15. Mirza N, Naqvi B, Rahat B, Rizvi SKA (2020). Price reaction, volatility timing and funds' performance during Covid-19. Finance Res Lett 36:101657
16. Albulescu C (2020) Coronavirus and financial volatility: 40 days of fasting and fear. arXiv:2003.04005
17. Albulescu CT (2021) COVID-19 and the United States financial markets' volatility. Finance Res Lett 38:101699
18. Giot P (2002). Implied volatility indices as leading indicators of stock index returns? Working Paper. CORE University of Leuvain
19. Zhang D, Hu M, Ji Q (2020). Financial markets under the global pandemic of COVID-19. Finance Res Lett 36:101528
20. Binder JJ, Merges MJ (2001) Stock market volatility and economic factors. Rev Quant Finance Acc 17(1):5–26
21. Prokopczuk M, Stancu A, Symeonidis L (2019). The economic drivers of commodity market volatility. J Int Money Finance 98:102063

Asymmetric Cryptosystem for Color Images Based on Unequal Modulus Decomposition in Chirp-Z Domain

Sachin, Phool Singh, Ravi Kumar, and A. K. Yadav

Abstract In this paper, we have presented an asymmetric encryption algorithm for color images in Chirp-Z domain. In the proposed cryptosystem, unequal modulus decomposition provides the private keys that are different from the encryption keys. Chaotic logistic map is utilized for pixel scrambling and provides additional layer of security to the cryptosystem. To evaluate the security of proposed cryptosystem, various statistical metrics such as correlation coefficient, mean squared error, and peak signal noise ratio are used. The encryption algorithm is also tested against statistical cryptographic attacks such as histogram, mesh plots, and correlation distribution analysis of adjacent pixels. The results show that the proposed cryptosystem resists the cryptographic attacks and exhibits a high level of security.

Keywords Asymmetric cryptosystem · Color image · Unequal modulus decomposition · Chirp-Z domain

1 Introduction

Nowadays, digital channels have become the most potent tool for data transmission. Since visual information has more impact, most of the data remain in the form of images or videos. Among the transmitted data, some data contain secret information such as one's banking details, government classified documents, military secret

Sachin (✉)
Department of Mathematics, Central University of Haryana, Mahendergarh, India
e-mail: sachin191219@cuh.ac.in

P. Singh
Department of Mathematics, SoET, Central University of Haryana, Mahendergarh, India

R. Kumar
School of Electrical and Computer Engineering, Ben-Gurion University of the Negev, P.O. Box 653, 8410501 Beer-Sheva, Israel

A. K. Yadav
Department of Mathematics, Amity School of Applied Sciences, Amity University Haryana, Gurugram, India

G. Gupta et al. (eds.), *Proceedings of Academia-Industry Consortium for Data Science*, Advances in Intelligent Systems and Computing 1411,
https://doi.org/10.1007/978-981-16-6887-6_27

missions, space projects, research projects of an institution, and many more. So, the security of data is very crucial. Data hiding, watermarking, and encryption [1–4] are some of the conventional methods to protect the image data. Among them, image encryption is a popular technique to secure images. Image encryption algorithm converts an input image to an unfamiliar noisy ciphertext. In a well-designed encryption algorithm, ciphertext can be converted back to the original image only with the help of correct decryption keys. Advanced encryption standard (AES), and data encryption standard (DES) are traditional algorithms of image encryption. However, these algorithms involve complex computations requiring more time and processing power, and therefore, and these methods are difficult to perform in real time [5].

To overcome these problems in 1995, Refregier and Javidi proposed an optical image encryption algorithm which is known as double random phase encoding (DRPE) algorithm [6]. In DRPE, they have used random phase mask in spatial and Fourier domain. The optical image encryption algorithm has features such as parallel processing, high speed, and low degree of complexity. So, a variety of optical, digital, and optoelectronic image encryption algorithms have been developed based on double-random phase encoding algorithm. To enhance the security of encryption algorithms, DRPE is also proposed with several integral transforms such as fractional Fourier, Fresnel, fractional Mellin, gyrator, and fractional Hartley to develop secure image encryption algorithms [7–11]. However, DRPE was found to be susceptible to basic cryptographic attacks such as chosen plaintext attack and known plaintext attacks [12, 13]. The image encryption algorithms based on DRPE are linear and symmetrical, which make them vulnerable to cryptographic attacks.

To combat cryptographic attacks, advanced algorithms such as asymmetric encryption algorithms have been proposed in which decryption keys are different from encryption keys [14, 15]. Many asymmetric encryption algorithms were developed on the principle of phase reservation and phase truncation in the Fourier domain, commonly known as PTFT [16–18]. Later, researcher demonstrated that PTFT-based image encryption schemes are vulnerable to specific iterative attack algorithms [19, 20]. In order to resist these cryptographic attacks, researchers proposed another one-way trapdoor function known as equal modulus decomposition (EMD). Thereafter, many image encryption algorithms were proposed based on EMD [3, 21]. In EMD, the plaintext image is decomposed into two complex masks of equal modulus in which one mask is used as a private key, and the other is the ciphertext. Since the modulus of the private key is same as that of the ciphertext, this property can easily be exploited by the attacker to establish a relationship between private key and ciphertext. An iterative algorithm is then deployed by the attacker to retrieve the private key [22]. EMD was further improved by taking different moduli, resulting in random modulus decomposition (RMD) [4, 23]. Later on, it was found that RMD also is vulnerable to iterative cryptographic attacks [24]. Random modulus decomposition is further improved by choosing unequal modulus, and thus known as unequal modulus decomposition (UMD) [25]. The security of UMD is also suspect as it is vulnerable to iterative cryptographic attacks [26]. Chaotic maps are used for enhancement of security of image encryption algorithms, as they are characterized by non-periodicity, sensitivity to initial values, uncertainty in prediction, and invertible. A chaotic map enlarges the key space and provides the additional encryption keys to the cryptosystem [26, 27].

In this paper, a secure image encryption algorithm for color images is proposed in the Chirp-Z domain using logistic map and cascaded unequal modulus decomposition. The logistic map is utilized for pixel scrambling and cascaded UMD provided the two private keys. The frequency parameters of Chirp-Z transform enlarge the key space of the cryptosystem. In Sect. 2, fundamentals of the proposed encryption scheme are discussed such as the principle of UMD, logistic map, and Chirp-Z transform. The encryption and decryption process of the cryptosystem are explained in Sect. 3. The simulation results are presented and discussed in Sect. 4. In the last section, the present work is summarized.

2 Fundamental Principle

2.1 *Logistic Map*

Logistic map is a one-dimensional nonlinear polynomial map of degree 2. In a cryptosystem, logistic map [28] is used to produce a random sequence. The sequence generated by logistic map is non-periodic and not convergent. Mathematically, logistic map is written as

$$x_{n+1} = \mu x_n(1 - x_n) \tag{1}$$

where μ is a parameter, x_0 is initial condition, $x_n(0, 1)$ and $\mu(2.5, 4)$. The sequence is sensitive to initial value of parameter μ and x_0 such that a small change in parameters alters the complete logistic sequence. The bifurcation diagram of logistic map is presented in Fig. 1.

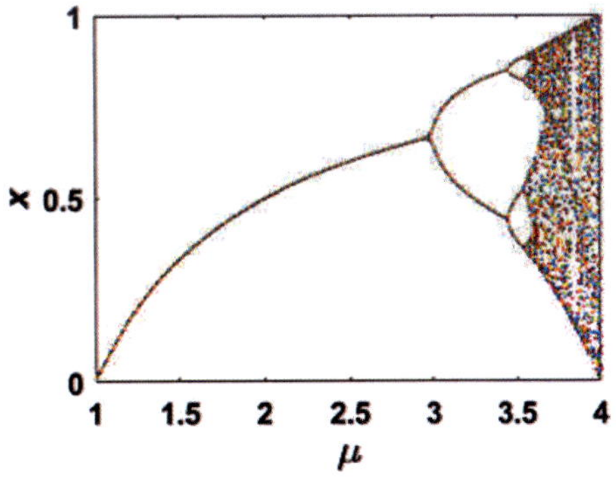

Fig. 1 Bifurcation diagram of logistic map

2.2 Unequal Modulus Decomposition

The method of decomposing a two-dimensional signal into unequal phase and amplitude mask is known as principle of unequal modulus decomposition [25]. $F(u, v)$ is assumed to represent two-dimensional signal of the input image $I(x, y)$ in transformed domain. The argument and phase of input signal is calculated by $\theta = \arg(F(u, v))$and $A = |F(u, v)|$. The transformed signal $F(u, v)$ is decomposed into two functions F_1 and F_2 by using simple geometrical representation demonstrated in Fig. 2, and mathematically, F_1 and F_2 are calculated from Eq. 2 and 3.

$$F_1 = \frac{A \sin(\beta - \theta)}{\sin(\beta - \alpha)} e^{i\alpha} \tag{2}$$

$$F_2 = \frac{A \sin(\theta - \alpha)}{\sin(\beta - \alpha)} e^{i\beta} \tag{3}$$

Here, $\theta(u, v)$, $\alpha(u, v)$, and $\beta(u, v)$ are arbitrary functions having the same size of input vector, and value varies in the interval $[0, 2\pi]$.

The compact form of UMD is expressed as:

$$[F_1(u, v), F_2(u, v)] \equiv \mathrm{UMD}_{\alpha,\beta}[F(u, v)] \tag{4}$$

where $\mathrm{UMD}_{\alpha,\beta}$ represents the unequal modulus decomposition operator along with $\alpha(u, v)$ and $\beta(u, v)$ functions.

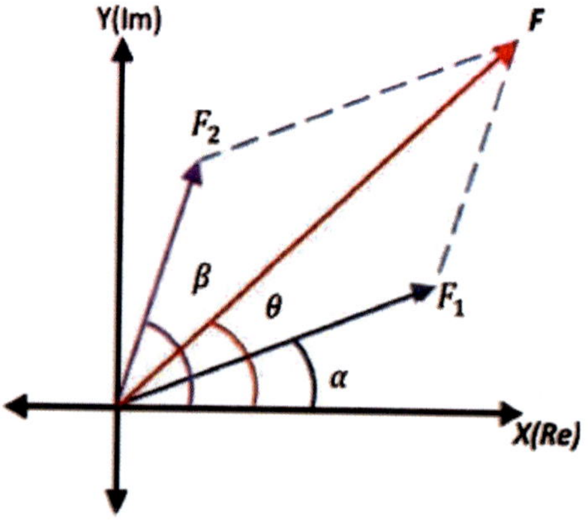

Fig. 2 Geometrical representation of unequal modulus decomposition [25]

2.3 *Chirp-Z Transform*

In 1969, Rabiner et al. [29] proposed the generalization of discrete Fourier transform known as Chirp-Z transform (CZT). However, Chirp-Z transform has difficulties in inverse calculation. In 2019, Vladimir Sukhov and Alexander Stoychev proposed an algorithm to compute the inverse of Chirp-Z transform [30]. The one-dimensional Chirp-Z transform is described by the following equation:

$$X_k = \sum_{l=0}^{N-1} x(l) Z_k^{-l} \tag{5}$$

where $Z_k = AW^{-k}$, $k = 0, 1, 2, 3, \ldots, M-1$, $x(l)$ denotes the input signal, W is complex matrix, and M and N represent the lengths of an input signal and an output signal, respectively.

The two-dimensional representation of the Chirp-Z transform is given by Eq. (6).

$$X = WAx \tag{6}$$

Here $A = \text{diag}\left(a^{-0}, a^{-1}, a^{-2}, a^{-3}, \ldots, a^{-(N-1)}\right)$ and $0.9a^{\frac{2\pi f_1}{f_s}}$ and A is a diagonal matrix of size $N \times N$.

Here, W is a matrix of size $M \times N$ and represented as

$$\begin{bmatrix} w^{0-0} & w^{0-1} & w^{0-2} & \cdots & w^{0-(N-1)} \\ w^{1-0} & w^{1-1} & w^{1-2} & \cdots & w^{1-(N-1)} \\ w^{2-0} & w^{2-1} & w^{2-2} & \cdots & w^{2-(N-1)} \\ \vdots & & & \ddots & \vdots \\ w^{(M-1)-0} & w^{(M-1)-1} & w^{(M-1)-2} & \cdots & w^{(M-1)-(N-1)} \end{bmatrix}$$

here $w = 0.995\,\mathrm{e}^{-2\pi i \frac{f_2 - f_s}{m f_s}}$, here f_1, f_2, and f_s are initial frequencies.

The inverse of Chirp-Z transform exists only for a square matrix and is calculated by Eq. (7).

$$x = A^{-1} W^{-1} X \tag{7}$$

3 Image Encryption Algorithm

In this paper, an encryption algorithm is proposed for color images in Chirp-Z domain. Logistic map is used for pixel permutation, and cascaded UMD is implemented to generate the private keys. The encryption process of the proposed algorithm is as follow:

Step 1: Split the color plaintext into three channels and store them as R, G, and B, respectively, for red, green, and blue channel. Here, we discuss encryption process of the red component only, and the other two components can also be encrypted by applying same encryption process.

Step 2: A random logistic sequence of length $3(m \times n)$ is generated by iterating the logistic map. Here, m and n are related to the size of the input image. The logistic sequence is stored as vector Z. Mathematically, Z is denoted as

$$Z(1, r; (m \times n) + r) \tag{8}$$

It should be noted that r may be assigned as an additional encryption key.

Step 3: R is bonded with a random phase mask (RPM1) and the result is store as R_2. Mathematically, R_2 is represented by Eq. (9).

$$R_2 = R \times \text{RPM1} \tag{9}$$

Step 4: Chirp-Z transform (czt) is performed on R_2 and result is stored as R_{32}.

$$R_3 = \text{czt}(R_2) \tag{10}$$

Step 5: By applying principle of unequal modulus decomposition, R_3 is decomposed into two parts say PR_1 and PR_2. The mathematical representation of UMD is briefed in Eq. (11).

$$[\text{PR}_1(u, v), \text{PR}_2(u, v)] \equiv \text{UMD}_{\alpha_1, \beta_1}[R_3(u, v)] \tag{11}$$

Here, PR_1 will serve as a private key and PR_2 will further participate in the encryption process. $\alpha_1(u, v)$ and $\beta_1(u, v)$ are randomly generated functions in interval $[0, 2\pi]$ and used in decomposition process.

Step 6: PR_2 is converted into row vector and sorted according to Z, and the result is stored as P_2.

Step 7: P_2 is further processed in inverse Chirp-Z (iczt) domain, and the result is stored in R_4.

$$R_4 = \text{iczt}(P_2) \tag{12}$$

Step 8: Again R_4 is decomposed into two parts PR_3 and CR under unequal modulus decomposition principle.

$$[\text{PR}_3(u, v), \text{CR}(u, v)] \equiv \text{UMD}_{\alpha_1, \beta_1}[R_3(u, v)] \tag{13}$$

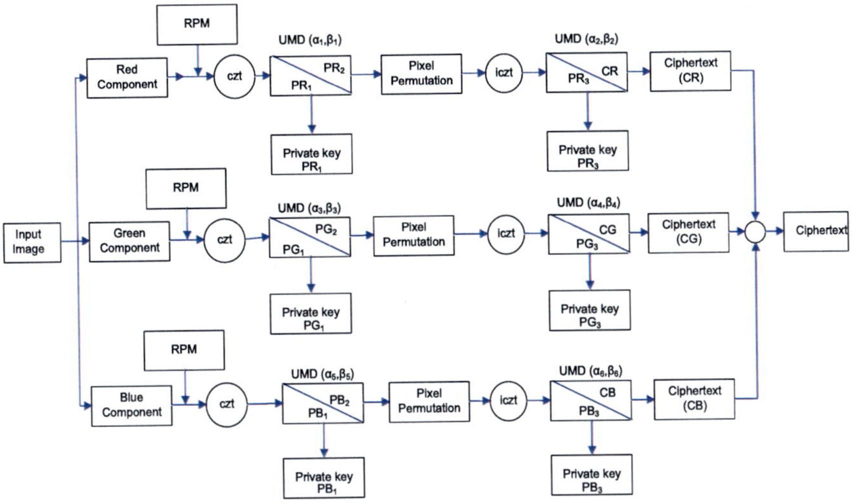

Fig. 3 Schematic flowchart for encryption process of the proposed cryptosystem

Here, PR_3 is the private key and CR is the ciphertext for the red component in encryption process. $\alpha_2(u, v)$ and $\beta_2(u, v)$ are randomly generated functions in interval $[0, 2\pi]$ and used in decomposition process.

Similarly, the encryption process is applied for green and blue components and encryption results are stored as CG and CB. The ciphertext of the encryption is obtained by mixing the ciphertext of different channels and the result is stored as C. The schematic flowchart of the encryption process is presented in Fig. 3.

The decryption process of the proposed algorithm is as follow:

The ciphertext (C) has red, green, and blue components. The decryption process for all the channels is similar. The decryption process for the red component is described in the following steps:

Step 1: The ciphertext of red component is combined with the private key (PR_3) and the output is subjected to Chirp-Z transform (czt) and resulting image is stored as RD_1. Mathematical representation of the Step 1 is shown in Eq. (14).

$$RD_1 = czt(CR + PR_3) \tag{14}$$

Step 2: The reverse pixel permutation is performed by applying a logistic map on the result obtained in Step1 and the output denoted as DR_{23} is combined with PR_1, and then inverse Chirp-Z transform (iczt) is performed on it. The output is decrypted image of red component. Step 2 is mathematically expressed in Eq. (15).

$$DR = czt(DR_{23} + PR_1) \tag{15}$$

Here, DR represents the decrypted image of the red component.

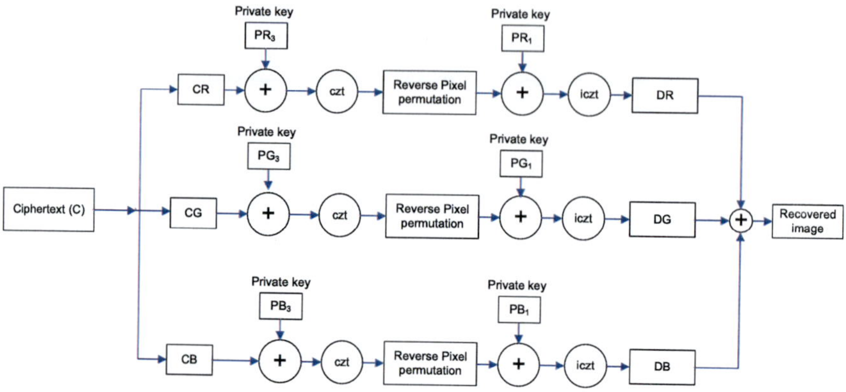

Fig. 4 Schematic flowchart for decryption process of the proposed cryptosystem

A similar process is applied to decrypt the green and blue components. The recovered image is obtained by combining the decrypted result of all the three components.

A schematic flowchart of the decryption process is presented in Fig. 4.

4 Numerical Simulations and Result Analysis

The efficacy of the proposed cryptosystem based on pixel permutation and UMD in Chirp-Z domain is analyzed in this section. Various simulations have been conducted for color images to test the feasibility and security of the proposed cryptosystem. Here, the simulation results for the 'Building' image are presented. In the simulation, we chose Chirp-Z transform frequency parameters such as $f_1 = 1024$, $f_2 = 480$, $f_s = 960$, and logistic map parameters are $x_0 = 0.3948$ and $\mu = 2.2859$. These parameters are also employed as encryption keys. The validation results of the proposed cryptosystem are presented in Fig. 5. The efficacy of the proposed

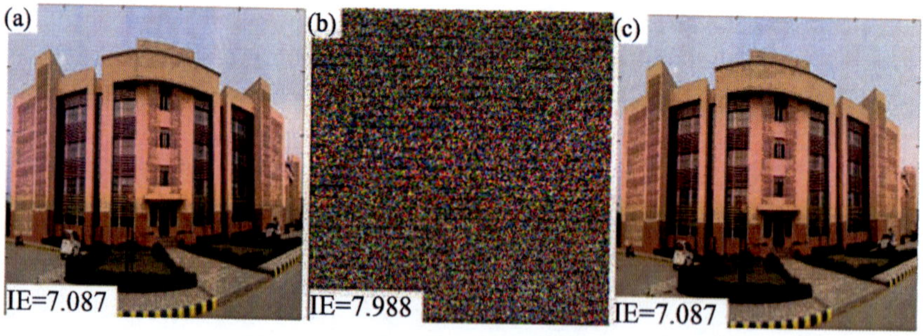

Fig. 5 **a** Plaintext; **b** ciphertext; **c** recovered image of 'Building'

cryptosystem is also analyzed for statistical attacks through histograms, mesh plots (3-D plots), and correlation distribution analysis [23]. The vulnerability of the cryptosystem is also tested against the noise attacks. The strength of the cryptosystem is also evaluated using statistical metrics such as correlation coefficient (CC), mean squared error (MSE), peak signal–noise ratio (PSNR), and information entropy (IE) [25]. The CC, MSE, and PSNR between plaintext (Fig. 5a) and the recovered image (Fig. 5c) are 1, 1.0892e^{-8}, and 79.6291, respectively.

4.1 Histogram and Mesh Plot Analysis

Histograms of an image possess critical statistical information about pixels of the image and mesh plots have information about position of the pixels. So, for a good image cryptosystem, the histogram and mesh plots of ciphertext should be different from plaintext and should be nearly flat. The histogram and mesh plots of plaintext and recovered image should be quite similar. The results for histogram analysis are depicted in Fig. 6a–c. From Fig. 6b, it is clear that histogram plot of ciphertext is completely from that of the plaintext and is nearly flat. The results for mesh plots are presented in Fig. 7a–c. From Fig. 7b, it is notable that the mesh plot of ciphertext is different from that of the plaintext. Thus, results of histogram and mesh plot analysis

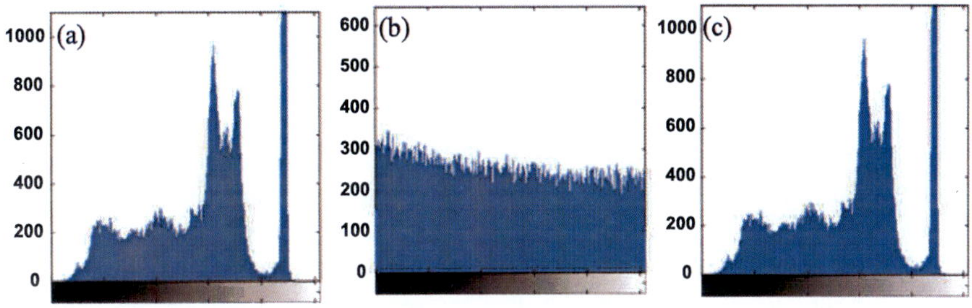

Fig. 6 Histogram plot for **a** plaintext; **b** ciphertext; **c** recovered image of 'Building'

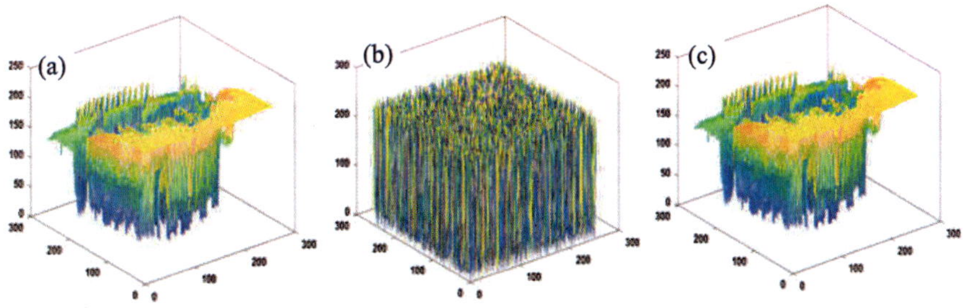

Fig. 7 Mesh plots for **a** plaintext; **b** ciphertext; **c** recovered image of 'Building'

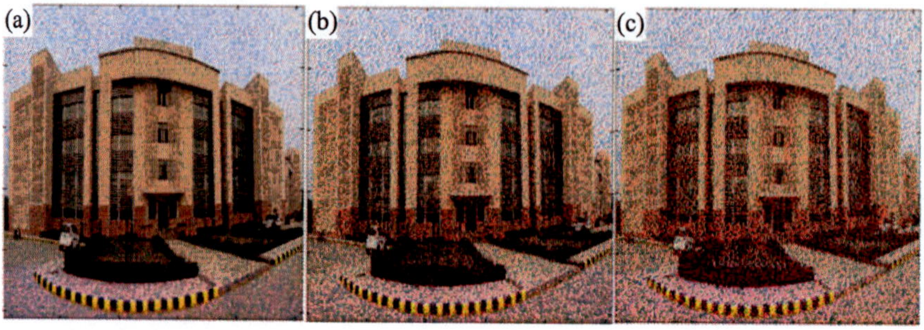

Fig. 8 Recovered image corresponding to noisy ciphertext with noise strength. **a** $n = 40$; **b** $n = 70$; **c** $n = 100$

establish that no statistical attack could be mounted successfully on ciphertext to reveal plaintext information.

4.2 Noise Attack Analysis

In real-time data transmission, sometimes unwanted frequencies get mixed, which is known as noise. The proposed cryptosystem resists various kinds of noise attacks. The simulation results for normal random noise with zero mean and unity variance are presented. The normal random noise is mixed using a formula provided in Eq. (16).

$$C^* = C + \alpha N \tag{16}$$

Here, C is ciphertext, N is normal random noise with zero mean, and variance unity, α is strength of noise, and C^* is noise affected ciphertext.

The results of noise attack simulation on red component are depicted in Fig. 7a–c, and results validate that the proposed cryptosystem resists the noise attack (Fig. 8).

4.3 Bruteforce Attack Analysis

The proposed cryptosystem has two parameters for logistic map, three frequency parameters of Chirp-Z transform, and two private keys of image size in the encryption process of each component. The parameters of Logistic maps are sensitive up to 10^{-15}, and Chirp-Z transform is sensitive up to 10^{-3}. So, the keyspace for the proposed cryptosystem is $10^{54} \times 4^{m\times n} \times 4^{m\times n} \times 4^{m\times n}$, which is large enough and cannot be breached in a foreseeable time. The results of brute force attack for proposed cryptosystem are presented in Fig. 9.

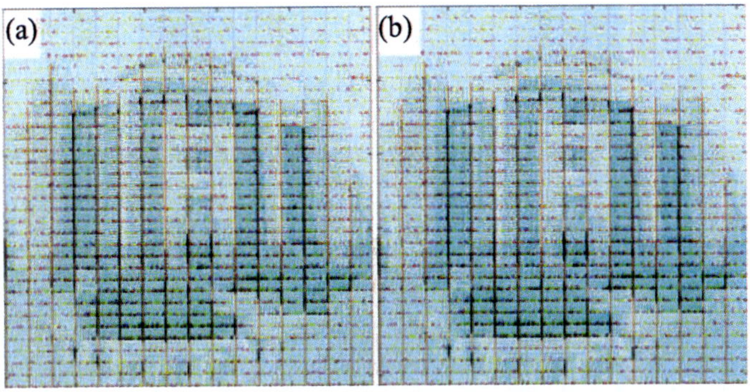

Fig. 9 Recovered image for wrong value of parameters. **a** $\mu = 2.2859 + 10^{-15}$; **b** $x_0 = 0.3948 + 10^{-15}$

4.4 Correlation Distribution Analysis

The correlation distribution of 7000 adjacent pixels of the plaintext and the ciphertext in diagonal direction for the red component is analyzed to see if any statistical information can be obtained for the proposed cryptosystem. The results of correlation distribution analysis for red component are presented in Fig. 10. The correlation distribution analysis results for green and blue component are similar.

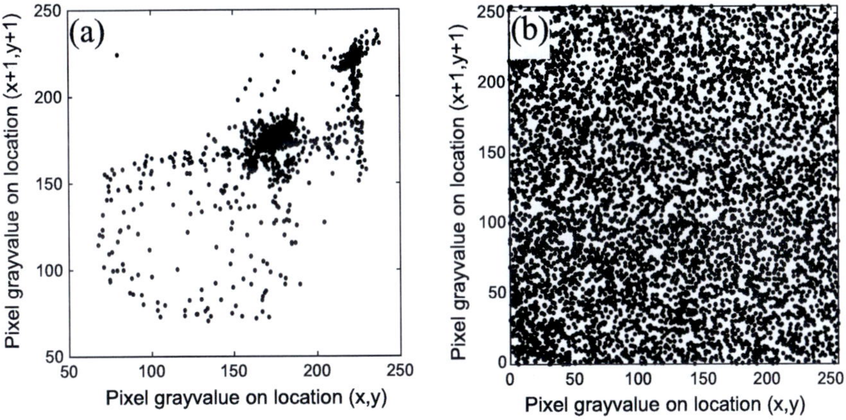

Fig. 10 Correlation distribution analysis of 7000 pixels of **a** red component of plaintext and **b** corresponding ciphertext in diagonal direction

5 Conclusion

In this paper, an asymmetric image encryption algorithm for color images is proposed. The proposed cryptosystem has used an unequal modulus decomposition and logistic map in the Chirp-Z domain. To establish the security of the cryptosystem, various statistical measures such as histogram, 3-D plots, and correlation distribution of adjacent pixels are analyzed. The statistical metrics such as correlation coefficient, mean square error, and peak signal noise ratio are analyzed. Results indicate that the proposed scheme resists statistical attacks. The proposed cryptosystem is also seen resisting noise attack and endures brute force attack for considerable duration. The validation results indicated that the proposed cryptosystem has a high level of security.

References

1. Biryukov A (2005) The boomerang attack on 5 and 6-round reduced AES. Dobbertin H, Rijmen V, Sowa A (eds) In: Advanced encryption standard—AES, vol 3373. Springer, Berlin, Heidelberg, pp 11–15
2. Kardas G, Tunali ET (2006) Design and implementation of a smart card based healthcare information system. Comput Methods Programs Biomed 81(1):66–78. https://doi.org/10.1016/j.cmpb.2005.10.006
3. Rakheja P, Vig R, Singh P (2019) Optical asymmetric watermarking using 4D hyperchaotic system and modified equal modulus decomposition in hybrid multi resolution wavelet domain. Optik 176:425–437. https://doi.org/10.1016/j.ijleo.2018.09.088
4. Singh P, Yadav AK, Singh K, Saini I. Asymmetric watermarking scheme in fractional Hartley domain using modified equal modulus decomposition. J Optoelectron Adv Mater 21:484–491
5. Hasib AA, Haque AAMM (2008) A comparative study of the performance and security issues of AES and RSA cryptography. In: 2008 Third international conference on convergence and hybrid information technology, Busan, Korea, Nov 2008, pp 505–510. https://doi.org/10.1109/ICCIT.2008.179
6. Refregier P, Javidi B (1995) Optical image encryption based on input plane and Fourier plane random encoding. Opt Lett 20(7):767. https://doi.org/10.1364/OL.20.000767
7. Yadav AK, Singh P, Saini I, Singh K (2019) Asymmetric encryption algorithm for colour images based on fractional Hartley transform. J Mod Opt 66(6):629–642. https://doi.org/10.1080/09500340.2018.1559951
8. Rakheja P, Vig R, Singh P (2020) Double image encryption using 3D Lorenz chaotic system, 2D non-separable linear canonical transform and QR decomposition. Opt Quant Electron 52(2):103. https://doi.org/10.1007/s11082-020-2219-8
9. Kumar J, Singh P, Yadav AK, Kumar A (2019) Asymmetric image encryption using Gyrator transform with singular value decomposition. In: Ray K, Sharan SN, Rawat S, Jain SK, Srivastava S, Bandyopadhyay A (eds) Engineering vibration, communication and information processing, vol. 478. Springer, Singapore, pp 375–383
10. Singh P, Yadav AK, Singh K (2019) Known-plaintext attack on cryptosystem based on fractional Hartley transform using particle swarm optimization algorithm. In: Ray K, Sharan SN, Rawat S, Jain SK, Srivastava S, Bandyopadhyay A (eds) Engineering vibration, communication and information processing, vol 478. Springer, Singapore, pp 317–327
11. Rakheja P, Vig R, Singh P (2019) A hybrid multiresolution wavelet transform based encryption scheme, Noida, India, p 020008. https://doi.org/10.1063/1.5086630

12. Peng X, Wei H, Zhang P (2006) Chosen-plaintext attack on lensless double-random phase encoding in the Fresnel domain. Opt Lett 31(22):3261. https://doi.org/10.1364/OL.31.003261
13. Tashima H, Takeda M, Suzuki H, Obi T, Yamaguchi M, Ohyama N (2010) Known plaintext attack on double random phase encoding using fingerprint as key and a method for avoiding the attack. Opt Express 18(13):13772. https://doi.org/10.1364/OE.18.013772
14. Anjana S, Saini I, Singh P, Yadav AK (2018) Asymmetric cryptosystem using affine transform in Fourier domain. In: Bhattacharyya S, Chaki N, Konar D, Kr. Chakraborty U, Singh CT (eds) Advanced computational and communication paradigms, vol 706. Springer, Singapore, pp 29–37
15. Kumar R, Quan C (2019) Asymmetric multi-user optical cryptosystem based on polar decomposition and Shearlet transform. Opt Lasers Eng 120:118–126. https://doi.org/10.1016/j.optlaseng.2019.03.024
16. Sharma N, Saini I, Yadav A, Singh P (2017) Phase image encryption based on 3D-Lorenz chaotic system and double random phase encoding. 3D Res 8(4):39. https://doi.org/10.1007/s13319-017-0149-4
17. Xiong Y, Kumar R, Quan C (2019) Security analysis on an optical encryption and authentication scheme based on phase-truncation and phase-retrieval algorithm. IEEE Photonics J 11(5):1–14. https://doi.org/10.1109/JPHOT.2019.2936236
18. Deng X, Zhao D (2012) Multiple-image encryption using phase retrieve algorithm and intermodulation in Fourier domain. Opt Laser Technol 44(2):374–377. https://doi.org/10.1016/j.optlastec.2011.07.019
19. Wang Y, Quan C, Tay CJ (2015) Improved method of attack on an asymmetric cryptosystem based on phase-truncated Fourier transform. Appl Opt 54(22):6874. https://doi.org/10.1364/AO.54.006874
20. Wang X, Zhao D (2012) A special attack on the asymmetric cryptosystem based on phase-truncated Fourier transforms. Opt Commun 285(6):1078–1081. https://doi.org/10.1016/j.optcom.2011.12.017
21. Deng X (2015) Asymmetric optical cryptosystem based on coherent superposition and equal modulus decomposition: comment. Opt Lett 40(16):3913. https://doi.org/10.1364/OL.40.003913
22. Wang Y, Quan C, Tay CJ (2016) New method of attack and security enhancement on an asymmetric cryptosystem based on equal modulus decomposition. Appl Opt 55(4):679. https://doi.org/10.1364/AO.55.000679
23. Rakheja P, Vig R, Singh P (2019) Asymmetric hybrid encryption scheme based on modified equal modulus decomposition in hybrid multi-resolution wavelet domain. J Mod Opt 66(7):799–811. https://doi.org/10.1080/09500340.2019.1574037
24. Rakheja P, Vig R, Singh P (2019) An asymmetric watermarking scheme based on random decomposition in hybrid multi-resolution wavelet domain using 3D Lorenz chaotic system. Optik 198:163289. https://doi.org/10.1016/j.ijleo.2019.163289
25. Archana, Sachin, Singh P (2021) Cascaded unequal modulus decomposition in Fresnel domain based cryptosystem to enhance the image security. Opt Lasers Eng 137: 10639912. https://doi.org/10.1016/j.optlaseng.2020.106399
26. Archana, Sachin, Singh P (2021) Cryptosystem based on triple random phase encoding with chaotic Henon map. In: Ray K, Roy KC, Toshniwal SK, Sharma H, Bandyopadhyay A (eds) Proceedings of international conference on data science and applications, vol 148. Springer, Singapore, pp 73–84
27. Sachin, Archana, Singh P (2021) Optical image encryption algorithm based on chaotic Tinkerbell map with random phase masks in Fourier domain. In: Ray K, Roy KC, Toshniwal SK, Sharma H, Bandyopadhyay A (eds) Proceedings of international conference on data science and applications, vol 148. Springer, Singapore, pp 249–262
28. Zhu A-h, Li L (2010) Improving for chaotic image encryption algorithm based on logistic map. In: 2010 The 2nd conference on environmental science and information application technology, Wuhan, China, Jul 2010, pp 211–214. https://doi.org/10.1109/ESIAT.2010.5568374

29. Rabiner LR, Schafer RW, Rader CM (1969) The chirp z-transform algorithm and its application. Bell Syst Tech J 48(5):1249–1292. https://doi.org/10.1002/j.1538-7305.1969.tb04268.x
30. Sukhoy V, Stoytchev A (2019) Generalizing the inverse FFT off the unit circle. Sci Rep 9(1):14443. https://doi.org/10.1038/s41598-019-50234-9

A Blockchain-Based Approach to Track Traffic Messages in Vehicular Networks

El-Hacen Diallo, Omar Dib, and Khaldoun Al Agha

Abstract Sharing traffic information on the vehicular network is essential to implement an efficient transportation system. Earlier communication of road closure warnings, car accident notices, and driver route changes is crucial to reduce traffic congestion. Unfortunately, tracking the exchanged traffic data and ensuring their correctness are not easy tasks. Vehicles can act maliciously by sending erroneous information or being subject to cyber-attacks, causing incorrect messages in the system. To efficiently identify the incorrect traffic information and track the exchanged data, we use a blockchain due to its decentralization, data immutability, security, and transparency features. We consider both the well-secured proof of work (PoW) and the practical Byzantine fault tolerance (PBFT) consensus algorithms. We study the adaptation of those two algorithms to enable an efficient, secure, and transparent exchange of data in vehicular networks. We also discuss their empirical performance under real-world vehicular network settings.

Keywords Blockchain · Vehicular networks · Consensus algorithms

1 Introduction

With the maturity of the Internet of Things (IoT) and the radio communication technologies, collecting and exchanging vehicles' data have become easier than before. Modern vehicles are now equipped with on-board units (OBUs), whereby lots of

E.-H. Diallo (✉) · K. Al Agha
Laboratoire Interdisciplinaire des Sciences du Numérique, CNRS, Université Paris-Saclay, 91190 Gif-sur-Yvette, France
e-mail: diallo@lisn.fr

K. Al Agha
e-mail: alagha@lisn.fr

O. Dib (✉)
Department of Computer Science, Wenzhou-Kean University, Wenzhou, China
e-mail: odib@kean.edu

G. Gupta et al. (eds.), *Proceedings of Academia-Industry Consortium for Data Science*, Advances in Intelligent Systems and Computing 1411,
https://doi.org/10.1007/978-981-16-6887-6_28

data are generated such as the interior temperature of the vehicle, its speed, GPS position, and the driving style. Not only vehicles' data are collected, environmental information can also be gathered such as the behavior of neighbor vehicles, weather and traffic conditions, and road features. Such data are usually transmitted to other vehicles by using vehicle-to-vehicle (V2V) communications and to the infrastructure through road-side units (RSUs).

Based on such information, critical driving decisions are made such as acceleration, deceleration, lane change, and route choice. The RSUs also provide driving recommendations to all vehicles based on their exchanged data and other environmental information such as weather forecasts, social media analysis, and road closure alerts. Despite the availability of traffic data, building an efficient vehicular network remains a challenge, that is mainly due to the good quality of the circulated data that must be guaranteed, and the high transparency and immutability that must be considered when storing the data for further analysis. Indeed, vehicles, as well as a portion of RSUs, might be malicious and can act against the system. For example, in V2I communications, malicious vehicles can inject incorrect messages that contain erroneous information, which may cause undesirable traffic redirection.

To address such behaviors, it is crucial to implement protocols that ensure the trustworthiness, integrity, immutability, and transparency of the traffic data, as well as the privacy of the nodes involved in the traffic data exploitation life cycle. The emergence of blockchain technology and its exciting capacity to maintain a distributed and consistent database between untrusted entities without relying on a central authority have attracted many researchers' interest. Recently, this technology was relied on to design robust vehicle network systems [1].

Essentially, a blockchain is a decentralized and distributed database that contains all transactions that have been executed between the participating nodes. This technology has recently gained increasing attention mainly due to its relevant features that include the network's decentralization, actions' transparency, data's immutability once written in the ledger, as well as its suitability for trust-less environments without relying on a single central authority.

From a network's participation point of view, the blockchain solutions can be classified under two main categories, public and private. A public blockchain is an open network; any external entity can create its proper blockchain node to join the network, as well as interact with its components without requiring any permission from a centralized authority. An example of public blockchain solutions is the famous Bitcoin network, which relies on the usage of the proof of work (PoW) consensus protocol to reach the necessary agreement on the network's state. On the contrary, the private blockchain solutions, also called consortium blockchains, usually implement an access control mechanism to limit the participation in the network, and consequently the interaction with the blockchain's components. More precisely, a subgroup of nodes is designated to execute a consensus protocol, whereby transactions are validated and new blocks are agreed upon and added to the blockchain ledger. Examples of consortium blockchain solutions include those relying on the Practical Byzantine Fault Tolerance (PBFT) consensus algorithm [2].

Based upon the blockchain's promising features, this paper proposes a blockchain-based solution to efficiently track the exchanged messages between vehicles and road-side units (RSUs) in the vehicular network. We believe that the proposed framework will securely, immutably, and transparently track transportation events without relying upon a centralized authority. RSUs will play the role of blockchain nodes that collectively validate traffic events messages' correctness, while vehicles will be the main source to generate the traffic data.

Technically speaking, a simple adoption of existing blockchain protocols is not directly applicable to ensure the correctness and immutability of messages in the vehicular network, that is mainly due to the nature of the traffic transactions and the way they must be validated in the vehicular network, which are quite different than the currency systems such as the bitcoin one [3].

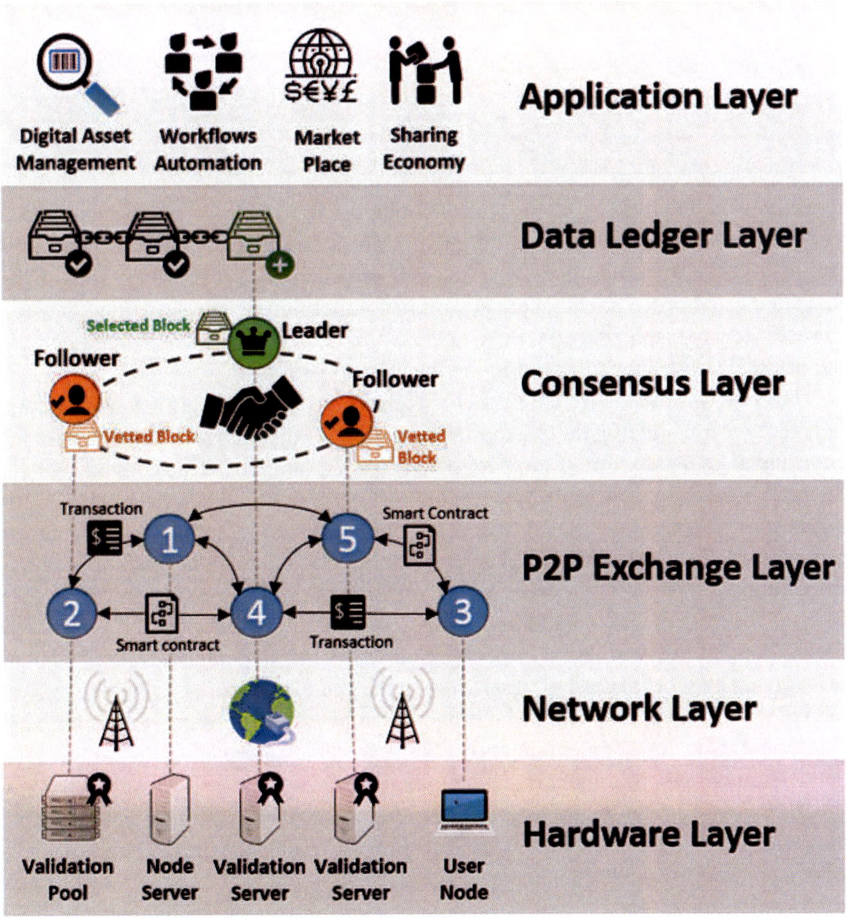

Fig. 1 Blockchain high-level architecture

Figure 1 shows the typical high- level architecture of blockchain and its main layers. However, it should be noted that blockchain architecture is not yet standardized, and other representations exist in the literature, such as [4].

We study in this paper the adaptation of both the PoW and PBFT consensus mechanisms. We mainly focus on the exchanged messages between the vehicles and the RSUs. We also empirically assess and compare the performance of both consensus mechanisms in the context of a real-world vehicular network context.

The remainder of this paper is organized as follows. Section 2 presents the related works. Section 3 gives a brief overview about the vehicular network architecture. Section 4 describes the adaptation of both PoW and PBFT consensus protocols. Section 5 introduces the experimental setup as well as the simulator that has been implemented to evaluate the proposed framework. In Sect. 6, the performance of the proposed protocols is assessed, and the empirical results are presented and discussed. Finally, Sect. 7 concludes this work and highlights future directions.

2 Related Works

Using a blockchain to deal with various challenges in the vehicular network has recently gained increasing attention from both researchers and practitioners. To start with, many works have focused on using PoW-based blockchains. The PoW that was introduced with Bitcoin [5], the first blockchain application, is essentially based upon dedicated processors that are always in competition to solve a difficult but easy to verify cryptographic puzzle. PoW is known to be very secure, resource intensive, but has scalability limits depending on the studied use case.

The PoW was used in [6] to build a blockchain-based distributed key management in heterogeneous intelligent transportation systems. Also, in [7], basic blockchain's concepts were used in order to share data among intelligent vehicles while relying on the PoW as a consensus. Moreover, in [8], a PoW featured blockchain was introduced for building a distributed trust management system. Similarly, in [9], authors proposed a PoW blockchain-based scheme for distributed trust management, where a fixed number of RSUs participate in the validation of traffic events using fuzzy logic. In [10], the PoW latency is reduced to speed up the block interval that might raise security issues and worsen the system performance, due to potential forks. Although there are many blockchain-based schemes in the vehicle network context that rely on the PoW [5], the state-of-the-art lacks works on how such consensus algorithm can be adapted to track fraudulent messages between vehicles and RSUs in the vehicular network, and how preferment it can be in real-world settings.

Alternatively, other blockchain-based solutions have been built using consortium blockchains. Practical Byzantine fault tolerance (PBFT) is a typical example of consensus protocols that are used in consortium-based blockchains. Using such protocol is motivated by its low latency in comparison with PoW, high throughput, and the acceptable security level.

For example, in [11], authors relied on a consortium blockchain for secure data sharing in vehicular edge computing; a new communication pattern for the consensus process was presented. The protocol seems attractive and simplified; nonetheless, more studies about its security model are needed. Differently, in [12], authors presented a consortium blockchain based on PBFT for secure data sharing and storage in vehicular networks [13].

Similarly, in another work [14], authors have proposed a consortium blockchain framework for traffic-related events management, built on the PBFT algorithm; a new protocol was introduced to control the block replication factor, relying on the concept of micro-transactions. In more recent work [15], a blockchain scheme aiming to secure data sharing in vehicular edge computing was introduced. A PBFT between RSUs was relyed on to reach an agreement between RSUs. A network of 10 RSUs was simulated, and the evaluated metrics were more related to packets delivery ratio over RSUs, packets receiving rate, and MAC/PHY layer overhead.

Most of those consortium-based blockchain schemes do not provide performance details about the consensus protocol nor how it can be adapted to deal with traffic messages in the vehicular network. From a theoretical point of view, the PoW and PBFT performance and security principles have been extensively discussed in the literature [16, 17], but mostly for currency-based use cases such as Bitcoin; however, traffic-related events differ from Bitcoin transactions. The latter rely only on the history (i.e., committed data) for transaction validation, while traffic-related records need to be approved by the witnesses of the concerned events. It is essential thus to study the suitability of widely adopted protocols in crypto-currency platforms, such as PoW and PBFT, to implement a secure traffic-related records sharing system in the vehicular network.

Other works in the literature introduced new consensus mechanisms mainly dedicated to cope with the vehicular network context. Typical examples are proof of event (PoE) [18], proof of driving (PoD) [7], and proof of authority (PoA) [19]. However, further discussions are needed regarding their security models and empirical performance. Such protocols are usually not able to resist against malicious nodes; they thus sacrifice the security of the system to achieve higher throughput and low transaction validation latency.

Moreover, stake-based protocols as proof of stake (PoS) [20] was also introduced in the context of VANETs applications [21]. Even if relying on the PoS will mitigate the high CPU demands on the implies by the PoW, the proof of stake raises issues that were not present in the PoW consensus mechanism, such as *Nothing at stake attack* and *Grinding attack* [22].

Our contribution is therefore to study the adaptation of PoW and PBFT consensus mechanisms to enable a decentralized, secure, transparent, and immutable traffic-related data sharing system between vehicles and RSUs in vehicular networks. We also assess the empirical performance of those two algorithms in the context of real-world vehicular network settings.

3 VANETs Architecture

The system architecture in a vehicular network can be separated into three layers (see Fig. 2). The first layer consists of the ad-hoc domain; it mainly includes the intelligent vehicles that are connected and equipped with OBUs and dedicated sensors, which ease the recording of relevant traffic data such as accident notices, traffic congestion, and road safety warnings. Such traffic-related data are sent to RSUs in order to be verified, processed, and efficiently exploited to enhance the overall quality of the vehicular network. Based on that exploitation, vehicles receive routing recommendations and safety announcements from RSUs. Vehicles can also request dedicated transportation services such as the real-time optimal itinerary to reach a destination from a specific place. Usually, the cooperation of vehicles and their engagement in transmitting pertinent traffic data will give them priority to access those dedicated services as a counter-party, thereby encouraging vehicles to work side-by-side with the RSUs in a trusted and transparent manner. However, as already mentioned, vehicles can be malicious or subject to cyber-attacks; as such, fraudulent messages can be spread in the system leading to poor traffic decisions.

The second layer in the vehicular network consists of edge nodes; it mainly includes RSUs that exchange relevant traffic information between each other using a pre-established peer-to-peer network. Interestingly, a secure edge node is usually associated with each RSU providing the required resources in terms of processing tools, computation power, and data storage space. Like vehicles, RSUs are also connected; they use embedded sensors to collect traffic and other environmental data.

To ensure the correctness of traffic-related events, the transparency of all actions performed by vehicles and RSUs, as well as the immutability of data sent across the vehicular network entities, we build in this paper a blockchain-based framework. A blockchain will be maintained by RSUs, which will collectively verify and store traffic data in a secure and decentralized manner. RSUs will primarily maintain a full copy of the blockchain ledger; the consistent state of the ledger and its content trustworthiness level are ensured by a consensus protocol and a validation mechanism that are both executed between relevant RSUs in a distributed manner.

When a RSU receives traffic information from a given vehicle, the data will first be stored within the RSU edge node's environment in a local *mempool* under a pending state. Received traffic messages are grouped by subjects, that is, messages providing the same information are merged into one traffic event; the digital signatures of the messages' issuers (i.e., vehicles) are associated with the newly created event; doing so helps in the detection, tracking, and punishment of wrong behaviors. The state of pending events will turn to valid after receiving a threshold of confirmations from different vehicles. By doing so, we ensure that each RSU makes a first local assessment of the trustworthiness of the received traffic messages. Such local verification will then be coupled with another verification step resulting from the consensus protocol, executed among multiple RSUs, which is used to agree on the next block of traffic messages to be added to the ledger.

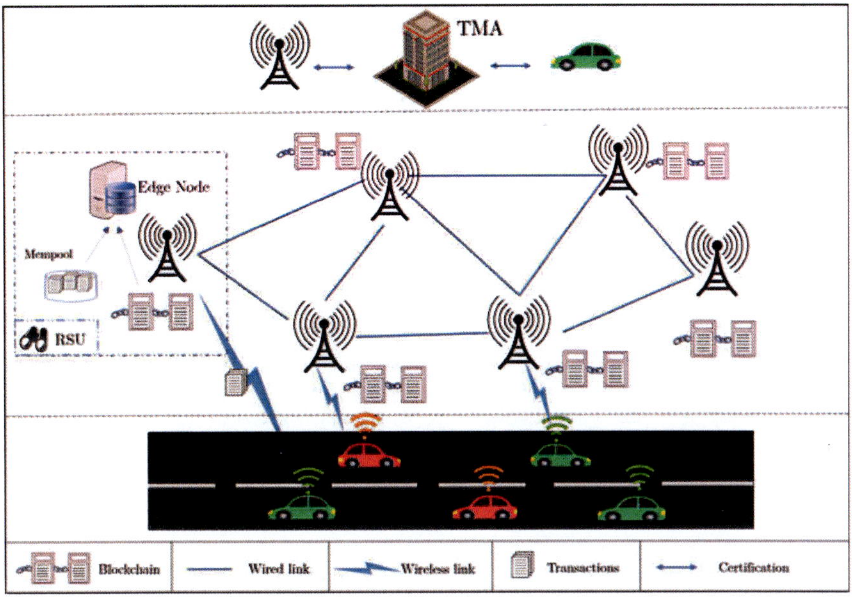

Fig. 2 Vehicular network system architecture

The last layer in the proposed vehicular network architecture comprises the traffic management authority (TMA) that latter is responsible for the generation of cryptographic materials such as digital certificates to the new entities that ask for joining the network such as vehicles and RSUs. In order to preserve the privacy of vehicles, the provided digital certificates must be anonymous; only the transportation authority is able to link a digital certificate to the real identity of the vehicle owner. It is worth mentioning that many TMAs can exist in the vehicular network [23], that is usually done to accelerate the issuance and revocation of digital certificates, as well as to increase the resilience of the network. The components that are involved in the vehicular network architecture as well as their interactions are shown in Fig. 2.

4 Evaluated Protocols

As discussed earlier, our goal is to enable a decentralized, secure, immutable, and transparent exchange of messages between the vehicles and RSUs. A blockchain is proposed for that purpose where traffic messages are turned into events (i.e., transactions) that will be verified and stored locally in RSUs. In this section, we will present how the RSUs will collectively agree on the traffic messages that must be added to the blockchain ledger. The PoW and the PBFT consensus mechanisms are considered.

4.1 PoW-like Protocols

The PoW consensus can be separated into two phases. The first one consists of decision making (who is going to append the next block?). To make a decision, the consensus participants (i.e., RSUs) also called miners allocate their dedicated CPUs to solve a difficult but easy to verify cryptographical puzzle. The difficulty level of the puzzle is parameterized. The whole process is completely random, but the more computation power a miner has, the higher its chance to try more possibilities and consequently solve the puzzle, and be the next block proposer.

Once a miner solves the puzzle, the second phase of the consensus starts. A set of transactions are collected from the *mempool* that contains all pending transactions (i.e., traffic events), put into a block, and sent to the rest of the network. When a miner receives a block, a verification process is performed to decide whether the block is valid before adding it to the blockchain ledger copy of the miner. Miners are therefore always in competition to solve the random challenge, so that they can add transactions to the blockchain and consequently receive some rewards in return.

Algorithm 1 Process of block validation

Input: A block received from another RSU;
Output: Check the validity of the block;
Workflow:

```
1: Check the signature of the block creator
2: Check the correctness of the PoW solving
3: Check the signature of the block creator
4: for every traffic event in the block do
5:     Check the threshold of events confirmation
6:     Check the signature of the vehicles
7:     Compare the content of the event with the local mempool information
8: end for
9: Return: Valid if all the above steps are correct
```

To ensure the validity of a block, the digital signature of its creator must be verified; its transactions must also be cryptographically ensured, and their content must be coherent with the ledger history. The content verification of a block is crucial and highly varies depending on the use case. For example, in the Bitcoin blockchain, the amount of the senders and receivers accounts must always be coherent across all the transactions which is relatively easy since it only consists of verifying numbers based on their initial state and the history of transactions. Unfortunately, easy verification process is not adapted to the vehicular network where traffic events must be verified according to the current traffic state.

To perform the verification process in the vehicular network context, we propose that after receiving a block, the RSU will verify the correctness of all its included traffic events (i.e., transactions). For example, for a road accident warning event to be considered as valid, the block proposer must prove the receipt of a certain number of messages from different vehicles confirming the occurrence of the event, that is done by including the signatures of all the vehicles that sent messages related to that event in the transaction. The number of signatures to be verified is a parameter. The validation process is illustrated in Algorithm 1.

Obviously, using a high number is essential to ensure a high level of events' trustworthiness. Interestingly, a RSU can act maliciously by not including all the signatures in the transaction. For instance, a RSU can only attach the signatures of vehicles confirming the occurrence of an event and ignore all the vehicles' messages confirming the opposite. To detect that, when receiving the block, the receiver will also compare the validity of events against its *mempool* of transactions. We assume in this work that all RSUs have synchronized *mempools* that contain all the traffic events circulated in the network. By doing so, we ensure double verifications, one based on receiving information from different vehicles, and another relied on including multiple RSUs in the process.

When implementing PoW, it is essential to balance between the difficulty (i.e., the average time for a new block to appear) and the wasted work. Usually, the easier the PoW challenge is, the more blocks are created that may lead to a situation where more than one miner finds the solution for the random puzzle at the same time. As a result, multiple new blocks will circulate in the network, which will cause different versions of the ledger (i.e., blockchain forks). Forks engender branches in the blockchain, hence causing inconsistency in the state of the ledger. To solve this, the ledger with the highest number of blocks will be considered as the valid ledger. When switching to the longest chain, the blocks in the forks, also known as "stale blocks," will be considered as wasted since an important amount of computational time was spent to create them, but they will ultimately not be added to the valid blockchain ledger. On the security aspect, forks in general and the high rate of stale blocks (rs), in particular, increase the advantage of the adversary in the network [17]. After forks resolving, transactions within stale blocks are put back into the RSUs' *mempool* for further inclusion in future blocks. Forks worsen the PoW performance since extra time is needed to ensure block's persistence, and additional computational power will be spend to add the same transactions. From a security point of view, the PoW security model relies on the principle that no miner controls more than 50% of the system's total computational power [16].

In this work, we evaluate PoW performance (i.e., throughput and latency) with different difficulties. We also measure the offset in terms of throughput and latency after forks resolving. Moreover, we assess the impact of the rate of stale blocks (rs) and the fork' size (i.e., the longest fork size) on the performance, always in the context of proposing a more efficient data sharing system in the vehicular network.

4.2 PBFT Based Blockchains

Due to the significant energy consumption required by miners to reach a consensus and validate transactions in PoW, alternative consensus mechanisms were proposed.

Authors in [2] first proposed PBFT as the first state machine replication algorithm to support Byzantine faults.

In this section, we explain how such algorithm can be adapted to the vehicular network context, and how different it is compared to PoW. Indeed, in the PBFT context, adding a new block to the blockchain ledger requires first the selection of a primary node that can be done either in a round-robin manner, or via a competition between the consensus participants like in the PoW.

Using a round-robin for the primary node election makes the system prone to DDoS attacks since the next leader is known in advance; therefore, the leader should be selected randomly using the PoW election mechanism. In addition to the primary node, other validator nodes should be selected to participate in the consensus and act as backups if needed. If the leader is compromised, the system backups impose a "change view," that will result in switching the current leader with one of the backups. The change view process is necessary to ensure the progress of the system.

After the selection of the leader and the backups, three communication phases (pre-prepare, prepare, and commit) are performed to reach a consensus on the next block. The elected primary node (i.e., an RSU) forges a block containing transactions that represent traffic-related events and forwards it to all the validator nodes (pre-prepare). Upon receiving a proposal, the validator node will verify it, and if it succeeds, a prepare message will be sent to all the other nodes (Prepare).

Upon receiving prepare messages from 2/3 of all validator nodes, the RSU will broadcast a commit message to all the validator nodes (commit). For a block to be committed and consequently added to the validator node blockhain ledger, it has to receive commit messages from 2/3 of the validator nodes. As can be noticed, two rounds of voting on a block are required to reach a consensus in PBFT. One is performed during the prepare phase, and another in the commit phase. After reaching consensus on the block, the block becomes publicly verifiable by other RSUs (non-consensus RSUs), thanks to cryptography signatures.

The workflow of the general PBFT consensus algorithm is illustrated in Fig. 3. As can be seen, three communication rounds are performed between the consensus participants to reach a consensus on the next block to be added to the ledger. As discussed earlier, after receiving a proposal, the validator node performs a verification process. As in the PoW, the RSU has to verify that each transaction in the block has the required number of vehicles signatures.

In addition, the digital signature of the block proposer is verified, as well as the coherence between the validator *mempool* and the content of the received block. Oppositely to PoW, the block is not added to the ledger copy, unless votes from other RSU validators are received as explained in the PBFT voting phases. As can be noticed, the verification process in PBFT is more thorough than PoW, but that is at the expense of additional communications between the validator nodes. Indeed, the PBFT suffers from communication overhead: $O(n^2)$ in the typical scenario and $O(n^3)$ for the change view, where n is the number of nodes participating in the consensus.

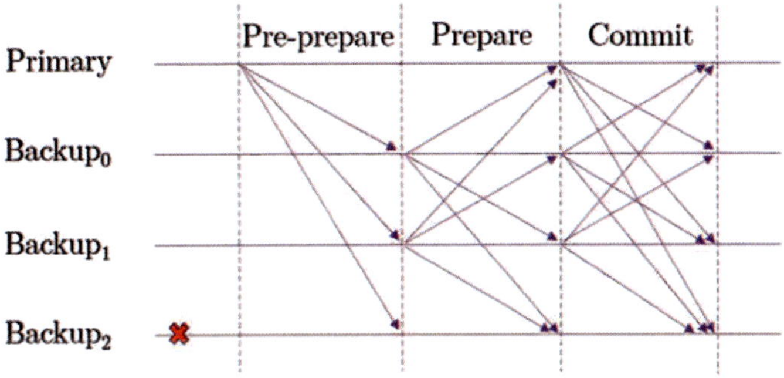

Fig. 3 Practical Byzantine fault tolerance (PBFT) three phases of communications

5 Simulator Description

To study the empirical performance of the PoW and PBFT to track traffic events, real-world settings have been simulated using NS3 vehicle network simulator [24]. Both consensus algorithms have been implemented in C++ and connected to NS3. Simulations are executed on a single machine having the following properties: Dell R640 server, Intel(R) Xeon(R) Silver 4112, CPU 2.60 GHz, 8 core CPU, 64 GB RAM, Ubuntu 18.04. The OpenSSL library has been used to implement the Schnorr signatures [25] for messages encryption and decryption; SHA-256 has been used for the hash function.

The simulated RSUs are connected by a point-to-point channel, forming a peer-to-peer connected network. To easily manage communications between RSUs without being bothered by the application routing layer, each RSU maintains a TCP connection with its peers. A vehicle is also represented as a node in the simulator and is connected to a group of RSUs. Traffic events transactions are generated according to a Poisson distribution. We assume that those events are generated by different vehicles, so that the limitation in terms of the number of vehicles the system can handle is assessed. The digital certificates of the vehicles and RSU are supposed to be well generated by a trusted CAs.

We assume also that vehicles messages are well received by RSUs, and the connectivity is not an issue in the system. Vehicles' primary function is traffic events generation, while RSUs focus on reaching an agreement on the next block to append to their maintained copy of the blockchain.

We assess in this work the throughput and latency of PoW and PBFT under the vehicular network context. Therefore, we have fixed some parameters to imitate real-world settings, such as the number of RSUs to 20; the average distance between two RSU to 50–500 m; the network connection speed to 100 Mbps; the block size (i.e., the maximum number of events in one block) to 1000 events; and event size to 800 bytes as fixed in [12, 21]. Besides, we vary the number of generated events; the difficulty to solve the mathematical puzzle in the PoW; and the number of consensus participants in the PBFT.

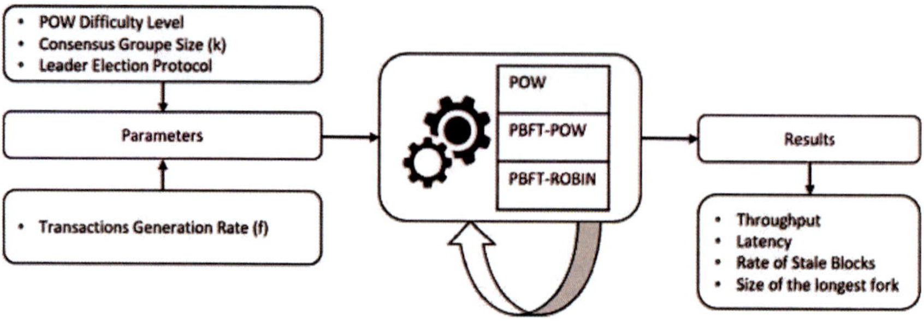

Fig. 4 Simulation workflow

The above settings include real-world VANETs settings. For instance, the network speed and the number of RSUs are justified by the possibility of relying on the WiMAX [26] technology instead of a wired network to connect RSUs. Doing so will increase the covered zone by RSUs, as so, 20 RSUs could cover a whole city since the WiMAX coverage rate can reach 15 km with a network speed that exceeds 100 Mbps [26]. Moreover, regarding the number and frequency of traffic-related events arrival, we have analyzed a dataset of the city San Francisco traffic-related warnings for the year 2019 [27]; we found out that the events arrival rate is in the order of 60 events per day. However, much more traffic records will be generated with the development of intelligent vehicles. Therefore, we vary events arrival rate from 200 event/s to 5000 event/s to represent as many possible real applications of the blockchain in VANETs context.

The workflow of the simulation is depicted in Fig. 4. As can be seen, PoW and PBFT related parameters are assessed by computing several performance indicators such as the throughput of the blockchain, its latency, and robustness.

6 Results and Discussions

This section presents the simulation of various instances of the implemented protocols. The assessed metrics are throughput and latency. And the plotted results represent an average of all end-to-end measurements from all RSUs.

6.1 *PoW-Like Blockchains*

In Fig. 5, we show the impact of the number of traffic events and the PoW difficulty on the throughput and latency of the PoW-based blockchain. By the throughput, we refer to the number of traffic events that the blockchain can verify per second; and the latency refers to the average amount of time that is required for an event to be

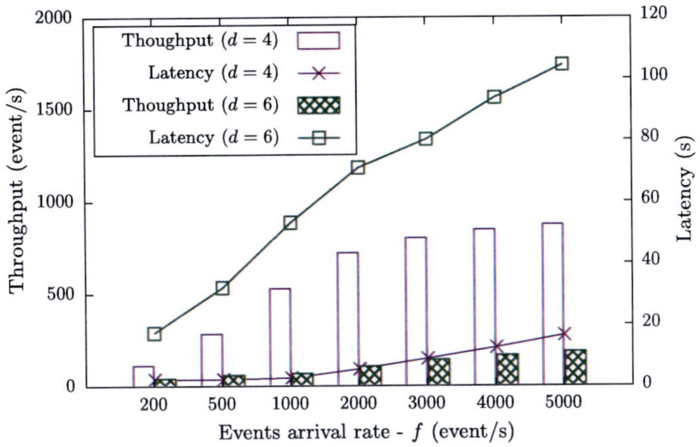

Fig. 5 PoW performance (throughput and latency) versus f (event/s)

validated and written as an immutable data in the blockchain ledger. As can be seen from Fig. 5, when the number of events generated per second (f) becomes more than 3 the block size (1000), the system becomes saturated: the throughput remains almost constant while the latency increases linearly with f. Fig. 5 also shows that with $d = 4$, the throughput can reach 870 event/s (i.e., traffic-related event per second), while for $d = 6$, the system does not process more than 186 event/s.

In terms of latency, if $f = 2000$ event/s for instance, it takes 71 s to confirm an event when $d = 6$. However, for $d = 4$, it requires less than 6 s. This huge difference comes from PoW solving when $d = 6$ compared with $d = 4$. Moreover, we can notice that in the worst case ($f = 5000$ event/s), with $d = 6$, the latency may exceed 104 s, which is not suitable for vehicular networks applications that require answers based on effective real-time valid traffic events. Therefore, it is essential to find the best trade-off between the security (PoW difficulty) and the required performance. It is also worth to mention that by including more RSUs in the system, the throughput will increase since more nodes will compete to generate new blocks, and consequently valid the correctness of traffic events.

Figure 6 depicts the rate of stale blocks (rs) against the traffic events generation rate (f). Results show that a lower difficulty leads to more stale blocks. For instance, for $d = 4$, the rs exceeds 38%. This means that RSUs waste more than 38% of their resources (i.e., computing and bandwidth) to process blocks that will not be part of the blockchain. The reason for this waste is that, with little difficulty, many RSUs will be likely to solve the puzzle simultaneously. This waste is because, with low difficulty, many RSUs will be likely to solve the puzzle at the same time. However, with $d = 6$, the wasted resources are much lower; they do not exceed 6%.

In Fig. 7, we measure the size of the longest fork when $d = 4$ and $d = 6$. The obtained results show that a low value of d increases the forks' size because the RSUs are likely to solve the PoW simultaneously. As can be seen, in case $d = 4$, the longest fork is equal to 2 blocks, while for $d = 6$, the fork size does not exceed one block. It

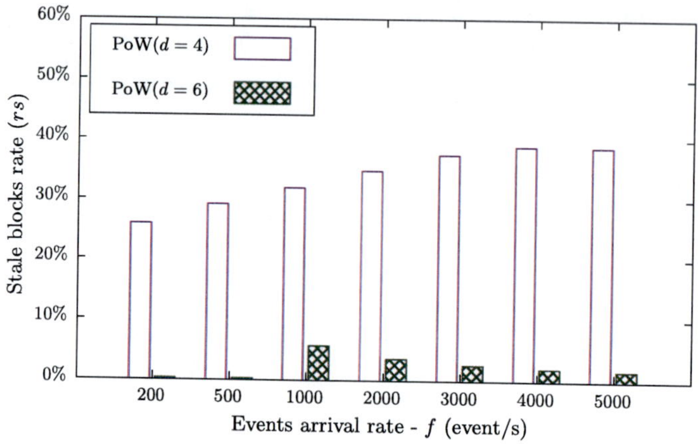

Fig. 6 PoW rate of stale blocks (%) versus f (event/s)

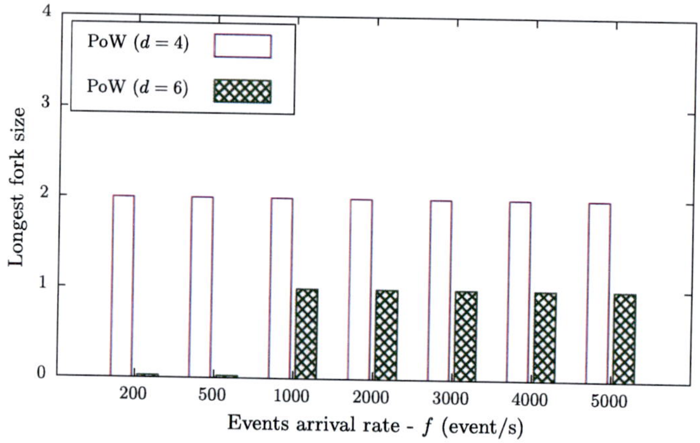

Fig. 7 PoW size of the longest fork versus f (event/s)

is important to mention that with $d = 6$, when $f = 200$ event/s or $f = 500$ event/s, there is no fork since there are no stale blocks (see Fig. 6).

Figure 8 depicts the offset of the throughput and latency after forks resolving. Results show that with low difficulty ($d = 4$), the offsets are more important. In the worst case ($f = 1000$ event/s), when $d = 4$, the throughput is reduced by 352 event/s and the latency is increased by 2 s. However, when $d = 6$, only 5 event/s and 0.4 s are the offsets regarding throughput and latency. We also can notice that when f becomes important, the offsets start decreasing; this is because the most extended fork size remains constant, while the number of blocks increases.

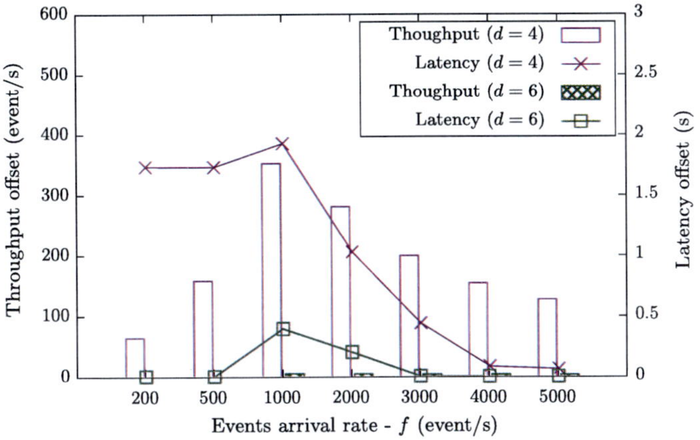

Fig. 8 PoW performance offset after fork resolving versus f (event/s)

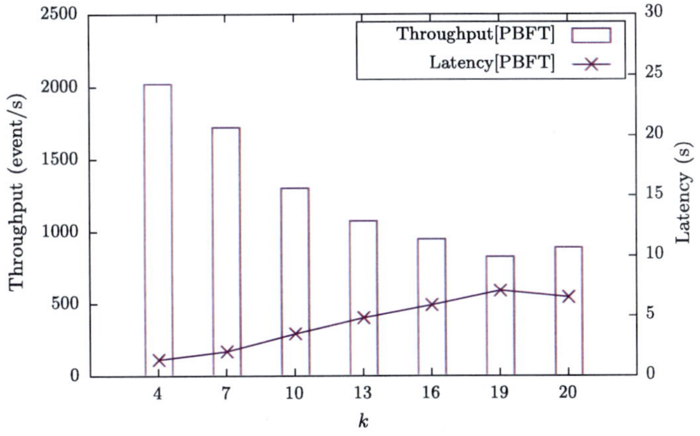

Fig. 9 PBFT performance (throughput and latency) versus k (consensus group size)

6.2 *PBFT Based Blockchains*

To study the performance of the PBFT based consortium blockchain, the throughput and the latency are assessed against the number of RSUs validating traffic events and participating in the consensus phase k. As shown in Fig. 9, we vary k from 4 to 20 with a step of 3; and we fix f the number of traffic events sent per second to 2000 That is equivalent to a vehicular network containing 2000 cars, each is generating one traffic event per second.

Figure 9 depicts the performance (i.e., throughput and latency) of the traffic events blockchain-based PBFT. In each round of the consensus, the PBFT leader (the primary RSU) is elected relying on a round-robin algorithm. The results show that the performance of the blockchain decreases with the consensus group size k. For

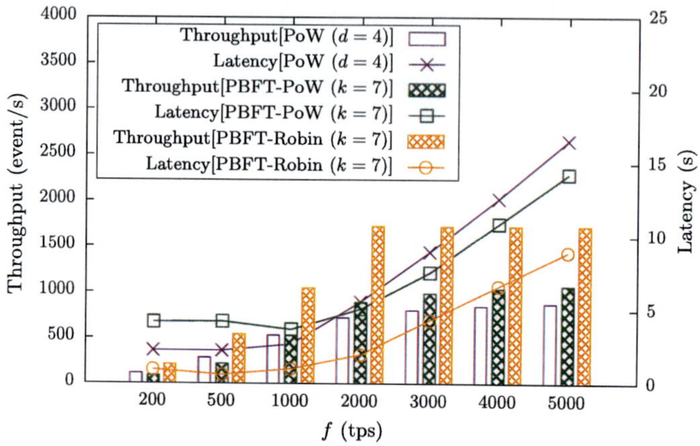

Fig. 10 Comparison between the performance of PBFT and PoW

instance, increasing the consensus participants' number from 4 to 19 decreases the throughput by 1196 event/s and increases the event confirmation latency by more than 5 s. Consequently, with such a high event confirmation delay, the relevance and efficiency of events sharing in the vehicular network will be affected negatively. This huge decrease in the blockchain's overall performance is due to the PBFT high communications cost, which is quadratic according to k ($O(k^2)$). Furthermore, results show that when $k = 4$, the system can confirm more than 2000 event/s in less than 1.3 s; this may be acceptable for some use cases. Nevertheless, such configuration does not support more than 1 ($k/3$) byzantine (malicious) RSU. It is therefore necessary to make a compromise between the level of security and the desired performance.

6.3 *PoW-Like Versus Consortium Blockchains*

In Fig. 10, we depict the throughput and latency of PoW and PBFT while varying the number of traffic events generated per second f. We compare between PoW of a difficulty $d = 4$, PBFT where $k = 7$ using PoW for leader election (PBFT-PoW), and another using round-robin (PBFT-Robin).

As can be seen, when f is greater than the block size 1000, the latency increases, while the throughput remains almost invariant for the three consensus versions. When f becomes greater than 2000, more transactions will be waiting in the *mempools* since the block size is saturated; thus, the latency increases while the throughout hits its maximum. It is worth to mention that increasing the block size will not linearly increase the throughput since the more transactions a block has, the more time will be required to reach a consensus on the validity of the corresponding traffic events.

Figure 10 shows that using a round-robin for the election of the PBFT leader increases the number of traffic events that can be validated and added to the blockchain per second. However, this compromises the blockchain's ability to withstand DDoS attacks against the leader. Therefore, it is essential to randomly select the leader, using a random mechanism that makes the next block proposer unknown, to minimize DDoS attacks on its resources.

Finally, as can be seen from Fig. 10, the PBFT has always better throughput and less latency in comparison with PoW. That is because in PoW, more time is required to solve the mathematical puzzle and reach a consensus on the next block. From a security point of view, the PoW performs better than PBFT. Indeed, the PoW cannot be compromised unless a malicious RSU controls more than 50% of the total computation power of all the RSUs participating in the consensus (i.e., 20 RSUs in our scenario) while the PBFT can resist up to $k/3$ of the total number of RSUs participating in the consensus work maliciously together to act against the system.

7 Conclusions and Perspectives

In this paper, we studied the adaptation of PoW and PBFT to validate and track messages in the vehicular network. We have shown how both protocols can solve the studied use case and how they perform when real-world parameters are considered. Experimental results have highlighted PoW limits regarding performance (i.e., throughput and latency). Also, reducing PoW difficulty causes many forks, which weakens the consistency of the blockchain. As such, the use case configuration decides the feasibility of the PoW usage. Furthermore, results have shown that using PoW for leader elections in PBFT induces a considerable delay in traffic-related events confirmation time compared to round-robin. To cover the V2V communications and enhance the events validation model, we have planned to integrate a credibility scoring module based on historical data exploitation from the blockchain. We also project to study the impact of vehicles' mobility on the validation of traffic events in the presence of malicious cars and RSUs.

References

1. Astarita V, Giofrè VP, Mirabelli G, Solina V (2020) A review of blockchain-based systems in transportation. Information 11(1):21
2. Castro M, Liskov B et al (1999) Practical byzantine fault tolerance. In: OSDI, vol 99, pp 173–186
3. Tomar R (2020) Maintaining trust in VANETs using blockchain. ACM SIGAda Ada Lett 40(1):91–96
4. Dib O, Brousmiche K-L, Durand A, Thea E, Hamida EB (2018) Consortium blockchains: overview, applications and challenges. Int J Adv Telecommun 11(1&2):51–64
5. Nakamoto S, Bitcoin A (2008) A peer-to-peer electronic cash system. Bitcoin 4. https://bitcoin.org/bitcoin.pdf

6. Lei A, Cruickshank H, Cao Y, Asuquo P, Anyigor Ogah CP, Sun Z (2017) Blockchain-based dynamic key management for heterogeneous intelligent transportation systems. IEEE Internet Things J 4(6):1832–1843
7. Singh M, Kim S (2017) Blockchain based intelligent vehicle data sharing framework. arXiv:1708.09721
8. Lu Z, Wang Q, Qu G, Liu Z (2018) Bars: a blockchain-based anonymous reputation system for trust management in VANETs. In: 2018 17th IEEE international conference on trust, security and privacy in computing and communications/12th IEEE international conference on big data and engineering (TrustCom/BigDataSE). IEEE, pp 98–103
9. Kchaou A, Abassi R, Guemara S (2018) Toward a distributed trust management scheme for VANET. In: Proceedings of the 13th international conference on availability, reliability and security, pp 1–6
10. Wagner M, McMillin B (2018) Cyber-physical transactions: a method for securing VANETs with blockchains. In: 2018 IEEE 23rd Pacific Rim International Symposium on Dependable Computing (PRDC). IEEE, pp 64–73
11. Kang J, Rong Y, Huang X, Maoqiang Wu, Maharjan S, Xie S, Zhang Y (2018) Blockchain for secure and efficient data sharing in vehicular edge computing and networks. IEEE Internet Things J 6(3):4660–4670
12. Zhang X, Chen X (2019) Data security sharing and storage based on a consortium blockchain in a vehicular ad-hoc network. IEEE Access 7:58241–58254
13. Guerraoui R, Knežević N, Quéma V, Vukolić M (2010) The next 700 BFT protocols. In: Proceedings of the 5th European conference on computer systems, pp 363–376
14. Diallo E-H, Al Agha K, Dib O, Laube A, Mohamed-Babou H (2020) Toward scalable blockchain for data management in VANETs. In: Workshops of the International Conference on Advanced Information Networking and Applications. Springer, pp 233–244
15. Firdaus M, Rhee K-H (2021) On blockchain-enhanced secure data storage and sharing in vehicular edge computing networks. App Sci 11(1):414
16. Vukolić M (2015) The quest for scalable blockchain fabric: proof-of-work versus BFT replication. In: International workshop on open problems in network security. Springer, pp 112–125
17. Gervais A, Karame GO, Wüst K, Glykantzis V, Ritzdorf H, Capkun S (2016) On the security and performance of proof of work blockchains. In: Proceedings of the 2016 ACM SIGSAC conference on computer and communications security, pp 3–16
18. Yang Y-T, Chou L-D, Tseng C-W, Tseng F-H, Liu C-C (2019) Blockchain-based traffic event validation and trust verification for VANETs. IEEE Access 7:30868–30877
19. De Angelis S, Aniello L, Baldoni R, Lombardi F, Margheri A, Sassone V (2018) Applying the cap theorem to permissioned blockchain, PBFT versus proof-of-authority
20. Vasin P (2014) Blackcoin's proof-of-stake protocol v2. 71. https://blackcoin.co/blackcoin-pos-protocol-v2-whitepaper.pdf
21. Yang Z, Yang K, Lei L, Zheng K, Leung VCM (2018) Blockchain-based decentralized trust management in vehicular networks. IEEE Internet Things J 6(2):1495–1505
22. Siim J (2017) Proof-of-stake. In: Research seminar in cryptography
23. Engoulou RG, Bellaïche M, Pierre S, Quintero A (2014) VANET security surveys. Comput Commun 44:10
24. Riley GF, Henderson TR (2010) The ns-3 network simulator. In: Modeling and tools for network simulation. Springer, pp 15–34
25. Schnorr C-P (1989) Efficient identification and signatures for smart cards. In: Conference on the theory and application of cryptology. Springer, pp 239–252
26. Vaughan-Nichols SJ (2004) Achieving wireless broadband with WiMax. IEEE Comput 37(6):10–13
27. Moosavi S, Samavatian MH, Nandi A, Parthasarathy S, Ramnath R (2019) Short and long-term pattern discovery over large-scale geo-spatiotemporal data. In: Proceedings of the 25th ACM SIGKDD international conference on knowledge discovery & data mining, pp 2905–2913

A Survey of Study Abroad Market for Chinese Student: Preliminary Data from a Pilot Study

John Chua, Haomiao Li, and Sujatha Krishnamoorthy

Abstract Prior to Covid-19, Chinese students spent US$30 billion on overseas tuition fees but the pandemic is unlikely to dampen future demand for international education. However, while China sends more students abroad than any other nation, this market's current landscape and its associated data are insufficiently studied. This paper outlines the scope of a study using primarily data from www.mygoodagency.com, a new free platform for Chinese students to share and get information about global education. Currently, in a soft launch pilot phase, the platform aims to help students in their application process, while we simultaneously gather and study data related to students' destination choices, concerns, and use of educational agencies. Two sources of data from the platform appear promising. Firstly, all registered users are asked to complete a questionnaire surveying their intentions, education background, as well as services required from, satisfaction with, and fees paid to agencies. Secondly, users can post questions and replies in our forums comprising 100 plus topics. With nearly 5000 questions and replies in the forums, we are using content analysis software to generate a word frequency count to evaluate what issues are of importance to users. Preliminary data suggest changing trends of interest to educators, agencies, and students. For example, there is almost two times more discussion about IELTS than TOEFL, and the test of English proficiency used in the US. There is evidence of increasing interest in pursuing further studies in the greater China area and Asia in general, including in Hong Kong and Singapore. Regardless of the destination, 67.43% of respondents said they would pay substantial sums, sometimes in excess of US$23,000 to agencies for services such as writing application essays,

J. Chua (✉)
Department of Communication, College of Liberal Arts, GEH B215, Wenzhou Kean University, 88 Daxue Rd, Ouhai, Wenzhou 325060, Zhejiang, China
e-mail: jchua@kean.edu

H. Li
Department of Computer Science and Technology, Wenzhou Kean University, 88 Daxue Rd, Ouhai, Wenzhou 325060, Zhejiang, China
e-mail: haomiaol@kean.edu

S. Krishnamoorthy
Department of Computer Science and Technology, Wenzhou Kean University, 88 Daxue Rd, Ouhai, Wenzhou 325060, Zhejiang, China

G. Gupta et al. (eds.), *Proceedings of Academia-Industry Consortium for Data Science*, Advances in Intelligent Systems and Computing 1411,
https://doi.org/10.1007/978-981-16-6887-6_29

securing internships, and selecting schools. The success of this pilot study will allow us to spread the platform's use across China, to support the continuing high demand for tertiary education, and help Chinese students make informed decisions, as well as enable us to conduct further research on their intentions and application processes, and possibly develop data-driven models to predict the results of their applications.

Keywords Study Abroad · Students across China · Discussion forum

1 Introduction

Prior to Covid-19, Chinese students spent US$30 billion on overseas tuition fees [1] In 2018, Chinese students contributed US$15 billion to the American economy [2]. However, while China sends more students abroad than any other nation [3], this current landscape and its associated data are insufficiently studied. The application process is a stressful time for students as many factors impact a student's success of getting admitted, including GPA, standardized test scores, individual university policies, and competition and popularity of a field in any given year. Coupled with the Covid-19 situation, the study abroad landscape and market for Chinese students is complex. Some countries like China and Iran have portals and websites where students share their application process and experience as well as their results. This paper proposes a pilot study to examine the shifting Chinese market for overseas studies in the post-Covid world. The scope of this current study is restricted to primarily using data from www.mygoodagency.com, a recently launched free platform for Chinese students to share and get information about global education. For now, the platform, with nearly 5000 forum messages and almost 1000 registrants, is mostly popular just with students at Wenzhou Kean University, a Sino-American cooperative institution. It is the intention of the platform's creators to spread its usage across China. This paper highlights some of the key results obtained from the site and its survey report. It is our hope that after a pilot stage and the platform's deployment across China, sufficient data will be collected in subsequent years to develop a predictive model of admissions success for Chinese students wishing to attend a top global university.

2 Related Work

Other researchers have conducted useful research related to our study. Ahmed et al. [4] mentioned graduate applications in a study. This research focuses on a system that uses data mining techniques to predict the application results of fully funded students [4]. They collected all the academic information of everyone with the support of a non-profit organization and used the academic data of successful students who have had the opportunity to study abroad to build a relational database containing all types of relevant information about students [4]. In the study, the authors found that most students who did not participate in the study abroad program believed that time and money were the main factors affecting their decision, and many people cannot get proper guidance [4]. These authors and their research provide us with suggestions and make us more committed to the integration of study abroad information and data in developing our project. Further, Ahmed, Hasan, Abdullah, and others [5] have continued their research to obtain the essential information about the students. Its purpose is to improve potential graduate applicants in Bangladesh and increase their application success rate [5]. They believed that it is not enough to process data through statistical analysis and observation, and further analysis must be done through data mining techniques [5]. This reference work provides good inspiration for our future data collection, data processing, and data application. Bhide et al. [6] also conducted similar research, but their research focused on exploring and defining patterns by analyzing and summarizing information from student profiles. The authors have mentioned a number of studies that are carried out in proposing a new recommendation system for higher education [6]. Their recommendation system examines various stages like enrolment, document auditing, tracking the process, and finally a score-based prediction [6]. In their study, the authors mentioned that summarizing all available information through data mining can effectively reduce time costs and improve system efficiency [6]. Since we are committed to researching and trying to predict the trend of Chinese students applying abroad, our current project is much smaller than a possible future research direction of predicting admission success; however, the possible difficulties in obtaining real massive data, including profiles and applications of successful students, is also a huge challenge for us.

3 Data Gathering

The respect and desire for higher education in China are well known. However, the nearly 3000 domestic universities cannot meet the demand from the 100 million Chinese citizens between 15–19 years old, many of them aspiring to tertiary studies (2). Middle-class Chinese families who can afford it look to overseas education for their children. (It might perhaps be ironic that it is often easier to get into one of the foreign QS 100 universities than into an elite Chinese institution.)

Our pilot study primarily uses data from students from undergraduate levels in a Sino-USA cooperative university with nearly 3000 students, which prides itself on its enviable record of preparing students for graduate studies in one of the QS global top 100 schools. Pre-Covid, in 2019 for example, 68% of its graduates pursued further studies upon completion of their bachelor's, with most heading to destinations in Western countries. This academic year because of the Covid-19 pandemic, many students are re-evaluating their plans and voicing their concerns on MyGoodAgency.com, allowing us to research the changing landscape of Chinese students intending to study abroad. This free platform (launched October 13, 2020) is the first in China to serve students in all majors and disciplines at all degree levels, to cover all the popular global destinations for studies, and to inform students in both Chinese and English.

4 Finding and Discussions

For many years now, many Chinese students rely on education agencies to process their applications for overseas studies. The agencies have become more than counselling and training centers but also serve as information gatekeepers. The education agency market within China, range from small one-shop firms to mega educational networks such as New Oriental Education, traded on the New York Stock Exchange (stock symbol EDU) with a market capitalization of US$30.77 billion (google.com/finance/quote/EDU:NYSE Feb. 24, 2021). These agencies market their services online, through their alumni networks, as well as presentations on campuses in China. On our platform, 67.43% of respondents said they would be willing to pay substantial sums, from US$1530 to in excess of US$23,000 to agencies for services such as writing application essays, securing internships, and selecting schools (MyGoodAgency.com). With such vast sums at play, our intention from the beginning was to encourage students to spend wisely by sharing information and experience, thereby avoiding poor service or possibly even the need for an agency.

However, of the 125 respondents who have already hired an agency, the majority replied they were satisfied with the service received, specifically 17.6 very satisfied and 56.8% satisfied, with the remaining replying they were unsatisfied (10.4%), very unsatisfied (8%), or felt it is still too early to tell (7.2%) (MyGoodAgency.com).

These 125 respondents hired agencies to write or edit their personal statements (86.4%), select schools for applications (76.8%), train them for admissions interviews (70.4%), complete application forms (67.2%), train them for admissions tests (57.6%), and remind and manage their deadlines (63.2%). Informally and outside this survey, however, students admitted to us that agencies can be used for more scandalous purposes, which can include allowing them to secure suspicious internships and fake recommendations, become "co-authors" of peer-reviewed publications, hire substitute test-takers for standardized entrance exams, and create portfolios to impress admission committees. It is unclear yet how the fallout from the Covid-19 pandemic impacting Chinese students going abroad will impact the long-term health of the educational agency market. What is clear from the content and textual analysis of our discussion forums is that there is still a demand for pursuing post-graduate studies abroad, but with shifting interests in desired destinations of study. With nearly 5000 topics or questions and replies in the forums, we are using content/textual analysis software to generate a word frequency count to evaluate what issues are of importance to users. For example, there is evidence of increasing interest in pursuing further studies in the Greater China area and Asia in general. There are more than twice the number of replies to questions about studying in Greater China than for studying in the US and the UK, specifically 448 replies in the Greater China section versus 205 replies for questions in the section for studying in the US and the UK (MyGoodAgency.com).

Examining all the posts reveals additional data. We created a web crawler program using Python to extract all available forum discussions, collected the posts as one file, and inserted the content into cloud-based platforms such as Weiciyun.com and Monkeylearn.com/Word-Cloud that rank the frequency of words used in a document. We did this for both the English and Chinese posts.

We found that "Hong Kong" is a frequently mentioned term, specifically 83 times for two months after the launch of the site on October 13, 2020. In contrast, "The UK" or "United Kingdom" is mentioned 60 times and 7 times, respectively. Meanwhile the "US" is mentioned 74 times.

All Hong Kong universities appear to be under discussion in the forums, revealing a curiosity about schools there. It is also possible that these students are simply more familiar with universities in the US and the UK, hence have more questions to ask about Hong Kong. To be sure, the UK is still the top potential destination for Chinese students. In our most current survey results, the UK currently edges the US in terms of popularity with 57.6% expressing a desire to study in the UK, versus 55.2% for the US (www.MyGoodAgency.com). That is a long climb up for the UK from pre-Covid times when 119,697 Chinese studied in the UK (gov.uk), while 369,548 studied in the US (statista.com) during 2018–19.

The Covid-19 pandemic and travel restrictions obviously played a role in the Chinese turning away from the US. One report indicated a 70% drop in student visas for Chinese citizens coming to study in the US for the 2020–21 academic year [2].

Another potential marker of shifting interest is the frequency of discussions about IELTS versus TOEFL. Traditionally, IELTS is the preferred test in the UK and Commonwealth countries, and TOEFL, while also accepted in many countries worldwide, is a test of American English. Test preparation is the third most popular category of discussion in our forums. Our analysis indicates IELTS is discussed or mentioned 175 times in the two-month period since our launch, while TOEFL only 120 times in the same period. When the discussions are in Chinese, IELTS (éZEæAI) was mentioned 169 times while TOEFL (æL'Ÿçe̡R) 85 times.

4.1 *Visualizing the Current Study Abroad Market with a Survey*

This study is a pilot study that has been conducted using student data primarily from one particular university. In this paper, we have listed seven sample questions that were presented to respondents via our platform and the analysis given according to prospective responses. Because this survey is ongoing and the results change whenever new platform registrants decide to answer the questionnaire, and the results here are captured on a particular day, specifically Feb. 24, 2021.

This site has been recommended to students across different universities through social media, but in the current pilot phase, most registrants are from Wenzhou Kean University. In future, we hope to gather more data from across Chinese universities for conclusions and generalization. Thus, it would be useful for the wider Chinese student community to post their experiences on this survey website or forum. Notably, this is preliminary work, and our results are deemed necessary for examination during this period of a global pandemic. In the following section, we present the survey questions and students' responses.

Question 1: How many prospective students plan to use an agency to help them apply for study in universities abroad?

Responses 1: A total of 919 respondents took part in this question. From their responses, 125 students are using an agency, 740 students would not use an agency while 54 respondents said they work for an agency.

Question 2: At what cost in RMB would prospective students pay agencies for help in applying to study abroad?

Responses 2: A number of 865 students responded with the range of fees (or charges) as quoted for them by different agencies. The charges vary from agency to agency according to the level of support that they render to applicants, and this is captured in percentage slice in a pie plot (Fig. 1). For further analysis, we categorized the range of charges into Low (less than 30,000 RMB), Moderate (between 30,000 and 50,000 RMB), High (between 30,001 and 80,000 RMB), and Extremely High (above 80,001 RMB) (Fig. 1).

Question 3: What level of education do prospective students prefer to pursue abroad?

Responses 3: From the survey, 865 students responded in which 138 (15.95%) students indicated they are pursuing the undergraduate program, 645 (74.57%) students for Master's degree, and 45 (5.2%) students for Ph.D. or research degree. 1.16% responded they want to pursue other types of degree or study (Fig. 2).

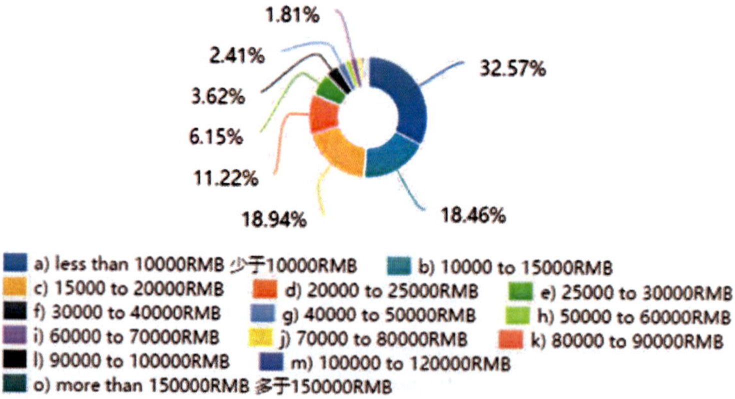

Fig. 1 The amount collected by the agency for their service

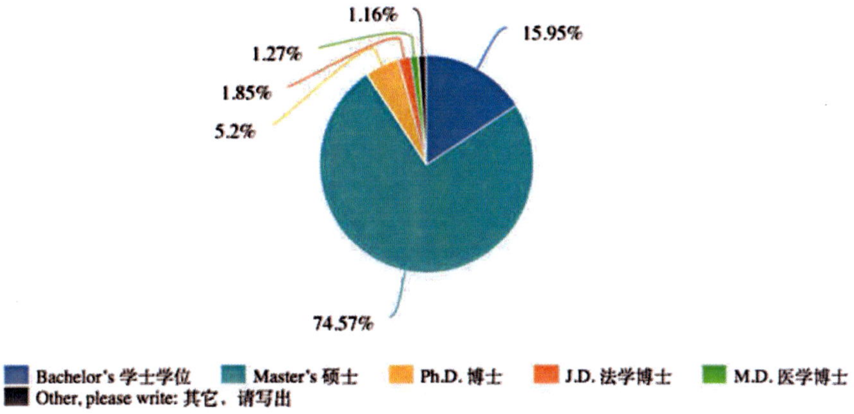

Fig. 2 The preference of the students for different level of education

Question 4: In which of the listed country would prospective students prefer to apply for their next level of education?

Responses 4: The survey listed 12 countries with an additional textbox for students to fill in a country that is not within the given list. From 125 responses, the majority of the students have the choice to study in the UK (57.6%), followed by the USA (55.2%), and thirdly in Australia (28.8%). While Taiwan (0.8%) is the country that is of least interest to pursue a study abroad, 1.6% percent, however, indicates that some students do not find the choice of country on the list (Fig. 3).

Question 5: What type of services do prospective students expect from the study abroad agencies?

Responses 5: 125 students stated different types of services they would expect from the agencies including personal statements, training, selection of schools, etc. Table 1 presents the percentage responses to the range of services expected (Table 1).

Question 6: How many universities are you applying to?

Responses 6: Of 865 students surveyed, the vast majority would apply to between one and 10 schools. Nine respondents indicated they would apply to more than 10. The breakdown is as follows (Table 2).

Question 7: What is your normal range of grades?

Responses 7: From the graph (Fig. 4), of 865 respondents, over 54.8% stated that their grade range would fall between 90 to 100%. The implication of this is that many of the prospective students seeking admissions for further studies abroad are likely to get a place (Fig. 4).

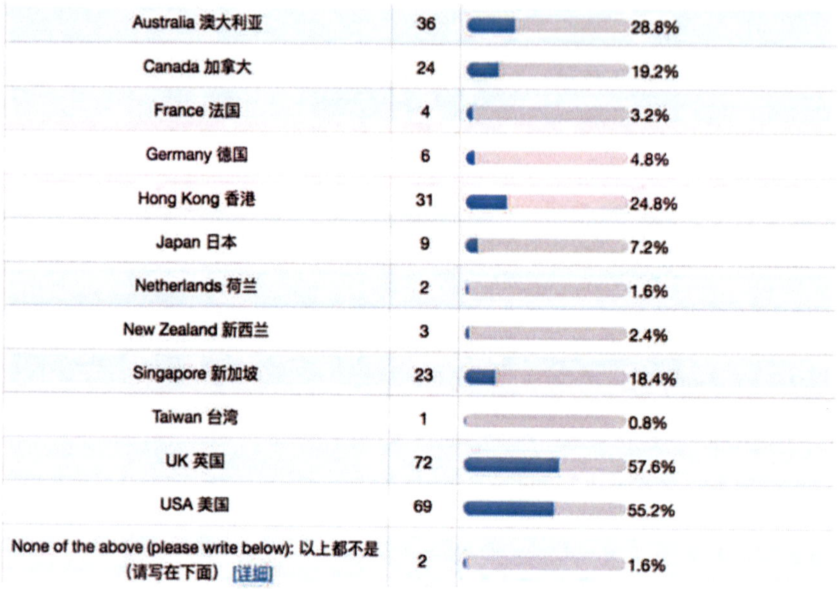

Fig. 3 Country preferred by students for their higher Education

Table 1 Type of Services seeking from an agency

No	Students	Type of services expected (%)
1	86.4	Writing or editing ray personal statement
2	70.4	Training for interviews
3	51.6	Training for admissions/Entrance tests
4	76.8	Selection of schools
5	63.2	Reminding and managing my deadlines
6	52	Arranging an internship
7	65.81	Completion of application form
8	58.97	Completing my visa application

5 Conclusion and Future Work

The completion of this soft launch pilot phase will allow us to be partner with interested parties and academic institutions to encourage Chinese students to use this platform to gather information to make educated and informed choices, simultaneously allowing us to study their decision-making patterns. In future, we hope to gather more data from across these Chinese universities for conclusions and generalization about the changing market for study abroad and eventually apply supervised learning algorithms to predict university placement for Chinese students within a number of parameters. In addition, when the platform's user numbers venture above a couple of

Table 2 Survey of No. of universities applied by the students

How many universities applied	Number of respondents	Number of respondents as percentage of total respondents
1	168	19.42
2	114	13.18
3	159	18.38
4	113	13.06
5	152	17.57
6	69	7.98
7	9	1.04
8	31	3.58
9	6	0.69
10	35	4.05

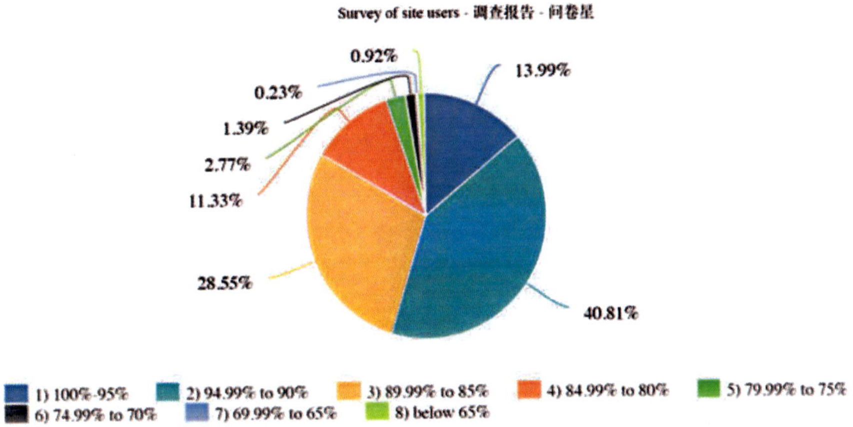

Fig. 4 Percentage of GPA (self-reported)

thousand registrants, we shall see if the current rising interest in Hong Kong, Singapore, and other Asian destinations would hold. However, the reality is that these other Asian countries do not yet have enough universities, especially in the QS100 top rankings to satisfy Chinese demand. Given that Chinese families invest more in education than any other nationality [7] our hypothesis is that Chinese citizens will continue to seek brand-name universities wherever they are. As the majority of these institutions are in the West, they would continue to benefit from the influx of Chinese students.

References

1. Qi Y, Wang S, Dai C (2020) Chinese students Abroad during the COVID crisis: challenges and opportunities. Glob Stud J 65:13–13
2. Kopf D (2020) https://www.yahoo.com/now/data-show-chinese-students-steering-181057877.html
3. Wang, S (2020) https://www.timeshighereducation.com/blog/covid-19-will-not-dent-chinese-demand-international-education
4. Hasan M, Ahmed S, Abdullah DM, Rahman MS (2016) Graduate school recommender system: assisting admission seekers to apply for graduate studies in appropriate graduate schools. In: 5th International conference on informatics, electronics and vision (ICIEV), pp 502–507
5. Ahmed S, Hoque ASML, Hasan M, Tasmin R, Abdullah DM, Tabassum A (2016) Discovering knowledge regarding academic profile of students pursuing graduate studies in world's top universities. In: International workshop on computational intelligence, pp 120–125
6. Bhide A, Karnik K, Badge P, Joshi V, Lomte VM. A review on recommender systems for university admissions. Int J Sci Res (IJSR) 5(12):141
7. Ischinger B (2020) Coronavirus challenges in the interconnected health and education sectors. Glob Stud J 13(29)

A Novel Image Encryption Algorithm Based on Circular Shift with Linear Canonical Transform

Poonam Yadav, Hukum Singh, and Kavita Khanna

Abstract This paper proposes a new image encryption based on circular shift on linear canonical transform, i.e., gyrator transform. We review various optical image encryption techniques proposed and inspired by the double random phase encoding (DRPE) system in the literature review. Also, we discussed optically inspired encryption techniques Fourier transform, fractional Fourier transform (FrFT), quadratic phase system (QPS), gyrator transform (GT), Fresnel transform (FrT), possible attacks and challenges in literature review and then explained various image scrambling techniques Jigsaw transform (JT) and Arnold transform (ART). Image encryption transforms are numerically implemented and compared with different DRPE schemes. Applied circular shift with Linear canonical transform (LCT) based transforms Fresnel transform, Gyrator transform, and compare the results of MSE, PSNR, and Histogram. The challenges faced in this area and the objectives of research have been discussed. This new circular shift approach provides adequate means security with acceptable complexity.

Keywords Optical image encryption · Linear canonical transform · Image scrambling · Circular shift · Gyrator transform

1 Introduction

To store and transmit the data securely, cryptographic methods or tools are used for ensuring all the goals of security. Cryptography is used to provide different security services in the electronic world. Cryptography is way of transforming a message into some other unintelligent form which is not understandable by an unauthorized person with the aim to make it immune from attacks. Modern cryptography involves different mechanisms such as symmetric key, asymmetric key, and hashing. In symmetric

P. Yadav (✉) · H. Singh
Applied Science Department, The NorthCap University, Gurugram, India
e-mail: poonam17csd011@ncuindia.edu

K. Khanna
CSE Department, The NorthCap University, Gurugram, India

G. Gupta et al. (eds.), *Proceedings of Academia-Industry Consortium for Data Science*, Advances in Intelligent Systems and Computing 1411,
https://doi.org/10.1007/978-981-16-6887-6_30

encryption, single secret key is used to both encryption and decryption. In symmetric encryption, communication entities must exchange the key because they are going to use the same key for decryption process, while in asymmetric key encryption scheme, the encryption key is published for all and encrypt messages via this public key. However, only the receiving entities has access to the decryption key that make message readable. This scheme is more secure as it does not face the problem of key distribution as in the case of symmetric schemes, but latter method is time-consuming and requires high computations. The security of information can be achieved by using multidimensional, high speed, and parallel processing optical systems [1]. Optical techniques have triggered interest of number of authors to work in this area. Many optical techniques such as encryption, authentication verification, watermarking, and data hiding have broadened the research interest of authors to maintain information security. Therefore, optical techniques have great potential as they possess several beneficial characteristics. To achieve the security, various encryption techniques are used which are basically rely on two methods.

Quadratic Phase-Based Encryption Techniques It includes DRPE-based techniques in FT, FrFT, FrT, or any linear canonical transformation domains.

Pixel Scrambling/permutation Techniques It includes technique or algorithms which scrambles the pixel like Arnold transform, jigsaw transform, logistic map, and affine transform.

1.1 Image Security Parameter

Generally, an excellent encryption system qualifies various security criteria, and some of them which we are using are described below:

- **Large key space**: A cryptosystem has considered to be secure if it has a large key space.
- **Key sensitivity**: A cryptosystem should be sensitive enough that with a small modification in key, and it should generate an absolutely contrary consequence.
- **Uniform histogram**: When the histogram of the encrypted image is uniform, then we say that cryptosystem is efficient
- **Correlation analysis**: An encrypted image should have low correlation between two adjoining pixels of the plane image and the encrypted image.
- **Sensitivity due to the presence of noise:** A cryptosystem should be able to produce recognizable results in the presence of noise as well.
- **Invulnerability against attacks**: A cryptosystem should be strong enough against various cyber-attacks.
- **Mean squared error**: The cumulative squared error is calculated between the obtained values and actual value. A cryptosystem should be able to minimize the MSE.

- **Peak signal-to-noise ratio**: A cryptosystem is said to be good if it provides a high PSNR.
- **Relative error**: A cryptosystem is said to be accurately define if it has RE close to zero. There are lot of other security parameters which can be considered while designing a security algorithm. All will depend on the level of security needed and type of algorithm (Fig. 1).

Introduction and overview of information security carried out in the field of optical image encryption. At first, we discussed various data encryption schemes for securing grayscale images. Next is the literature review of DRPE techniques and systems based on DRPE such as FrFT, FrT, and GT are described. Also, we listed wrong parameters and various attacks against DRPE-based techniques. We applied circular shift with Fresnel transform and gyrator transform and compared the results of MSE, PSNR, and Histogram. The challenges faced in this area and the objectives of research have been discussed.

Experimentation was carried out on the Lena, OTP, and barcode image of size 256 × 256 pixels. The MATLAB platform (R20019a) by Mathworks USA on an Intel i7 Processor is used to implement the proposed algorithm.

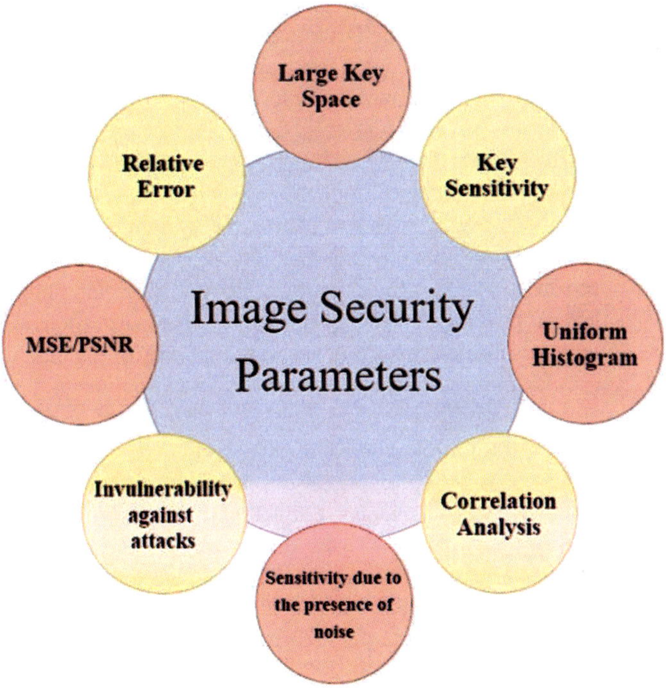

Fig. 1 Image security parameters

2 Observations Based on Literature Review on Image Encryption

In this area, first optical technology was developed by Javidi and Refregier [2] in 1995 known as double random phase encoding (DRPE) based on Fourier transform (FT). Such system or techniques is developed to improve the security, flexibility, and make it user-friendly. This DRPE technique encrypts the input image into the cipher image or a white noise which is not understandable by the third party.

The DRPE method in optical image encryption is due to inherent linearity, and it has been shown that this image encryption method is vulnerable to some attacks [3, 5] The attacks applied on traditional DRPE were mainly chosen-ciphertext attack (CCA) [3], chosen-plaintext attack (CPA) [4], known-plaintext attack [5, 6], and ciphertext-only attack [7, 8]. To render nonlinear in the DRPE technique, asymmetric image encryption scheme based on phase truncation Fourier transform (PTFT) was proposed [9]. Nonlinearity makes system robust against many existing attack; however, it is still vulnerable to known public key attack based on iterative amplitude-phase retrieval algorithm [10–12].

Some optical transforms are also used in the optical security system except the Fourier transform, such as fractional Fourier transform (FRT) [13–23], and the Fresnel transform (FST) in which the fractional order and the propagation distance are served as additional keys. Situ and Zhang proposed image encryption by using DRPE encoding in the Fresnel domain, and more degrees of freedom of the optical wave provides the higher security [24–27]. LCT has also been proposed for the optical encryption techniques using a quadratic phase system (QPS). The DRPE encrypted image is complex and unsuitable for many real applications. The gyrator transform (GT) has been used for optical encryption systems, which also belongs to the class of the linear canonical transforms [28, 30–38]. Since the color image is commonly used in daily life, the encryption method for color image is necessary. In [29], Chen proposed a pure random intensity mask as a filter with Hartley transforms to overcome this drawback. However, the security of this method is weak against the bare attack [39, 40]. With the help of chaotic mapping, Liu extended the Hartley transform to color image encryption [41]. Based on the wavelet decomposition [42] generates normal good-looking encrypted image instead of noise-like encrypted image. Recently, deoxyribonucleic acid (DNA) for parallelism and low-power consumption was applied to an image chaotic cryptosystem. The scrambling operations are used to enhance the security of the cryptosystems, such as Arnold transform (ART) [43–45], jigsaw transform (JT) [46], and baker mapping. Some parameters or some random elements, such as phase, amplitude, and position order, are used as additional key to enhance the security of encryption algorithm.

3 Optical Image Encryption by LCT-Based Encryption Technique (Fresnel and Gyrator Transform) with Pixel Scrambling Technique—Circular Shift

LCT can be implemented using quadratic-phase encryption system (QPES) by Unnikrishnan et al. All parameters of LCT are extra key in the algorithm which increases the degree of freedom except from two random phase mask and wavelength. Special cases of LCT are FT, QPS, FrFT, and FrT where fractional order of FrFT and propagation distance and wavelength for FrT act as additional keys. GT also belongs to LCT class where rotation angle acts as extra key.

Fresnel Transform (FrT): In 1999, Matoba and Javidi proposed encryption scheme based on FrT using random keys which added third dimension to the RPMs by shifting the RPM away from Fourier plane. This is mathematically computed through Fresnel–Kirchhoff formula as

$$Fz(u, v) = \mathrm{FrT}_{\lambda, z}\,\{I(x, y)\} = \int_{+\infty}^{-\infty}\int_{+\infty}^{-\infty} f(x, y)h_{\lambda, z}(u,\ v,\ x,\ y)\mathrm{d}x\mathrm{d}y$$

where the operator FrT λ, z denotes the (FrT) Fresnel transform with parameters λ and z, where h, λ, z are the kernel of the transform. The number of parameters for key domain is $R1$, $R2$, $Z1$, $Z2$, and λ which is large as compare to conventional DRPE.

Gyrator transform (GT) is one of the canonical transforms proposed by Rodrigo in the field of image processing, and it uses three lenses with fixed distance between them. The 2D function $f(x, y)$ for GT can be mathematically expressed as

$$\mathrm{GT}(u, v) = \mathrm{GT}_{\alpha}\{f(x\ y)(u, v)\} = \iint f(x, y)K_{\alpha}(x, y, u, v)\mathrm{d}x\mathrm{d}y$$

Let input image $I(x, y)$ be diffused with random phase key $R1(x, y)$ in input plane and propagates a rotation angle α in GT domain.

$$G(u, v) = \mathrm{GT}_{\alpha}[I(x, y) \times R1(x, y)]$$

Then, the output obtained $G(u, v)$ is diffused with second phase RPM $R2(x, y)$ and propagates a rotation angle б in the GT domain to get encrypted image.

$$\mathrm{E}(\mu,\mathrm{v}) = \mathrm{FrT}\text{-}б\ [\mathrm{G(u,v)\ x\ R2(u,\ v)}]$$

The number of parameters for key domain is $R1$, $R2$ (random phase mask) α, and б (rotation angel) which is large as compared to conventional DRPE that enhance the security.

3.1 Optical Image Encryption by Pixel Scrambling

The proposed hybrid encryption algorithm pixel scrambling is applied as first layer. The new algorithm performs image encryption with circular shift. A brute-force attack on this new algorithm is impossible. To increase the security of system and resistance toward attacks, repositioning of pixels is performed to confuse the relationship between the input image and encrypted image. Pseudorandom sequences are generated to shuffle the image pixel positions; there are multiple methods we can use for pixel scrambling, and here, we used circular shift. We implemented the circular shift with multiple LCT techniques, and statistical analysis using MSE, Histograms, PSNR showed the algorithm is not vulnerable to statistical attacks. Result analyses have shown that this new method has high security.

In pixel scrambling, we displace the position of pixels that is equal to image's division. Bit-shifting is done on the input image based on the key. Pixel scrambling is the interchange the position of gray or RGB pixel values. It means that the value of the pixel (x, y) is exchanged with the another pixel value (x_0, y_0). At the receiver end, the original image can be retrieved by reversing the scrambling according to the key. The pixel scrambling can be achieved by breaking up original image into small pixel blocks or 256 subsections, which are repositioned relative to one another according to some random permutation; here, we use circular shift (Fig. 2).

This new algorithm performs lossless encryption of an input image via adding circular shift as additional key. We applied circular shift that shifts the bytes via x-axis and y-axis, and we can change the value within size of image that is 256.

$$\text{shiftVariable} = [200, 100];$$

Each block in a set of blocks is circularly shifted by bits horizontally and vertically. Here, size of input image is 256, and we shift $x = 200$ and $y = 100$. The shifting is done bitwise circular motion, where the vacant bit positions are filled with the shifted-out bits in circular position. This makes input image more senseless, and now, we applied LCT transforms to encrypt the image and send to the receiver.

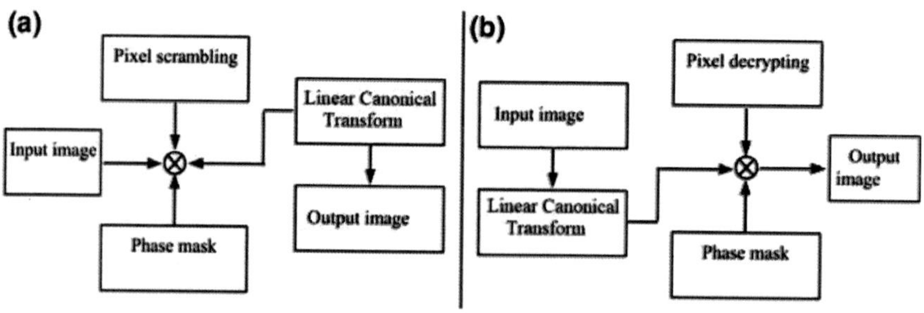

Fig. 2 Schematic diagram of the LCT encryption based on pixel scrambling: **a** encryption and **b** decryption

$$\text{reverseShiftVariable} = [-200, -100]$$

At receiver end, the decryption algorithm is performed by inverting the encryption algorithm without any loss of data. First, we performed inverse of LCT transform and then inverse of circular shift as shown in Fig. 3. The number of bits to circular shift is also performed at decryption end in the opposite direction like if left-shift was used in the encryption, right-shift must be used in the decryption (Figs. 4 and 5).

Here is another example of circular shift with OPT image.

The main aspect of circular shift security is performing key-dependent encryption at the bit level. Changing one or more bits in the key would cause a change in the block sizes and change the number of bits shifted. The large number of possible keys in circular shift makes a brute-force attack impossible.

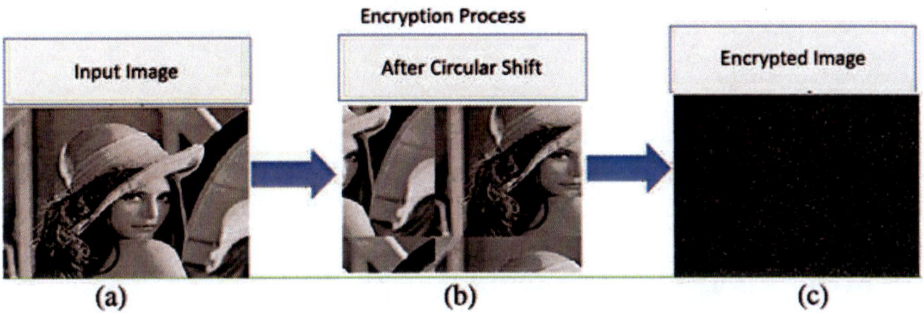

Fig. 3 **a** Input image lena, **b** image after circular shift, **c** encrypted image after applying gyrator transform

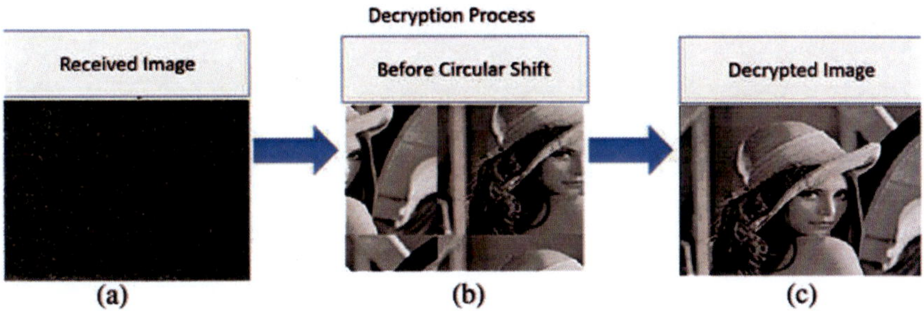

Fig. 4 **a** Received image encrypted form, **b** image after decryption process, **c** final image after applying reverse shift

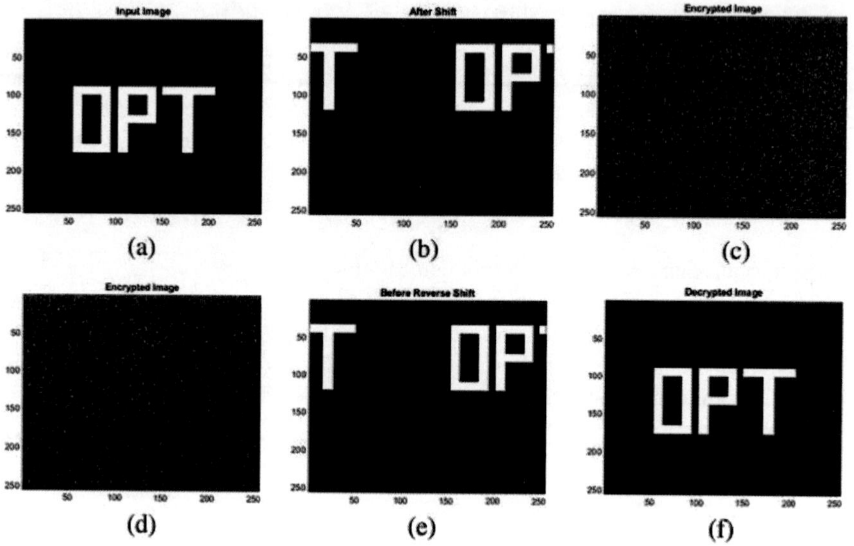

Fig. 5 OPT image encryption gyrator transform with circular shift and decryption. **a** input image, **b** image after shift, **c** encrypted image with GT, **d** image received, **e** image before reverse shift, **e** decrypted image

4 Result Comparison Histogram, MSE, PSNR

The test is carried out on several widely used standard images like lena, OTP, and barcode. Using different keys with the same image caused different encrypted results. In addition, analysis using histograms, MSE, peak signal-to-noise ratio (PSNR) showed effectiveness of the new encryption algorithm in resisting statistical attacks.

4.1 Histogram Comparison

The histogram represents the relative frequency of occurrence in the image at various gray levels (Fig. 6).

Histogram of input image is not uniformly distributed; it changes according to input image. On the other hand, the histogram of encrypted image is uniformly distributed and is similar for each input image which proves that there is no statistical similarity. They gave no indication that may help statistical attacks. So, attacker cannot able to detect input image information via encrypted image.

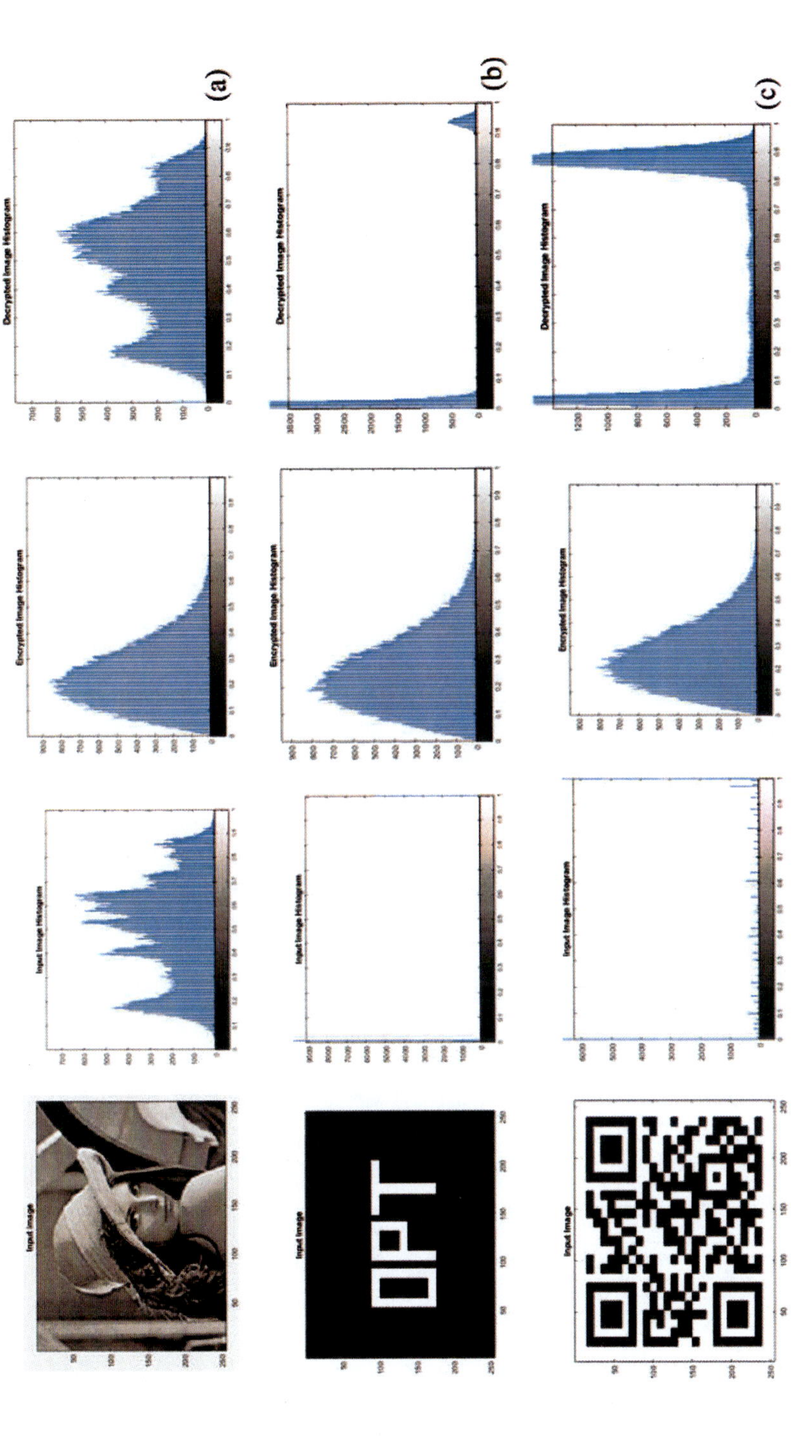

Fig. 6 Histogram comparison of gyrator transform between input image. **a** Lena image, **b** OPT image, **c** barcode image, encrypted image, and decrypted image

4.2 MSE

The iteration process convergence is evaluated by the mean square error (MSE) between the original image $P(x, y)$ and the recovered image $P'(x, y)$. The smaller the MSE is, the more precise the recovered signal becomes.

$$\mathrm{MSE} = \frac{\sum_{M,N} [I_1(m,n) - I_2(m,n)]^2}{M * N}$$

4.3 PSNR

The peak signal-to-noise ratio (PSNR): this ratio is used as a quality measurement between the input image and a received image. The higher the PSNR, the better the quality of the compressed or reconstructed image.

$$\mathrm{PSNR} = 10\log_{10}\left(\frac{R^2}{\mathrm{MSE}}\right)$$

We applied this circular shift on FrFT, gyrator transform, and Fresnel transform. Here is the comparison of MSE and PSNR with circular shift or without circular shift (Table 1).

In the comparison, we observed that value of MSE with circular shift is less as compared to MSE without circular shift and value of PSNR is more with circular shift, i.e., OPT image, and with other images also, there is very less difference between the other values. Each encryption algorithm has its own pros and cons, and it should not be vulnerable to various attacks such as statistical attacks, brute force, and differential attacks. In circular shift, we increase image security via adding shift key without increasing system complexity. This research proposes additional layers of encryption algorithms to make it more resistant to various attacks.

Table 1 Comparison of MSE and PSNR of GT and GT with circular shift—Lena, OPT, and barcode image

Gyrator transform	MSE	PSNR	MSE with circular shift	PSNR with circular shift
Lena	1.452×10^{-10}	336.543	1.463×10^{-10}	336.510
OPT	9.859×10^{-10}	338.226	9.815×10^{-10}	338.245
Barcode	2.809×10^{-10}	323.677	3.506×10^{-10}	322.716

5 Conclusions

Image scrambling algorithm based on circular shift is proposed, and related experiments are implemented and compared in MATLAB (R20019a). In literature review, we discussed various optical DRPE methods, its numerical simulation algorithm, vulnerabilities, or possible attack also explored some image scrambling technique—Arnold transforms and jigsaw transform. The image scrambling methods, including circular shift, are employed in the optical encryption systems to randomly scramble the input images. The proposed algorithm is used for position scrambling as well as pixel value scrambling, which enlarges the scope of application. Finally, the performance comparisons (MSE, PSNR, Histogram) of circular shift are presented in tabular format. In our future communication, we would like to further discuss the latest security application and data hiding techniques for practical applications.

References

1. Goodman, Joseph W (1996) Introduction to fourier optics, secondMcGraw-Hill. New York
2. Refregier P, Javidi B (1995) Optical image encryption based on input plane and Fourier plane random encoding. Opt Lett 20(7):767–769
3. Carnicer A, Montes-Usategui M, Arcos S, Juvells I (2005) Vulnerability to chosen-cyphertext attacks of optical encryption schemes based on double random phase keys. Opt Lett 30(13):1644–1646
4. Peng X, Wei H, Zhang P (2006) Chosen-plaintext attack on lensless double-random phase encoding in the Fresnel domain. Opt Lett 31(22):3261–3263
5. Peng X, Zhang P, Wei H, Yu B (2006) Known-plaintext attack on optical encryption based on double random phase keys. Opt Lett 31(8):1044–1046
6. Gopinathan U, Monaghan DS, Naughton TJ, Sheridan JT (2006) A known-plaintext heuristic attack on the Fourier plane encryption algorithm. Opt Express 14(8):3181–3186
7. Frauel Y, Castro A, Naughton TJ, Javidi B (2007) Resistance of the double random phase encryption against various attacks. Opt Express 15(16):10253–10265
8. Zhang C, Liao M, He W, Peng X (2013) Ciphertext-only attack on a joint transform correlator encryption system. Opt Exp 21(23):28523–28530
9. Wang X, Zhao D (2012) A special attack on the asymmetric cryptosystem based on phase-truncated Fourier transforms. Opt Commun 285(6):1078–1081
10. Wang Y, Quan C, Tay CJ (2015) Improved method of attack on an asymmetric cryptosystem based on phase-truncated Fourier transform. Appl Opt 54(22):6874–6881
11. Rajput SK, Nishchal NK (2013) Known plain-text attack on asymmetric cryptosystem. In: Optics and photonics for information processing VII, vol 8855. International Society for Optics and Photonics, p 88550U
12. Yadav AK, Vashisth S, Singh H, Singh K (2015) Optical cryptography and watermarking using some fractional canonical transforms, and structured masks. In: Advances in optical and engineering: proceedings of IEM Optronix 2014. Springer, pp 25–36
13. Vashisth S, Singh H, Yadav AK, Singh K (2014) Devil's vortex phase structure as frequency plane mask for image encryption using the fractional Mellin transform. Int J Opt 728056:1–9
14. Dahiya M, Sukhija S, Singh H (2014) Image encryption using quad masks in fractional Fourier domain and case study. In: IEEE international advance computing conference (IACC), pp 1048–53

15. Zhu B, Liu S (2001) Optical image encryption based on the generalized fractional convolution operation. Opt Commun 195(5–6):371–381
16. Liu S, Mi Q, Zhu B (2001) Optical image encryption with multistage and multi- channel fractional Fourier-domain filtering. Opt Lett 26(16):1242–1244
17. Singh H (2016) Optical cryptosystem of color images using random phase masks in the fractional wavelet transform domain. In: AIP conference proceedings, vol 1728, pp 020063–1/4
18. Zhang Y, Zheng C-H, Tanno N (2002) Optical encryption based on iterative fractional Fourier transform. Opt Commun 202(4–6):277–285
19. Girija R, Singh H (2018) A cryptosystem based on deterministic phase masks and fractional Fourier transform deploying singular value decomposition. Opt Quantum Electron 50(5):1–24
20. Yadav PL, Singh H (2018) Optical double image hiding in the fractional Hartley transform using structured phase filter and Arnold transform. 3D Res 9(20):1–20
21. Liu S, Sheridan JT (2013) Optical encryption by combining image scrambling techniques in fractional Fourier domains. Opt Commun 287:73–80
22. Maan P, Singh H (2018) Non-linear cryptosystem for image encryption using radial Hilbert mask in fractional Fourier transform domain. 3D Res 9(53):1–12
23. Singh H (2016) Devil's vortex Fresnel lens phase masks on an asymmetric cryptosystem based on phase-truncated in gyrator wavelet transform. Opt Lasers Eng 81:125–139
24. Singh H, Yadav AK, Vashisth S, Singh K (2015) Optical image encryption using devil's vortex toroidal lens in the Fresnel transform domain. Int J Opt (2015)
25. Singh H (2016) Cryptosystem for securing image encryption using structured phase masks in Fresnel wavelet transform domain, 3D Res 7:34
26. Unnikrishnan G, Singh K (2001) Optical encryption using quadratic phase systems. Opt Commun 193(1–6):51–67
27. Rodrigo JA, Alieva T, Calvo ML (2007) Gyrator transform: properties and applications. Opt Express 15(5):2190–2203
28. Rodrigo J, Alieva T, Calvo ML (2007) Applications of Gyrator transform for image processing. Opt Commun 278(2):279–284
29. Chen L, Zhao D (2006) Optical image encryption with Hartley transforms. Opt Lett 31(23):3438–3440
30. Singh H, Yadav AK, Vashisth S, Singh K (2015) Double phase-image encryption using gyrator transforms, and structured phase mask in the frequency plane. Opt Lasers Eng 67:145–156
31. Singh N, Sinha A (2009) Gyrator transform-based optical image encryption, using chaos. Opt Lasers Eng 47(5):539–546
32. Vashisth S, Yadav AK, Singh H, Singh K (2015) Watermarking in gyrator domain using an asymmetric cryptosystem. In: Proceedings of SPIE, vol 9654, pp 96542E-1/8
33. Liu Z, Xu L, Lin C, Dai J, Liu S (2011) Image encryption scheme by using iterative random phase encoding in Gyrator transform domains. Opt Lasers Eng 49(4):542–546
34. Li H, Wang Y (2008) Information security system based on iterative multiple-phase retrieval in Gyrator domain. Opt Lasers Eng 40(7):962–966
35. Liu Z, Xu L, Lin C, Liu S (2010) Image encryption by encoding with a nonuniform optical beam in Gyrator transform domains. Appl Opt 49(29):5632–5637
36. Singh H (2018) Hybrid structured phase mask in frequency plane for optical double image encryption in gyrator transform domain. J Mod Opt 65(18):2065–2078
37. Khurana M, Singh H (2018) Asymmetric optical image triple masking encryption based on gyrator and Fresnel transforms to remove Silhouette problem. 3D Res 9:38
38. Singh H (2018) Hybrid structured phase mask in frequency plane for optical double image encryption in gyrator transform domain. J Mod Opt 65(18):2065–2078
39. Khurana M, Singh H (2019) A spiral-phase rear mounted triple masking for secure optical image encryption based on gyrator transform. Recent Patents Comput Sci 12(2):80–94
40. Liu Z, Liu S (2007) Comment on Optical image encryption with Hartley transforms. Opt Lett 32(7):766
41. Liu Z, Zhang Y, Liu W et al (2013) Optical color image hiding scheme based on chaotic mapping and Hartley transform. Opt Lasers Eng 51(8):967–972

42. Bao L, Zhou Y, Philip Chen CL (2103) Image encryption in the wavelet domain. In: Mobile multimedia/image processing, security, and applications vol 8755. International Society for Optics and Photonics, p 875502
43. Liu Z, Chen H, Liu T, Li P, Xu L, Dai J et al (2011) Image encryption by using Gyrator transform and Arnold transform. J Electron Imaging 20(1):013020–013026
44. Guo Q, Liu Z, Liu S (2010) Color image encryption by using Arnold and discrete fractional random transforms in its space. Opt Lasers Eng 48(12):1174–1181
45. Liu Z, Xu L, Liu T, Chen H, Li P, Lin C, Liu S (2011) Color image encryption by using Arnold transform and color-blend operation in discrete cosine transform domains. Opt Commun 284(1):123–128
46. Sinha A, Singh K (2005) Image encryption by using fractional Fourier transform and jigsaw transform in image bit planes. Opt Eng 44(5):057001

Performance of Artificial Electric Field Algorithm on 100 Digit Challenge Benchmark Problems (CEC-2019)

Dikshit Chauhan and Anupam Yadav

Abstract The Artificial Electric Field Algorithm (AEFA) is a population-based stochastic optimization algorithm for solving continuous and discrete optimization problems, and it is based on Coulomb's law of electrostatic force and Newton's laws of motion. Over the years, AEFA has been used to solve many challenging optimization problems. In this article, AEFA is used to solve 100 digit challenge benchmark problems, and the experimental results of AEFA are compared with recently proposed algorithms such as dragonfly algorithm (DA), whale optimization algorithm (WOA), the salp swarm optimization (SSA), and fitness dependent optimizer (FDO). The performance of AEFA is found to be very competitive and satisfactory in comparison with other optimization algorithms chosen in the article.

Keywords Metaheuristic algorithms · Optimization · Swarm intelligent · Artificial electric field algorithm AEFA · 100 digit challenge

1 Introduction

Optimization is an essential tool in decision making and in analyzing physical systems. It is a process of making a trading system more effective by adjusting the variables used for technical analysis. An optimization problem is the problem of finding the *best* solution among the set of all feasible solutions. The mathematical formulation of optimization problem choices as decision variables and seeks values of maximum (minimum) objective functions of the decision variables subject to constraints.

D. Chauhan (✉) · A. Yadav
Department of mathematics, Dr. B.R. Ambedkar, National Institute of Technology, Jalandhar, Punjab 144011, India
e-mail: yadava@nitj.ac.in

G. Gupta et al. (eds.), *Proceedings of Academia-Industry Consortium for Data Science*, Advances in Intelligent Systems and Computing 1411,
https://doi.org/10.1007/978-981-16-6887-6_31

$$\begin{aligned}&\text{optimize } f(X)\\&\text{subject to}\\&h(x)=0\\&g(x)\leq 0\\&X=[x_1,x_2,\ldots,x_n]\end{aligned}$$

The optimization problems have an important role in real-world optimization. These optimization problems can be solved by using traditional and stochastic optimization techniques available in the literature and the active area of the research to find the global optimal solution of these optimization problems. The stochastic approach becomes more popular than the traditional approach in the last few decades. When performing optimization on complicated and complex nonlinear optimization problems, the optimum value can be obtained very fast using population-based algorithms (stochastic approaches) than the algorithm consider only one parameter of the search space at a time. Apart from fast solution, the traditional methods fail to solve NP-hard problems due to some limitations such as it fails to solve higher dimensional and highly nonlinear problems, fails to chase the best optimal solution. The population-based search method can be defined as:

$$S' = m(f(X)) \tag{1}$$

where X is a set of positions in the search space and X is called the population, f is a fitness function and m is a population modification function that returns a new populations [1].

AEFA algorithm is a newly designed and promising optimization technique to solve constraint and non-constraint optimization problems [2]. AEFA algorithm has many similarities with other meta heuristic techniques, such as differential evolution (DE) [3], particle swarm optimization (PSO) [4], artificial bee colony (ABC) [5], hybridization of PSO [6], ant colony optimization (ACO) [7], and gravitational search algorithm [8, 9].

In this paper, we solved 100 digit challenge unconstrained benchmark problems by AEFA algorithm and compared the performance with other existing state-of-the-art algorithms, namely fitness-dependent optimizer (FDO) [10], dragonfly algorithm (DA) [11], the whale optimization algorithm (WOA) [12], and the salp swarm algorithm (SSA) [13]. Section 2 describes brief explanation of AEFA algorithm; 100 digit challenge benchmark problems described in Sect. 3; Sect. 4 presented the experimental value and comparison with other state-of-the-art algorithms and convergence plots for each 10 benchmark problems. Conclusion is presented in Sect. 5.

2 Artificial Electric Field Algorithm (AEFA)

AEFA is a new and promising optimization technique to solve constraint and non-constraint optimization problems. AEFA follows swarm behavior and works in a collaborative manner [14].

Artificial Electric Field Algorithm (AEFA) [2] is a population-based meta-heuristic algorithm for solving multimodal complex or complicated optimization problems. This algorithm is based on two laws of the physics world. First, Coulomb's law of electrostatics, the electrostatic force is proportional to the product of charges and inversely proportional to the square of the distance between that charges. Second, in Newton's law of motion, the force applied to the mass agent of mass M is $F = Ma$, where a is the acceleration of the mass agent. In AEFA, consider each agent as a charged agent and charge is defined as a function of fitness value of that function which is optimizing. The inherent theory of this scheme is that if the charge of an agent is higher and opposite then it will attract the nearby lower charge agents. AEFA has used the history of previous personal and global fitness of each agent to update their position toward the promising positions.

(1) **Position Initialization**: We consider the swarm of the agents is randomly generated, and the position of any charged agent at any time in D dimensional search space is $X_i = (x_i^1, x_i^2, \ldots, x_i^D)$ for $i \in \{1, 2, 3, \ldots, N\}$

$$p_{id}(t+1) = \begin{cases} p_{id}(t), & \text{if} f(p_i(t)) < f(X_i(t+1)) \\ X_{id}(t+1), & \text{if} f(X_i(t+1)) \le f(p_i(t)) \end{cases} \tag{2}$$

(2) **Calculation of force**: The force acting from the jth agent to the ith agent is given by:

$$F_{ijd}(t) = K(t) \frac{Q_i * Q_j (P_{jd}(t) - X_{id}(t))}{R_{ij}(t) + \epsilon} \tag{3}$$

where X_i is the position vector, Q_i and Q_j are the charges of the ith and jth agent, respectively, $K(t)$ is the coulomb's constant, ϵ is the small positive constant and R_{ij} is the Euclidean distance between the ith and jth charge agents.

The coulomb's constant is defined as follows:

$$K(t) = K_0 \exp\left(-\alpha \frac{\text{iter}}{\text{maxiter}}\right), \tag{4}$$

where α and K_0 are the two initial parameters, and iter and maxiter are the current and maximum number of iterations, respectively.

The distance $R_{ij}(t)$ is given as follows:

$$R_{ij}(t) = \|X_i(t), X_j(t)\|_2, \tag{5}$$

The charge of the ith agent is defined as a function of fitness value of that function which is optimizing and updated as follows:

$$q_i(t) = \exp\left(\frac{\text{fit}p_i(t) - \text{worst}(t)}{\text{best}(t) - \text{worst}(t)}\right), \tag{6}$$

$$Q_i(t) = \frac{q_i(t)}{\sum_{i=1}^{N} q_i(t)}. \tag{7}$$

where $\text{fit}_i(t)$ is the current fitness value, N is the total number of agents in the search space, and worst(t) and best(t) is the maximum and minimum fitness for whole population, respectively, (for minimization) and is defined as follows:

$$\text{best}(t) = \min(\text{fit}_i(t)), \quad i \in (1, 2, \ldots, \text{ps}), \tag{8}$$

$$\text{worst}(t) = \max(\text{fit}_i(t)), \quad i \in (1, 2, \ldots, \text{ps}). \tag{9}$$

The total force of any agent by all other agents in dth dimension is defined as follows:

$$F_{id}(t) = \sum_{j=1, j \neq i}^{N} \text{rand}() * F_{ijd}(t). \tag{10}$$

where rand() is a uniformly distributed random number in lie between 0 and 1, N is the total number of agents. The electric field acting on ith agent is given as follows:

$$E_{id}(t) = \frac{F_{id}(t)}{Q_i(t)}. \tag{11}$$

Now, the acceleration of the ith agent is given by the Newton's law of motion, defined as follows:

$$\text{ac}_{id}(t) = \frac{E_{id}(t) Q_i(t)}{M_i(t)}. \tag{12}$$

where $M_i(t)$ is the unit mass of the ith agent.
(3) **Movement**: The ith agent update their velocity and position using equations:

$$\text{vel}_{id}(t+1) = \text{rand}() * vel_{id}(t) + \text{ac}_{id}(t) \tag{13}$$

$$X_{id}(t+1) = X_{id}(t) + \text{vel}_{id}(t+1). \tag{14}$$

After updating position X and velocity vel, the fitness values are updated using new positions X. For calculating the optimal value of given benchmark problems, repeat the process of AEFA until unless the stopping criteria are satisfied; stopping criteria may be the maximum number of iterations or error tolerance.

Table 1 Pseudocode of artificial electric field algorithm

Initialization
Randomly initialize $(X_1(t), X_2(t), \cdots, X_N(t))$ of population size N
Initialize the velocity
Evaluate the fitness values $(fit_1(t), fit_2(t), \cdots, fit_N(t))$ of agent X
set iteration $t = 0$
Reproduction and Updating
while
Stopping criterion is not satisfied
do calculate $K(t)$, $best(t)$ and $worst(t)$
for $i = 1 : N$ **do**
Evaluate the fitness values $fit_i(t)$
calculate the total force in each direction $F_i(t)$
calculate the acceleration $a_i(t)$
$vel_i(t+1) = rand() * vel_i(t) + ac_i(t)$
$X_i(t+1) = X_i(t) + vel_i(t+1)$
end for
end while

2.1 *Pseudocode of Artificial Electric Field Algorithm AEFA*

The procedure of AEFA is explained in the pseudocode, depicted in Table 1.

2.2 *Tuned Parameter for AEFA*

Tuning of parameter for AEFA is given in Table 2.

3 100 Digit Challenge Benchmark Test Functions

Table 3 have a group of 10 CEC benchmark problems, and these benchmark test functions were improved for a single objective optimization problem by professor Suganthan and his colleges [15] and are known as "The 100-digit challenge." All benchmark functions are highly scalable. The benchmarks C04 to C10 are rotated and shifted, whereas the benchmarks C01 to C03 are not but complicated. The benchmarks C01 to C03 are minimization problem of different dimensions and range; dimensions are 9, 16, 18 and range $[-8192, 8192]$, $[-16384, 16384]$, and $[-4, 4]$, respectively. The benchmark functions C04 to C10 are also minimization problems with 10 dimension in the range $[-100, 100]$. All benchmarks have a global optimum

Table 2 Tuned parameter

Benchmark functions	K_0	Population size (N)	Maximum iterations	α
C01	300	30	500	3
C02	300	30	500	3
C03	300	30	500	3
C04	500	30	500	30
C05	500	30	500	30
C06	500	30	500	30
C07	500	30	500	30
C08	300	30	500	3
C09	300	30	500	3
C10	300	30	500	3

Table 3 100-digit challenge problems

No.	Benchmark functions	Dimension	Range	Global optimal value
C01	Storn's Chebyshev polynomial fitting problem	9	[−8192, 8192]	1
C02	Inverse Hilbert matrix problem	16	[−16384, 16384]	1
C03	Lennhard-Jones minimum energy cluster	18	[−4, 4]	1
C04	Rastrigin's function	10	[−100, 100]	1
C05	Griewangk's function	10	[−100, 100]	1
C06	Weierstrass's function	10	[−100, 100]	1
C07	Modified Schwefel's function	10	[−100, 100]	1
C08	Expanded Schwefel's F6 functions	10	[−100, 100]	1
C09	Happy cat function	10	[−100, 100]	1
C10	Ackley function	10	[−100, 100]	1

at 1. AEFA competes with four new optimization algorithms: FDO, DA, WOA, and SSA. There are three selecting ideas of these algorithms: (1) All four algorithms FDO, DA, WOA, and SSA are population-based same as AEFA. (2) All algorithms are well-cited algorithms in most cited algorithms. (3) All algorithms have outstanding performance on selected benchmark problems.

4 Results and Discussion

4.1 *Evaluation of AEFA Over 100-Digit Challenge Benchmark Problems*

The results were obtained after 25 runs and evaluated for a 500 * N maximum number of function evaluations on each function. The recorded results are presented in Table 4, and it is observed that the AEFA algorithm gives better result for most of the benchmark problems in comparison with DA, WOA, FDO, and SSA. In the table, best optimum values obtained from the selected algorithms are highlighted using bold face.

4.2 *Convergence Plots for AEFA*

In this section, we compared the convergence of AEFA with existing algorithms. In this article, we plotted the best values against the iterations for all benchmark problems. The convergence plots of the comparative algorithms are given in Fig. 1. From these convergence plots, the convergence phenomenon of AEFA can be observed against other algorithms for each function.

Table 4 Comparative results of benchmark problem values for 100 digit challenge

Benchmark functions	Metric	DA	WOA	SSA	FDO	AEFA
C01	Best values	1.60e+09	1.01e+06	2.01e+08	2.89e+06	**1.00e+06**
	Mean values	543e+08	225e+08	876e+07	765.34e+05	235e+07
	Standard deviation	669e+08	410e+08	100e+08	638.84e+05	390e+07
	Worst values	212.0e+09	152.82e+09	408.17e+08	218.79e+06	802.41e+07
C02	Best values	17.3430	17.3439	17.3430	**17.3428**	941.2
	Mean values	78.0368	17.3502	17.3605	17.3428	4586
	Standard deviation	87.7888	0.0072	0.0060	1.9947e−09	1831
	Worst values	61.521e+01	17.3776	17.6328	17.3428	88.80e+02
C03	Best values	**12.7024**	**12.7024**	**12.7024**	**12.7024**	**12.7024**
	Mean values	13.0368	127.0240	127.0262	12.7024	127.0128
	Standard deviation	0.0007	0.0000	0.0006	3.9115e−12	0.0005
	Worst values	12.7049	12.7024	12.7049	12.7024	12.7052
C04	Best values	55.9331	72.3884	16.9142	9.2988	**2.9848**
	Mean values	344.3561	412.4570	35.0224	31.8094	52.1731
	Standard deviation	414.0982	263.3009	14.6001	11.4068	82.5916
	Worst values	7.1490e+02	1.3093e+03	6.8651e+01	6.0623e+01	4.1982e+02

(continued)

Table 4 (continued)

Benchmark functions	Metric	DA	WOA	SSA	FDO	AEFA
C05	Best values	1.0947	1.4021	1.0762	1.0245	**1.0000**
	Mean values	2.5572	1.7996	1.2263	1.1147	1.0065
	Standard deviation	0.3245	0.3561	0.0091	0.0045	0.0010
	Worst values	2.3819	2.6559	1.4111	1.1998	1.0368
C06	Best values	5.3309	1.4021	1.7930	7.2413	**1.0000**
	Mean values	9.8955	9.6929	5.2928	9.2099	1.3031
	Standard deviation	1.6404	1.3337	2.0725	0.0767	0.5798
	Worst values	1.1690e+01	1.1798e+01	9.9036e+00	11.007	2.7780
C07	Best values	3.9861e+01	1.6125e+01	−2.3575e+02	−50.2418	**1.4688**
	Mean values	578.9531	595.7386	298.1940	76.7688	171.8117
	Standard deviation	329.3983	321.7981	266.6329	81.5225	93.0219
	Worst values	1.0480e+03	1.3256e+03	1.0179e+03	3.2583e+02	3.5373e+02
C08	Best values	5.1274	5.0695	4.1580	3.4193	**1.4451**
	Mean values	6.8734	5.9145	5.4431	4.2295	2.6137
	Standard deviation	0.5015	0.4489	0.5569	0.05143	0.7687
	Worst values	6.8853	7.0181	6.0944	5.0380	3.9945
C09	Best values	3.0387	3.1453	2.4269	**2.3428**	2.3959
	Mean values	6.0467	5.0293	2.6204	2.3994	2.4822
	Standard deviation	2.871	1.0980	0.1770	0.0023	0.0673
	Worst values	6.8853	8.6430	3.0634	2.4640	2.6562
C10	Best values	20.0209	20.0474	4.4228e−05	2.2204e−14	**0.3009**
	Mean values	21.2604	20.2831	19.2485	19.2625	9.9676
	Standard deviation	0.1715	0.01039	4.0115	4.0141	10.0546
	Worst values	20.5292	20.4572	20.3641	20.0688	20.3696

5 Conclusion

In this article, we studied the of ability artificial electric field algorithm (AEFA) on 100-digit challenge problems. A set of 10 problems (100-digit challenge) is solved using the AEFA algorithm, and the optimization results are compared with recent optimization algorithms of the same category. The results obtained ensure the superiority of AEFA over other selected algorithms. Further, the convergence plots for all the 10 problems shows the promising nature of the convergence of AEFA.

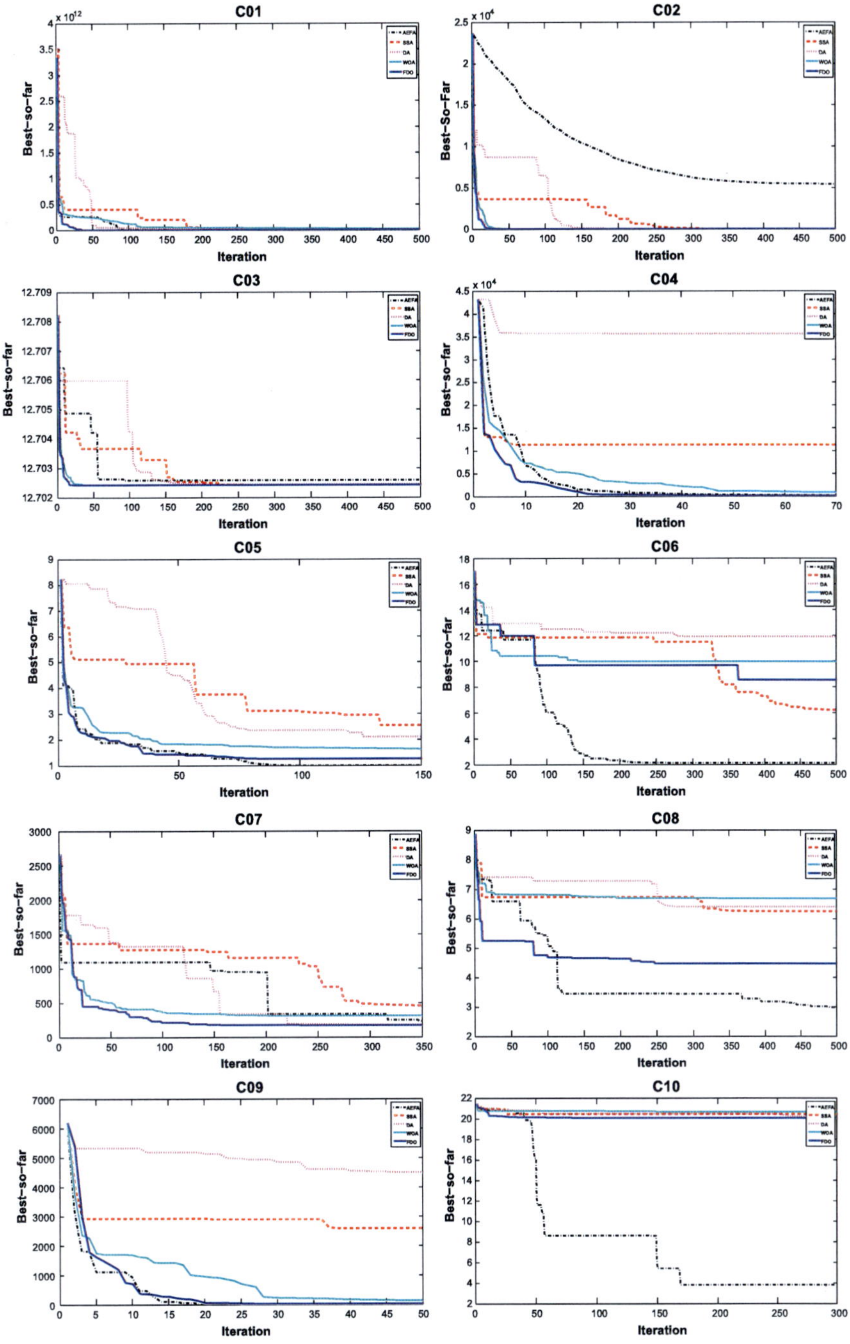

Fig. 1 Convergence plot of for IEEE CEC-C06, 2019

References

1. Angeline PJ (1998) Using selection to improve particle swarm optimization. In: 1998 IEEE International conference on evolutionary computation proceedings. IEEE World Congress on Computational Intelligence (Cat. No. 98TH8360). IEEE, pp 84–89
2. Yadav A et al (2019) AEFA: artificial electric field algorithm for global optimization. Swarm Evol Comput 48:93–108
3. Storn R, Price K (1997) Differential evolution—A simple and efficient heuristic for global optimization over continuous spaces. J Glob Optim 11(4):341–359
4. Kennedy J, Eberhart R (1995) Particle swarm optimization. In: Proceedings of ICNN'95-International Conference on Neural Networks, vol 4. IEEE, pp 1942–1948
5. Karaboga D (2010) Artificial bee colony algorithm. Scholarpedia 5(3):6915
6. Huang C-L, Huang W-C, Chang H-Y, Yeh Y-C, Tsai C-Y (2013) Hybridization strategies for continuous ant colony optimization and particle swarm optimization applied to data clustering. Appl Soft Comput 13(9):3864–3872
7. Dorigo M, Gambardella LM (1997) Ant colony system: a cooperative learning approach to the traveling salesman problem. IEEE Trans Evol Comput 1(1):53–66
8. Venkata Rao R (2016) Teaching-learning-based optimization algorithm. In: Teaching learning based optimization algorithm. Springer, pp 9–39
9. Yadav A, Deep K, Kim JH, Nagar AK (2016) Gravitational swarm optimizer for global optimization. Swarm Evol Comput 31:64–89
10. Abdullah JM, Ahmed T (2019) Fitness dependent optimizer: inspired by the bee swarming reproductive process. IEEE Access 7:43473–43486
11. Mirjalili Seyedali (2016) Dragonfly algorithm: a new meta-heuristic optimization technique for solving single-objective, discrete, and multi-objective problems. Neural Comput Appl 27(4):1053–1073
12. Mirjalili S, Lewis A (2016) The whale optimization algorithm. Adv Eng Softw 95:51–67
13. Mirjalii SZ, Saremia C, Farisd H, Mirjalilia S, Gandomibf AH, Mirjalilie SM (2017) Salp swarm algorithm: a bio-inspired optimizer for engineering design problems. 114:163–191
14. Yadav Anupam, Kumar Nitin et al (2020) Artificial electric field algorithm for engineering optimization problems. Expert Systems Appl 149:113308
15. Ali MZ, Sunganthan PN, Price KV, Awad NH. The 2019 100-digit challenge om real-parameter, single objective optimization: analysis of results

Stock Investment Strategies and Portfolio Analysis

Zheng Tao and Gaurav Gupta

Abstract With the advent of the era of big data, the application of new technologies in the financial field has continued to deepen. Based on Markowitz's portfolio theory, this paper uses Python to select 5 from 11 representative stocks in different industries for portfolio investment analysis. Through empirical research, obtain the optimal solution of the investment portfolio: (1) the sharpe ratio is the largest; (2) the variance is the smallest. Additionally, their expected return, standard deviation, and sharpe ratio are compared and analyzed, and Monte Carlo simulation is used to calculate the efficient frontier of the asset portfolio. Finally, create two independent simple moving averages as a trading strategy. Through back-testing, it is found that under this method, the return rate of the selected optimal investment portfolio is close to 29%, the highest is close to 40%.

Keywords Markowitz's portfolio theory · Sharpe ratio · Monte Carlo simulation · Back-testing

1 Introduction

The uncertainty of the stock market causes investors to face risks and may lead to losses. Therefore, investment needs to be cautious. There are many factors that cause risks, such as relevant laws and regulations, and some emergencies. Moreover, financial crisis, national policies, and the general environment also affect the financial market. If these problems are encountered, the rate of return on investment will decrease. Everyone likes a high-return, low-risk investment model, and everyone wants to be rich. As the common people have an urgent need for a reasonable wealth

---

Z. Tao · G. Gupta (✉)
College of Science and Technology, Wenzhou-Kean University, Wenzhou, China
e-mail: ggupta@kean.edu

Z. Tao
e-mail: taozhe@kean.edu

G. Gupta et al. (eds.), *Proceedings of Academia-Industry Consortium for Data Science*, Advances in Intelligent Systems and Computing 1411,
https://doi.org/10.1007/978-981-16-6887-6_32

distribution theory, learning how to optimize investment and financial management and achieve rational investment is the most concerned issue of current investors.

The construction of investment portfolio is a vital aspect of investment management. Although the expected return of assets is an important factor, investors also need to consider the interdependence of investment risks and the rate of return on assets. Starting from Markowitz's theory, Kedi [4] gave a combination of cut-points of n types of risk asset allocation on the effective front by maximizing the Sharpe ratio. After a given investor's risk aversion level, he determines that the investor is risk-free. The investment ratio between assets and n types of risk assets is thus constructing investment portfolios suitable for investors with different risk aversion levels. Chakrabarty and Biswas [1] used the strategic Markowitz portfolio optimization method to build a portfolio based on the performance of eight companies' stocks and achieved good investment income. Lipo [2] selected 5 of 20 stocks from different industries for portfolio investment analysis, obtained the optimal investment portfolio with the largest Sharpe ratio through empirical analysis, and gave the efficient frontier of the asset portfolio. Nawrocka [3] used neural networks for portfolio creation and trading strategies. The performance of this method is based on the S&P 500 index, Markowitz portfolio's buy-and-hold strategy and rebalancing. It turns out that the deep learning-based portfolio outperforms the market.

Through literature review, it is found that although they constructed the optimal investment portfolio, they did not elaborate on how to choose stocks and formulate investment strategies. This paper will select 11 representative stocks and use the covariance and correlation coefficients between the stocks to select five stocks with smaller correlation coefficients. Then, the two optimization methods of maximizing Sharpe ratio and minimizing variance are used to find the optimal portfolio weight parameters and calculate the expected return, expected volatility, and Sharpe ratio of these two optimal portfolios. Next, visualize the efficient frontier of the portfolio. Through the comparison of the four investment portfolios, find the combination with the highest cumulative return and use the simple moving averages strategy for it. Finally, a back-test is used to confirm the rate of return under this method.

2 Method

2.1 Markowitz's Portfolio Theory

In 1952, Harry Markowitz conducted research on probability theory and solving quadratic programming, trying to find high-quality investment models to achieve low-risk and high-return results. This is the modern portfolio management theory. Assuming a premise: investors hope to maximize target returns while having zero risk. This set of theories by Harry Markowitz consists of two parts: (1) mean variance model; (2) efficiency boundary theory. Combine the two values to quantify the existing risks and target benefits. The ultimate goal is to find the optimal solution: (1)

Under the condition of equal target returns, find the investment model with the least risk; (2) Under the condition of equal risk, find the project with the highest return. A portfolio on the efficient frontier will meet these two conditions together.

Mean-variance model Make an assumption: the investor selects n projects (risky) at the same time within a certain period of time, r_i represents the target rate of return of the ith project, the following formula: the target rate of return of each project

$$E(r_{\mathrm{p}}) = \sum_{n=1}^{n} x_i E(r_i) \tag{1}$$

where x represents the investment proportion of the ith asset. Let ${\sigma_i}^2$ denote the variance of the ith asset, then the variance of the portfolio consisting of n assets is:

$$\sigma_{\mathrm{p}}^2 = \sum_{i=1}^{n}\sum_{j=1}^{n} x_i x_j \mathrm{cov}(r_i, r_j), i \neq j \tag{2}$$

$$\sigma_{\mathrm{p}}^2 = \sum_{i=1}^{n} {x_i}^2 {\sigma_i}^2 + \sum_{i=1}^{n}\sum_{j=1}^{n} x_i x_j \rho_{ij} \sigma_i \sigma_j \tag{3}$$

where i and j represent different assets, and $\mathrm{cov}(r_i, r_j)$ is the covariance between asset i and asset j, which is an index used to measure the linkage between the return rates of the two assets. ρ_{ij} represents the coefficient between two items i and j, and the function is to compare the degree of correlation between the two.

According to formula (3), the proportion of the amount of investment projects determines the degree of risk in the investment portfolio. Different projects have different standard deviations, and different securities have different corresponding coefficients. Therefore, priority is given to selecting assets with a small variance and a low correlation coefficient between the two to construct an investment portfolio to reduce investment risks. In practical applications, the sample average and sample variance of past return data are usually used to estimate future returns and risks.

The efficient frontier of the asset portfolio All possible combinations of each asset in the portfolio constitute a feasible set, whose shape is similar to a left-convex solid area shown in Fig. 1.

Among them, the investment portfolio located on the edge of the upper half meets the conditions of the risk is the smallest in the equivalent amount of return, and the return is the highest in the equivalent risk, which is called the efficient frontier of the asset portfolio.

Fig. 1 Efficient frontier

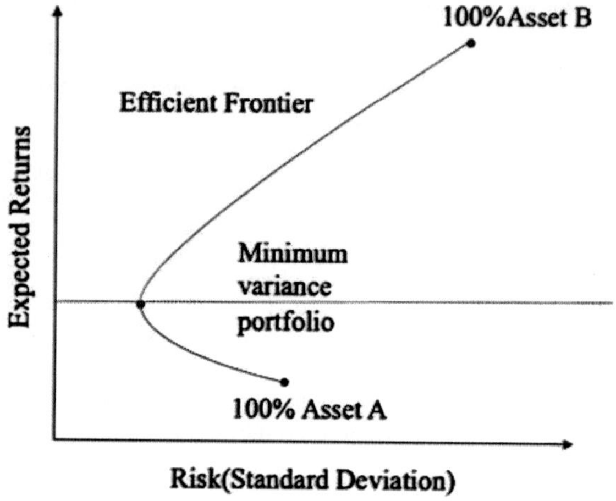

2.2 *Sharpe Ratio*

In 1966, Sharp invented the Sharpe ratio, which is used to judge the performance of the fund and use the risk-adjusted index for analysis. The ratio of the excess target return of the portfolio to the overall standard deviation is the Sharpe ratio. The calculation formula is:

$$S_p = \frac{E(r_p) - r_f}{\sigma_p} \tag{4}$$

where S_p represents the Sharpe ratio, σ_p is the overall standard deviation of the portfolio, and R_f is the risk-free interest rate. S_p represents how much excess return the asset portfolio can have for each additional unit of risk. S_p comprehensively considers returns and risks. The empirical research part of this paper will also use this indicator as a criterion for judging the strengths and weaknesses of investment portfolios.

3 Empirical Research

3.1 *Selection of Sample Stocks and Data Acquisition*

Select 11 representative stocks in the market (Table 1) and use the quandl package in python to obtain historical data of their daily closing prices for a total of 252 trading days from December 30, 2016 to December 30, 2017.

From the above 11 stocks, 5 stocks with smaller correlation coefficients are selected (Fig. 2). They are Apple, Exxon Mobil, JP Morgan, General Electric, and Tesla.

Table 1 Eleven representative stocks

Name	Code name
Apple	AAPL
Microsoft	MSFT
Exxon Mobil	XOM
Johnson & Johnson	JNJ
JP Morgan	JPM
Amazon	AMZN
General Electric	GE
Facebook	FB
AT & T	T
Tesla	TSLA
WalMart	WMT

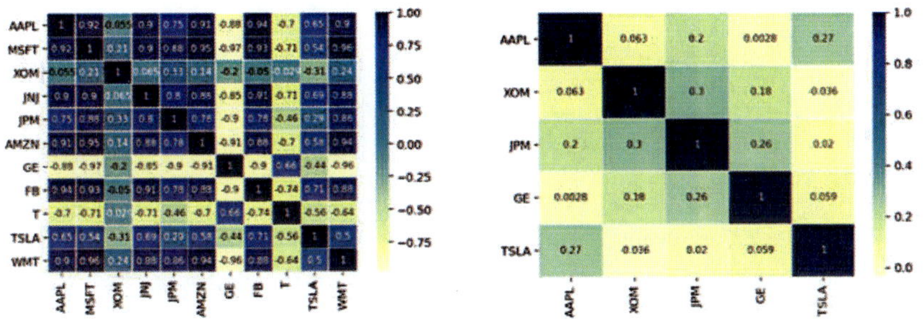

(a) Correlation coefficients of 11 stocks

(b) Correlation coefficients of 5 stocks

Fig. 2 Correlation coefficients

3.2 *The Expected Return Rate, Mutual Covariance, and Correlation Coefficient of Five Stocks*

Recalculate the covariance matrix and correlation coefficient of the selected five stocks. The results are given in Table 2 and Fig. 2b.

According to Table 2 and Fig. 2b, it can be seen that the covariance and correlation coefficients among the five stocks are small, which is suitable for constructing investment portfolios.

3.3 *Simulate a Large Number of Random Investment Portfolios*

To find the best investment portfolio and effective frontier that meets your needs, we first obtain 10,000 sets of random weight vectors through a Monte Carlo simulation

Table 2 Covariance matrix of five stocks

Codename	AAPL	XOM	JPM	GE	TSLA
AAPL	0.031577	0.001268	0.005721	0.000099	0.016972
XOM	0.001268	0.012688	0.005522	0.004045	−0.001425
JPM	0.005721	0.005522	0.026546	0.008451	0.001127
JPM	0.005721	0.005522	0.026546	0.008451	0.001127
GE	0.000099	0.004045	0.008451	0.039601	0.004192
TSLA	0.016972	-0.001425	0.001127	0.004192	0.125785

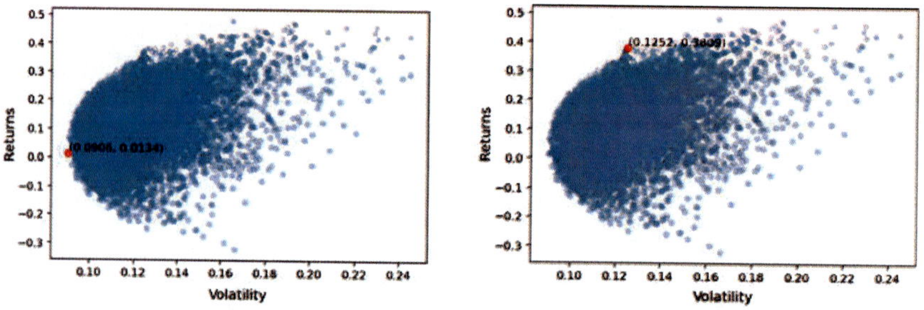

(a) The optimal portfolio with the smallest variance

(b) The optimal portfolio with the largest Sharpe ratio

Fig. 3 Use Monte Carlo simulation to find the optimal portfolio

and record the expected returns, standard deviations, and Sharpe ratios of each combination through an array. The for loop statement, append function, random function, and array function in the numpy package are mainly used here. To consolidate all the target returns, you need to add the weights of the target returns of each stock. The risk-free interest rate is set at 0% when calculating the Sharpe ratio, and we do not consider short-selling situations. Therefore, it is also necessary to limit the weight value of each equity between 0 and 1. The distribution of these random combinations can be seen in Fig. 3. They constitute the feasible set of asset portfolios.

3.4 Find the Optimal Portfolio with the Smallest Variance

First define a stat function to record the expected return, standard deviation, and Sharpe ratio of the portfolio, and then, introduce the optimze module in the scipy package. Under the constraint that the sum of weights is equal to 1, using the minimize function to obtain the optimal portfolio weight vector with the smallest variance is [0.177, 0.496, 0.109, 0.168, 0.050]. The expected return rate, standard deviation, and Sharpe ratio vector of the combination is: [0.0134, 0.0906, 0.143].

Table 3 Comparison of the two optimal investment portfolios

Portfolio	Weight of each equity	Standard deviation	Sharpe ratio
The optimal portfolio with the largest Sharpe ratio	[0.472, 0.087, 0.325, 0.001, 0.115]	0.1252	3.042
The optimal portfolio with the smallest variance	[0.177, 0.496, 0.109, 0.168, 0.050]	0.0906	0.143

Table 4 Comparison of the two optimal investment portfolios

Portfolio	Weight of each equity	Standard deviation	Sharpe ratio
The optimal portfolio with the largest Sharpe ratio	[0.472, 0.087, 0.325, 0.001, 0.115]	0.1252	3.042
The optimal portfolio with the smallest variance	[0.177, 0.496, 0.109, 0.168, 0.050]	0.0906	0.143

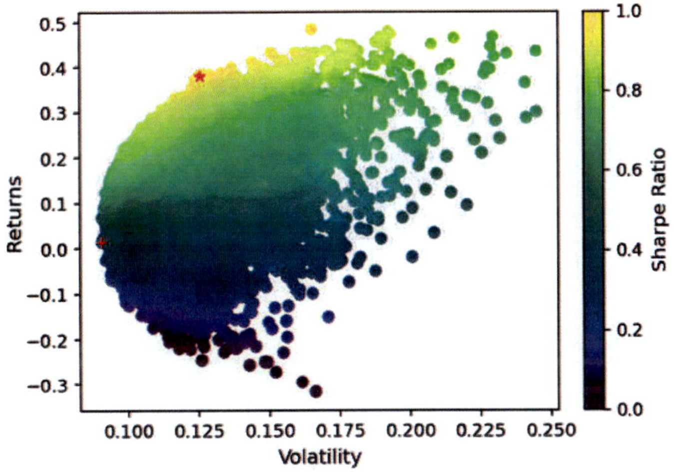

Fig. 4 Visualization of the optimal portfolio

3.5 *Find the Optimal Portfolio with the Largest Sharpe Ratio*

Use the minimize function to minimize the negative value of the Sharpe ratio, and the starting parameter list is uniformly distributed, that is, the weights of five stocks are all 0.2. Finally, the weight vector of the portfolio with the largest Sharpe ratio is [0.472, 0.087, 0.325, 0.001, 0.115]. The vector of expected return, standard deviation, and Sharpe ratio is [0.3809, 0.1252, 3.042] (Table 3).

Use the correlation function in matplotlib.pyplot to visualize the position of the above optimal investment portfolio in the feasible concentration, and get Fig. 4.

In Fig. 4, the five-pointed star marks the portfolio with the largest Sharpe ratio (risk–return equilibrium point), and the plus sign marks the portfolio with the smallest variance. Taking the minimum variance portfolio as the boundary, the feasible set can be divided into upper and lower parts, and the edge of the upper half is the efficient frontier. Moreover, the Sharpe ratios of all investment portfolios are positive, and the investment portfolio on the efficient when the return results are equal, frontier has the lowest degree of risk; when the risk degree is equal, frontier's target return is the highest.

3.6 Comparative Rate of Return

The first option is to evenly distribute the weight of each stock, so that they are all equal. This is the simplest investment method.

The second option needs to consider the market value of the stock and plan the investment weight according to a certain ratio. In other words, invest a larger proportion of the weight in stocks with more market value. If these stocks have risen, the return on the portfolio will be higher. The third option is when the stock weight is in the minimum risk portfolio (global minimum variance). The fourth option is the portfolio with the largest Sharpe ratio. Figure 5 shows the cumulative rate of return under these four method.

It can be seen from Table 4 that among the four portfolios, the best performance is the maximum Sharpe ratio portfolio, followed by the equal weighted portfolio, and the worst is the portfolio weighted by market value.

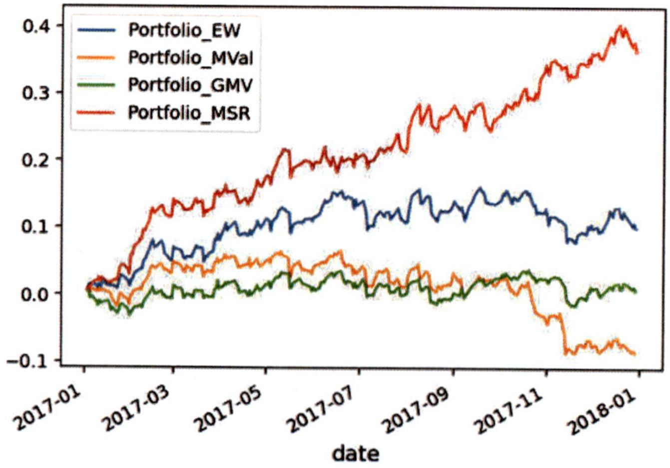

Fig. 5 Visualization of the optimal portfolio

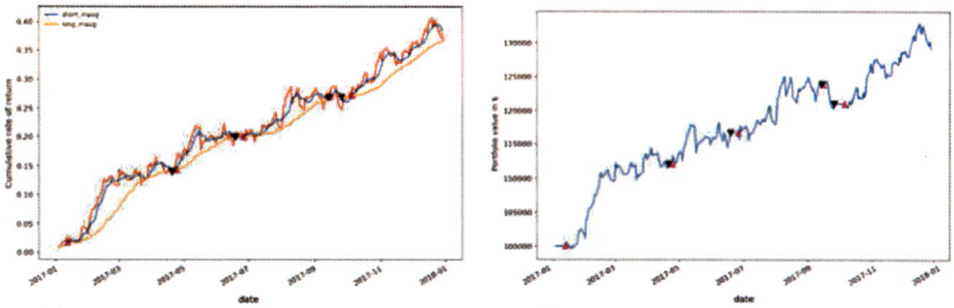

(a) Trading strategy for simple moving averages

(b) Maximum Sharpe ratio portfolio in the backtest

Fig. 6 Trading strategy for the maximum Sharpe ratio portfolio

Next, set up a simple trading strategy and create two simple moving averages for the time series, they exist independently. They have different lookback periods, they are 7 days and 30 days. When the short-term moving average is longer than the long-term moving average, operate: buy; when the long-term moving average is shorter than the short-term moving average, operate: sell. Figure 6 shows the trading strategy for the maximum Sharpe ratio portfolio. The upward arrow represents buying, and the downward arrow represents selling.

Finally, create the maximum Sharpe ratio portfolio in the back-test. First, set the initial capital to 100,000 $. If the short-term moving average is longer than the long-term moving average (for periods greater than the shortest moving average window), then all funds are bought into the maximum Sharpe ratio portfolio. Then calculate the total value of cash and stocks held and the rate of return obtained by storage. Figure 6 shows the fluctuations in the cash and stock value of the maximum Sharpe ratio portfolio. Because in the later period, short-term moving average exceeds the long-term moving average. Therefore, there is no option to sell and still hold the portfolio. In the end, the total value of cash and stocks can be obtained through the back-test as 129008.69 $, and the yield is almost 29%.

4 Conclusion

In this paper, we find the portfolio with the smallest risk or the largest Sharpe ratio and the effective boundary in a portfolio composed of multiple assets using Markowitz's portfolio theory. Investors can make rational investments based on their actual capabilities and risk preferences, reduce risks as much as possible while ensuring certain returns, or obtain the highest returns at the same risk level. Python greatly facilitates the calculation of the expected rate of return and portfolio variance in Markowitz's portfolio theory and can quickly find the optimal portfolio.

Results shows that that not all securities portfolios are located on the efficient boundary, and risk and return are not completely positively correlated. There is a certain degree of difference. Therefore, investors must use a cautious attitude and scientific concepts to invest in the selection of securities, the judgment of investment proportions, and the analysis of the market environment, and they must not blindly pursue returns while ignoring the existence of risks. Meanwhile, due to the limitations of Markowitz's theory itself, investors should not blindly apply this theory to make investment decisions. They should also make appropriate corrections and adjustments based on the actual situation in order to avoid unnecessary investment losses and give full play to investment diversification.

Acknowledgements The authors would like to thank the anonymous reviewers and editors for their valuable comments and suggestions to improve the quality of this paper.

References

1. Chakrabarty N, Biswas S (2019) Strategic Markowitz portfolio optimization (SMPO): a portfolio return booster. In: 2019 9th Annual information technology, electromechanical engineering and microelectronics conference (IEMECON). IEEE, pp 196–200
2. Libo Sun (2020) An empirical study of Markowitz's portfolio theory based on Python. Times Finance 25:46–47
3. Nawrocka D (2018) Machine learning for trading and portfolio management using Python. Doctoral dissertation, Hochschulbibliothek HWR, Berlin
4. Kedi L (2018) Markowitz theory constructs investment portfolio. Mod Bus 36:44–45. https://doi.org/10.14097/j.cnki.5392/2018.36.020
5. Markowitz H (1952) Portfolio selection. J Finance 77-91
6. Markowitz H (1959) Portfolio selection, efficient diversification of investments. Wiley
7. Sharpe WF (1966) Mutual fund performance. J Bus 39(1):119–138
8. Sharpe WF (1994) The sharpe ratio. J Portfolio Manage 21(1):49–58
9. Maller RA, Durand RB, Jafarpour H (2010) Optimal portfolio choice using the maximum Sharpe ratio. J Risk 12(4):49
10. Reinganum MR (1983) Portfolio strategies based on market capitalization '. Ariel 134:60–119
11. Banz RW (1981) The relationship between return and market value of common stocks. J Financial Econ 9(1):3–18
12. Raudys A, Lenčiauskas V, Malčius E (2013) Moving averages for financial data smoothing. In: International conference on information and software technologies, pp 34–45. Springer, Berlin, Heidelberg
13. Ming-Ming L, Siok-Hwa L (2006) The profitability of the simple moving averages and trading range breakout in the Asian stock markets. J Asian Econ 17(1):144–170
14. Harvey CR, Liu Y (2015) Backtesting. J Portfolio Manage 42(1):13–28

Automatic Keyword Extraction in Economic with Co-occurrence Matrix

Bingxu Han and Gaurav Gupta

Abstract As the financial market is very sensitive to the sudden news of economic events, it is very important to identify the events in the news items accurately and timely. Keywords can effectively provide information in this aspect. Keywords contain a certain amount of information; people can easily select relative news through keywords. In addition, keyword extraction is an important technology in the field of literature retrieval and Web search. This paper proposes a keyword extraction method based on a co-occurrence matrix. First of all, based on the dictionary method, we collect more comprehensive information of Chinese characters and automatically segment Chinese words according to the bidirectional maximum matching method. At the same time, we can extract the words not included in the dictionary as a separate unit. Then, the co-occurrence matrix is constructed to obtain the co-occurrence frequency of different phrases. Finally, we get keywords according to the co-occurrence frequency of phrases. Compared with the traditional TD-IDF method, this method has better accuracy and extracts more effective keywords in the face of a single paragraph. The experimental results show that this method can extract more effective keywords in some single paragraphs, and the experimental results are satisfactory.

Additional Keywords and Phrases Chinese NLP · Co-occurrence matrix · Keyword extraction

1 Introduction

Economy plays an important role in today's society. It is the main support to maintain the normal operation of society. Economic information is gradually informationized, and decision makers are faced with a large amount of economic information and data.

B. Han · G. Gupta (✉)
College of Science and Technology, Wenzhou-Kean University, Wenzhou, China
e-mail: ggupta@kean.edu

B. Han
e-mail: hanbi@kean.edu

G. Gupta et al. (eds.), *Proceedings of Academia-Industry Consortium for Data Science*, Advances in Intelligent Systems and Computing 1411,
https://doi.org/10.1007/978-981-16-6887-6_33

Therefore, it is very important to pick up the key information quickly and effectively. Often, keywords can effectively explain the key information [1]. Keyword extraction is an important technology in document retrieval, Web page retrieval, text index, abstract, and classification. By keyword extraction, people can easily choose which documents are needed, and keywords can also show the relationship between documents [2, 3]. In the field of text information processing, keywords are the most concise description of the text and can also effectively reflect the theme of the text [4]. For example, through a person's high-frequency search for certain keywords, the Web page can provide the corresponding articles or information. At the same time, the first problem of Chinese keyword extraction is word segmentation. Word segmentation is the process of recombining continuous word sequences into word sequences according to certain norms. In English, spaces are used as natural boundaries between words, while there is no formal one in Chinese, so Chinese is much more complicated and difficult than English. At the same time, because of the particularity of Chinese characters, most of the word segmentation methods and keyword extraction methods used in English are not suitable for Chinese [5]. The current Chinese word segmentation methods include string matching method, understanding method, and statistics method [6]. A popular keyword extraction method for Chinese is TF-IDF, which extracts keywords that often appear in documents but not in the rest of corpus. The main idea of this method is that if a word appears more frequently in one article and less frequently in other articles, it can better represent the meaning of the current article. That is to say, the importance of a word is directly proportional to the number of times it appears in the document and inversely proportional to the frequency it appears in the corpus. At the same time, TF-IDF needs a large number of corpora and keyword extraction based on the comparison of multiple documents [2, 7].

This paper explains a keyword extraction method based on a co-occurrence matrix for a single paragraph in economic. Firstly, according to the user-defined dictionary, the bidirectional maximum matching method is used to segment the text, and the words not included in the dictionary are extracted. Then, the co-occurrence matrix is made according to the frequency of different phrases. If a pair of terms often appear together, it may be of great significance to the terms. We extract keywords according to the co-occurrence matrix. Finally, it is compared with the TD-IDF method.

The organizational structure of this paper is as follows. The second part will introduce some related work. The third part describes the method. The fourth part describes the experimental results. Finally, we summarize our contributions and expectations for future development.

2 Related Work

Although this is a keyword for the economic field, it still belongs to the task of natural language processing. In the task of Chinese natural language processing, word segmentation is the most basic and important part. In the past two decades, Chinese word segmentation is divided into dictionary-based method, statistical method, and other methods.

Dictionary-based word segmentation often relies on large dictionaries. At the same time, according to the dictionary string matching, according to the maximum matching method for word segmentation. However, without a good comprehensive dictionary, the success rate of this method will be greatly reduced [8]. At the same time, dictionary-based word segmentation tools often make mistakes when facing new word combinations or strings. In some specific word segmentation tools, words that do not match are treated as single characters [8]. These methods represent domain knowledge, which is the mainstream method of Chinese word segmentation for specific domain [9]. In addition, according to different fields, there will be corresponding dictionaries [10].

The word segmentation method based on statistics is generally implemented according to statistical models or other machine learning algorithms. For example, Chinese word segmentation based on conditional random field (CRF) is implemented by Gui et al. [11]. Natural language task processing is based on support vector machine implemented by Nawroth et al. But the method based on statistics is more flexible, which can be changed according to a different training or test set, and improves the segmentation performance automatically [8]. But they still rely heavily on large amounts of hard-to-process, high-quality, manually segmented data. In addition, they are domain-specific and involve a lot of time and resource-intensive retraining when converting domains.

The other method is a mixture of dictionary-based method and statistical method to improve the performance of Chinese word segmentation. It is usually based on a neural network model and widely used in NLP tasks [8]. For example, the bidirectional LSTM recurrent neural network with word embedding function is proposed by Wang P et al., the greedy neural-based word segmenter is proposed by Cai et al., and the Chinese word segmentation method is based on wordhood memory networks proposed by Tian et al. [12–14]. Hogenboom et al. detect economic events by semantics-based information extraction [1].

In natural language processing for a single paragraph, we prefer to use dictionary-based word segmentation. Although it is not a new method, it can effectively increase the performance of word segmentation according to the expansion of the dictionary. At the same time, we make our own dictionary, which can contain multiple fields and be competent for word segmentation in specific fields.

At the same time, there are several main methods for keyword extraction:

1. Based on TF-IDF: It weighted the frequency of a given word in the file and the frequency in other files and finally got different TF-IDF values and got the keywords according to the ranking [15].
2. Based on Textrank: The algorithm divides the text into words as network nodes and forms a network graph model of words. The similarity between words is regarded as a recommendation or voting relationship, which makes it possible to calculate the importance of each word [16].

3. Based on Word2vec. The main idea of this algorithm is to cluster words in the article by K-means algorithm for text words represented by word vectors. The clustering center is selected as a keyword of the article, and the distance between other words and cluster centers is calculated, namely similarity. The words closest to the cluster center are selected as text keywords, and the similarity between words can be calculated by the vector generated by Word2vec [17].

Compared with these methods, TF-IDF is the most widely used, in which the idea of word frequency is used [15]. At the same time, another way to express word frequency is the co-occurrence matrix. It can show the co-occurrence frequency of *different phrases. In ad*dition, if a pair of terms often appear together, it may be of great significance to the terminology [2, 6]. The co-occurrence matrix can show the relationship between different phrases, but it faces the problem of sparsity, and the amount of computation will grow geometrically in the face of a large number of texts. Then, we proposed a method for a single paragraph, so it does not require a huge amount of computation.

3 Method

In this section, we introduce our main methods. The first step is the making of dictionaries, which is based on a large number of corpora. Then, Chinese automatic word segmentation is based on bidirectional maximum matching method. Finally, the co-occurrence frequency of different word combinations is obtained by constructing the co-occurrence matrix, and the keywords are extracted. Our workflow is shown in Fig. 1.

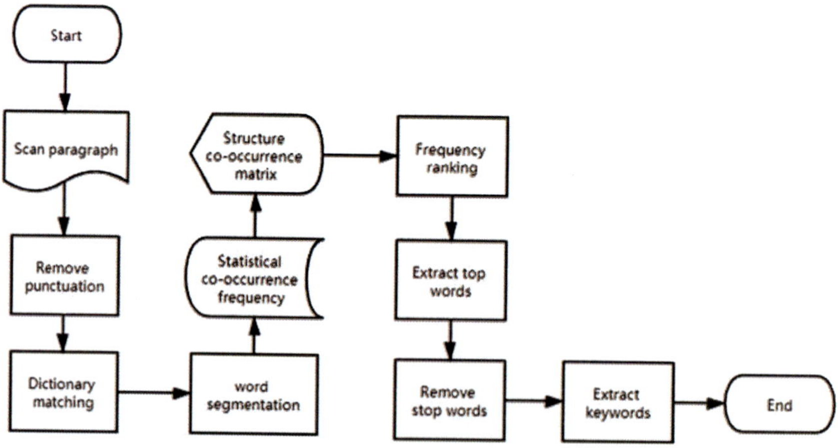

Fig. 1 Flowchart

3.1 Text Preprocessing

After scanning the text, we need to preprocess the text. Because the text from the network contains different kinds of formats, we need to format the text. We translate all the texts into Chinese and remove punctuation.

3.2 Dictionary

For segmentation, we will use Chinese maximum matching segmentation method (CMMS) which is a dictionary-based matching method [3]. Therefore, we need to make an effective dictionary while achieving the maximum matching. According to the Internet, we searched a lot of relevant corpora, and the corpora we finally used included Thu open Chinese lexicon (THUOCL) from Tsinghua University, SougouW (Internet Thesaurus) from Sogou Laboratory, and modern Chinese common words from Baidu Library.

THUOCL contains a variety of thesaurus in different fields, including IT, finance, idioms, place names, historical celebrities, poetry, medicine, food, law, cars, animals, a total of 11 different fields. THUOCL contains statistical information of word frequency, and after several rounds of manual screening, it can effectively ensure the accuracy of thesaurus collection, and constantly update, including more categories of thesaurus [18].

Sogou lab's corpus is from the Internet thesaurus, and its data is from the statistical analysis of the Chinese Internet corpus indexed by Sogou search engine. The statistics were conducted in October 2006, and the scale of the internet corpus involved is more than 100 million pages. The number of words counted is about 150,000 high-frequency words. In addition to the word frequency information of these words, the commonly used part of speech information is also marked [19].

Finally, we use the commonly used modern Chinese words from Baidu Library, including more than 23,000 two-character words from daily life. Through sorting out, we merged them together and added some common words of single word. The final dictionary has 306, 911 words, including common words, Internet thesaurus and 11 professional words in different fields [20].

At the same time, we also made another dictionary called ignore dictionary. The dictionary is used to eliminate the negligible words in the final keywords, such as "的", "和", and so on. At present, there are 47 entries in the dictionary, which will be kept up-to-date to add more negligible words.

3.3 Chinese Automatic Word Segmentation

Chinese automatic word segmentation is the most basic step in Chinese natural language processing. This step is used to input natural language articles or paragraphs into the computer and output them in words. This step provides a prerequisite for subsequent processing. Chinese word segmentation is quite different from English. Because of the habit of using the English language, people can easily split the words by using the space as a space symbol. However, in Chinese sentences, it is often difficult to distinguish which characters can form words and which are individual characters. At the same time, there are many difficulties in Chinese automatic word segmentation, such as the combination of different characters that can form different words; the same word has different meanings; there are many kinds of ambiguity, and so on. Therefore, we use the forward maximum matching in dictionary-based matching.

Forward maximum matching is a basic word segmentation method, which requires that the statement has been preprocessed; that is, the special punctuation has been removed, and then, the method will match based on a pre-set dictionary. The forward maximum matching method is based on the dictionary, taking the longest word in the dictionary as the first scanning string of the number of words, and scanning in the dictionary according to the order from front to back. At the same time, the reverse maximum matching method is the opposite of the forward maximum matching method, which gets the final result through the reverse scanning order.

Combining the two maximum matching methods, the bidirectional maximum matching method can be obtained. It is based on the comparison of the above two maximum matching methods. According to the minimum number of single words, it decides which maximum matching method to choose. At the same time, in our algorithm, we add the function of taking a single string as the smallest unit for the words that cannot match the dictionary.

3.4 Co-occurrence Matrix

Co-occurrence matrix is a kind of matrix that can be used to express the number of co-occurrence of two elements in a component. In natural language processing, using co-occurrence matrix to construct high-frequency keyword analysis is a hot spot. Keywords usually have the characteristics of high frequency, but not all the high-frequency words are keywords, and at the same time, there is not only one keyword. Therefore, the co-occurrence matrix can well show the correlation between high-frequency words. Often some common words can be used as the keywords of the text.

We need to process the text, so we need to input it first. After word segmentation, we get a new vector text by removing the repeated elements. Then, we use vector text as the row vector and vertical quantity of the matrix, and i and j represent the row and column of the matrix, respectively.

Through calculation and comparison, the largest number of word combinations in the matrix is obtained. If there are multiple word combinations with the same frequency, all the combinations are selected as the most important keywords. Generally, it is not enough to select the combination with the highest frequency as the keyword in a paragraph, because it may lead to errors if the combination does not have very effective keyword information. Therefore, we chose the combination with the highest extraction frequency and the second one. Although some conditions may lead to such a situation, extract more keyword combination, but this can effectively reduce the occurrence of error. At the same time, for the extracted keyword list, further processing is carried out, that is, to remove duplicate words and invalid words (such as "的", "和"). Finally, we can get the keyword list through the co-occurrence matrix.

4 Experiment and Results

In this part, we carry out keyword extraction experiment based on China finance and Economics Web site. Some paragraphs are selected for keyword extraction, and the experimental results and evaluation are given. The algorithm is as follows.

1. Text preprocessing: Remove punctuation
2. Word segmentation: Dictionary matching
3. Structure co-occurrence matrix
4. Remove stop words
5. Output keywords.

At the same time, we use other random paragraphs on the Internet for experiments. Under the condition of pure text, the keywords obtained by keyword extraction based on co-occurrence matrix are effective and can summarize the content of the test paragraph. However, for a paragraph or an article, keywords are not unique, and there are no same keywords for different people.

The accuracy of keyword extraction is difficult to define. There are two reasons. The first is quantity. In the process of manual keyword extraction, people can choose the number of keywords freely. In other words, we only extract all the keywords in this paper, and there is no limit on the number of keywords. The second problem is the content of keywords. In the process of manually extracting keywords, different people may have different results, including which are keywords and how many words can be used as keywords. In the evaluation of keyword extraction, Jiao h and others used the method of comparing with the results extracted by experts [5]. Therefore, we use the keywords extracted from TF-IDF as the comparison results and give the recall as the evaluation criteria in Table 1.

The evaluation results in economics of this experiment are shown in Table 1. For different texts, the accuracy is very different. This is because in these texts, there may be a big gap in their features, and the co-occurrence frequency of some texts

Table 1 Comparison with TF-IDF

Experiment objects	Total words	Our method	TF-IDF
Text 1	36	0.667	0.600
Text 2	133	0.571	0.800
Text 3	137	0.800	0.600
Text 4	191	0.800	0.600
Text 5	436	0.333	0.333

is very low and the same, so our method is difficult to extract keywords. At the same time, we are not satisfied with the result of TF-IDF method. However, for text 3, the keywords extracted by our method can better summarize the content of the text, while TF-IDF method does not extract some effective keywords for the text. For paragraphs with obvious features, our method has good performance, but with the decrease and increase of text features, the amount of calculation and error will gradually increase.

5 Conclusions and Future Work

In this paper, we use co-occurrence matrix to achieve keyword extraction in Chinese natural language processing and carry out many experiments in economics field. This method can calculate the number of common occurrences of different words in Chinese sentences and obtain the keywords of the text according to the frequency. At the same time, it has its own dictionary, including finance and other 11 different fields of professional vocabulary, and at the same time, it can filter the final results and remove meaningless words. In addition, we also discuss the accuracy rate of keyword extraction in Chinese natural language processing. According to its significance and complexity, we give the recall rate by comparing with TF-IDF method. The final experiment shows that this method can effectively extract keywords in a certain range and has great development potential.

For the future development, this method has the ability to extract the words that do not exist in the dictionary, but we hope it can automatically identify the word combination and add it to the dictionary by the common frequency of the front and back words. In addition, we need to constantly update the dictionary to include more words. Finally, we hope this method can help the existing keyword extraction and promote the development of Chinese natural language processing.

References

1. Luftman J, Zadeh HS, Derksen B, Santana M, Rigoni EH, Huang Z (2012) Key information technology and management issues 2011–2012: an international study. J Inf Technol 27(3):198–212
2. Matsuo Y, Ishizuka M (2004) Keyword extraction from a single document using word co-occurrence statistical information. Int J Artif Intell Tools 13(01):157–169
3. Liu F, Huang X, Huang W, Duan SX (2020) Performance evaluation of keyword extraction methods and visualization for student online comments. Symmetry 12(11):1923
4. Liang Y (2017) Chinese keyword extraction based on weighted complex network. In: 2017 12th international conference on intelligent systems and knowledge engineering (ISKE). IEEE, pp 1–5
5. Jiao H, Liu Q, Jia HB (2007) Chinese keyword extraction based on N-gram and word co-occurrence. In: 2007 International conference on computational intelligence and security workshops (CISW 2007). IEEE, pp 152–155
6. Long SQ, Zhao ZW, Tang H (2009) Overview on Chinese segmentation algorithm. Comput Knowl Technol 5(10):2605–2607
7. Hong B, Zhen D (2012) An extended keyword extraction method. Phys Procedia 24:1120–1127
8. Qiu Q, Xie Z, Wu L, Li W (2018) DGeoSegmenter: A dictionary-based Chinese word segmenter for the geoscience domain. Comput Geosci 121:1–11
9. Zeng D, Wei D, Chau M, Wang F (2011) Domain-specific Chinese word segmentation using suffix tree and mutual information. Inf Syst Front 13(1):115–125
10. Junfeng H, Shiwen Y (2000) The multi-layer language knowledge base of Chinese NLP. In: Proceedings of the second international conference on language resources and evaluation (LREC'00)
11. Gui KZ, Ren Y, Peng ZM (2014) CRFs based Chinese word segmentation. In: Applied mechanics and materials, vol 556. Trans Tech Publications Ltd., pp 4376–4379
12. Wang P, Qian Y, Soong FK, He L, Zhao H (2015) A unified tagging solution: Bidirectional LSTM recurrent neural network with word embedding. arXiv:1511.00215
13. Cai D, Zhao H, Zhang Z, Xin Y, Wu Y, Huang F (2017) Fast and accurate neural word segmentation for Chinese. arXiv:1704.07047
14. Tian Y, Song Y, Xia F, Zhang T, Wang Y (2020). Improving Chinese word segmentation with wordhood memory networks. In: Proceedings of the 58th annual meeting of the association for computational linguistics, pp 8274–8285
15. Zhang W, Yoshida T, Tang X (2011) A comparative study of TF* IDF, LSI and multi-words for text classification. Expert Syst Appl 38(3):2758–2765
16. Mihalcea R, Tarau P (2004) Textrank: bringing order into text. In: Proceedings of the 2004 conference on empirical methods in natural language processing, pp 404–411
17. Lilleberg J, Zhu Y, Zhang Y (2015) Support vector machines and word2vec for text classification with semantic features. In: 2015 IEEE 14th international conference on cognitive informatics & cognitive computing (ICCI* CC) IEEE, pp 136–140
18. Han S, Zhang Y, Ma Y, Tu C, Guo Z, Liu Z, Maosong Sun. (2016). THUOCL: Tsinghua Open Chinese Lexicon.
19. SogouLab (2006). SogouW
20. Vocabulary in modern Chinese (2018). Baidu Wenku. https://wenku.baidu.com/view/0ea3b0fd0129bd64783e0912a216147917117e81.html

Author Index

G. Gupta et al. (eds.), *Proceedings of Academia-Industry Consortium for Data Science*, Advances in Intelligent Systems and Computing 1411,
https://doi.org/10.1007/978-981-16-6887-6

Precision Forming Technology of Large Superalloy Castings for Aircraft Engines